Optimal Learning

T0305740

Optimal Learning

Warren B. Powell
Operations Research and Financial Engineering
Princeton University

Ilya O. Ryzhov
Robert H. Smith School of Business
University of Maryland

A JOHN WILEY & SONS, INC., PUBLICATION

Library of Congress Cataloging-in-Publication Data:

Powell, Warren B., 1955–
  Optimal learning / Warren B. Powell, Operations Research and Financial Engineering, Princeton University, Ilya O. Ryzhov, Robert H. Smith School of Business, University of Maryland.
      pages cm. — (Wiley series in probability and statistics)
  Includes bibliographical references and index.
    ISBN 978-0-470-59669-2 (hardback)
  1. Machine learning. I. Ryzhov, Ilya Olegovich, 1985– II. Title.
  Q325.5.P69 2012
  006.3'1—dc23   2011047629

Printed in the United States of America.

10 9 8 7 6 5 4 3 2 1

To our families

CONTENTS

PREFACE

This book emerged from a stream of research conducted in CASTLE Laboratory at Princeton University during the period 2006 - 2011. Initially, the work was motivated by the "exploration vs. exploitation" problem that arises in the design of algorithms for approximate dynamic programming, where it may be necessary to visit a state to learn the value of being in the state. However, we quickly became aware that this basic question had many applications outside of dynamic programming.

The results of this research were made possible by the efforts and contributions of numerous colleagues. The work was conducted under the guidance and supervision of Warren Powell, founder and director of CASTLE Lab. Key contributors include Peter Frazier, Ilya Ryzhov, Warren Scott, and Emre Barut, all graduate students at the time; Martijn Mes, a post-doctoral associate from University Twente in the Netherlands; Diana Negoescu, Gerald van den Berg, and Will Manning, then undergraduate students. The earliest work by Peter Frazier recognized the power of a one-step look-ahead policy, which was named the knowledge gradient, in offline (ranking and selection) problems. The true potential of this idea, however, was realized later with two developments. The first, by Peter Frazier, adapted the knowledge gradient concept to an offline problem with correlated beliefs about different alternatives. This result made it possible to learn about thousands of discrete alternatives with very small measurement budgets.

The second development, by Ilya Ryzhov, made a connection between the knowledge gradient for offline problems and the knowledge gradient for online problems. This relationship links two distinct communities: ranking and selection in statistics and simulation, and the multi-armed bandit problem in applied probability and computer science. This link provides a tractable approach to multi-armed bandit problems with correlated beliefs, a relatively new extension of the well-known bandit model.

This foundation led to a number of additional results. Peter Frazier and Diana Negoescu created a version of the algorithm for problems where the belief structure is given by a linear model, which made it possible to tackle a problem in drug discovery, sorting through 87,000 possible drug combinations with just a few hundred experiments. Martijn Mes adapted the knowledge gradient approach to create an algorithm for a nonparametric model where the value of an alternative was represented through a hierarchical aggregation model. Emre Barut adapted this result for a nonparametric belief model using kernel regression. Warren Scott derived a very difficult but powerful algorithm for when the choice of what to measure is a multidimensional vector of continuous parameters, which was applied to calibrate an industrial simulator for airline operations. Ilya Ryzhov then used the knowledge gradient idea to connect optimal learning with classical mathematical programming, making it possible to incorporate learning issues into fundamental optimization models such as linear programs. As of this writing, we are developing (with Gerald van den Berg) a method to handle the exploration issue in approximate dynamic programming – the very problem that we originally set out to solve.

The work inspired an undergraduate course at Princeton University called "Optimal Learning." Over several years, it has repeatedly attracted talented and enthusiastic students who have produced a creative collection of projects. This book evolved out of lecture notes written the first year that the course was offered. Indeed, the course covers roughly the first seven chapters, with other topics selected from the second half as time permits. The book is designed to be accessible to an advanced undergraduate audience, and presents an overview of the extensive body of research we compiled around the idea of the knowledge gradient. However, we also kept another goal in mind: to recognize the important contributions that have been made by a number of different communities such as economics, computer science, applied probability, simulation optimization, stochastic search, and ranking and selection.

The languages of these different communities have posed a particular challenge. For example, we use the term "online learning" to refer to learning where we have to live with the rewards we receive while also learning to improve decisions in the future, while some use this same term to refer to any sequential learning policy. Different communities are guided by specific classes of applications with characteristics that guide the choice of algorithms. The application setting is rarely transparent in the mathematics, complicating the evaluation of competing methods which have been designed with specific issues in mind. In the multi-armed bandit literature, an alternative x is always referred to as a "bandit," even if x is continuous and vector-valued. Even within this literature, the preferred techniques for solving these problems are quite different in applied probability (Gittins indices) and computer science (upper confidence bounding).

Audience

The book is aimed primarily at an advanced undergraduate or masters-level audience with a course in statistics and a full course in probability. The course can easily be taught at a Ph.D. level by putting more emphasis on the derivations and supporting theory, which is quite deep. However, the book was written for students and practitioners who are interested in practical tools for real problems. For this reason, the core of each chapter focuses on a specific learning problem and presents practical, implementable algorithms.

The later chapters cover material that is more advanced, including (a) learning on graphs and linear programs and (b) learning where the alternatives are continuous (and possibly vector-valued). We have provided chapters designed to bridge with communities such as simulation optimization and machine learning. This material is designed to help Ph.D. students and researchers to understand the many communities that have contributed to the general area of optimal learning.

While every effort has been made to make this material as accessible as possible, the theory supporting this field can be quite subtle. More advanced material is indicated with an * in the section title. Some derivations and proofs are provided in sections called "Why does it work." These sections provide more advanced students with a more thorough foundation, but they can be skipped without loss of continuity.

Organization of the Book

The book is roughly organized into three parts:

Part I: Fundamentals

      Chapter 1 - The Challenges of Learning

      Chapter 2 - Adaptive Learning

      Chapter 3 - The Economics of Information

      Chapter 4 - Ranking and Selection

      Chapter 5 - The Knowledge Gradient

      Chapter 6 - Bandit Problems

      Chapter 7 - Elements of a Learning Problem

Part II: Extensions and Applications

      Chapter 8 - Linear Belief Models

      Chapter 9 - Subset Selection Problems

      Chapter 10 - Optimizing a Scalar Function

      Chapter 11 - Optimal Bidding

      Chapter 12 - Stopping Problems

Part III: Advanced Topics

Chapter 13 - Active Learning in Statistics

Chapter 14 - Simulation Optimization

Chapter 15 - Learning in Mathematical Programming

Chapter 16 - Optimizing over Continuous Measurements

Chapter 17 - Learning with a Physical State

The book is used as a textbook for an undergraduate course at Princeton. In this setting, Part I covers the foundations of the course. This material is supplemented with traditional weekly problem sets and a midterm exam (two hourly exams would also work well here). Each of these chapters have a relatively large number of exercises to help students develop their understanding of the material.

After this foundational material is covered, students work in teams of two to design a project which involves the efficient collection of information. In the initial problem definition, it is important for students to clearly identify the information that is being collected, the implementation decision (which may be different, but not always), and the metric used to evaluate the quality of the implementation decision.

While the students work on their projects, the course continues to work through most of the topics in Part II. The material on linear belief models is particularly useful in many of the student projects, as is the subset selection chapter. Sometimes it is useful to prioritize the material being presented based on the topics that the students have chosen. The chapters in Part II have a small number of exercises, many of which require the use of downloadable MATLAB software to help with the implementation of these more difficult algorithms.

Part III of the book is advanced material and is intended primarily for researchers and professionals interested in using the book as a reference volume. These chapters are not accompanied by exercises, since the material here is more difficult and would require the use of fairly sophisticated software packages.

Additional material for the book is available at the website:

http://optimallearning.princeton.edu/

Downloadable software is provided for several of the most important algorithms, and a fairly elaborate implementation, the Optimal Learning Calculator, is available as a spreadsheet interface calling a sophisticated Java library. Further reading, software, sample projects, and additional thoughts about the field will be made available here.

WARREN B. POWELL AND ILYA O. RYZHOV

Princeton University
University of Maryland
October 2011

ACKNOWLEDGMENTS

This book reflects the contributions of a number of students at Princeton University. Special thanks go to Peter Frazier, who developed the initial framework that shaped our research in this area. Warren Scott, Diana Negoescu, Martijn Mes, and Emre Barut all made important contributions, and their enthusiastic participation in all stages of this research is gratefully acknowledged.

The Optimal Learning Calculator was co-written by Gerald van den Berg and Will Manning, working as undergraduate interns. The spreadsheet-based interface has provided valuable insights into the behavior of different learning policies.

We are very grateful to the students of ORF 418, Optimal Learning, who put up with the formative development of these ideas and who provided an extensive list of creative projects to highlight potential applications of information collection.

The research was supported primarily by the Air Force Office of Scientific Research from the discrete mathematics program headed by Don Hearn, with additional support from the Department of Homeland Security through the CICCADA Center at Rutgers under the leadership of Fred Roberts and the National Science Foundation. Special thanks also goes to SAP and the enthusiastic support of Bill McDermott and Paul Hofmann, along with the many corporate partners of CASTLE Laboratory and PENSA who have provided the challenging problems which motivated this research.

W. B. P. and I. O. R.

CHAPTER 1

THE CHALLENGES OF LEARNING

We are surrounded by situations where we need to make a decision or solve a problem, but where we do not know some or all of the relevant information for the problem perfectly. Will the path recommended by my navigation system get me to my appointment on time? Am I charging the right price for my product, and do I have the best set of features? Will a new material make batteries last longer? Will a molecular compound help reduce a cancer tumor? If I turn my retirement fund over to this investment manager, will I be able to outperform the market? Sometimes the decisions have a simple structure (which investment advisor should I use), while others require complex planning (how do I deploy a team of security agents to assess the safety of a set of food processing plants). Sometimes we have to learn while we are doing (the sales of a book at a particular price), while in other cases we may have a budget to collect information before making a final decision.

There are some decision problems that are hard even if we have access to perfectly accurate information about our environment: planning routes for aircraft and pilots, optimizing the movements of vehicles to pick up and deliver goods, or scheduling machines to finish a set of jobs on time. This is known as deterministic optimization. Then there are other situations where we have to make decisions under uncertainty, but where we assume we know the probability distributions of the uncertain quantities:

Optimal Learning. By Warren B. Powell and Ilya O. Ryzhov
Copyright © 2012 John Wiley & Sons, Inc.

How do I allocate investments to minimize risk while maintaining a satisfactory return, or how do I optimize the storage of energy given uncertainties about demands from consumers? This is known as stochastic optimization.

In this book, we introduce problems where the probability distributions themselves are unknown, but where we have the opportunity to collect new information to improve our understanding of what they are. We are primarily interested in problems where the cost of the information is considered "significant," which is to say that we are willing to spend some time thinking about how to collect the information in an effective way. What this means, however, is highly problem-dependent. We are willing to spend quite a bit before we drill a \$10 million hole hoping to find oil, but we may be willing to invest only a small effort before determining the next measurement inside a search algorithm running on a computer.

The modeling of learning problems, which might be described as "learning how to learn," can be fairly difficult. While expectations are at least well-defined for stochastic optimization problems, they take on subtle interpretations when we are actively changing the underlying probability distributions. For this reason, we tend to work on what might otherwise look like very simple problems. Fortunately, there are very many "simple problems" which would be trivial if we only knew the values of all the parameters, but which pose unexpected challenges when we lack information.

1.1 LEARNING THE BEST PATH

Consider the problem of finding the fastest way to get from your new apartment to your new job in Manhattan. We can find a set of routes from the Internet or from our GPS device, but we do not know anything about traffic congestion or subway delays. The only way we can get data to estimate actual delays on a path is to travel the path. We wish to devise a strategy that governs how we choose paths so that we strike a balance between experimenting with new paths and getting to work on time every day.

Assume that our network is as depicted in Figure 1.1. Let p be a specific path, and let $x_p = 1$ if we choose to take path p. After we traverse the path, we observe a cost $\hat{c}_p$. Let μ_p denote the true mean value of $\hat{c}_p$, which is of course unknown to us. After n trials, we can compute a sample mean $\bar{\theta}_p^n$ of the cost of traversing path p along with a sample variance $\hat{\sigma}_p^{2,n}$ using our observations of path p. Of course, we only observe path p if $x_p^n = 1$, so we might compute these statistics using

$$N_p^n = \sum_{k=1}^{n} x_p^k, \tag{1.1}$$

$$\bar{\theta}_p^n = \frac{1}{N_p^n} \sum_{k=1}^{n} x_p^k \hat{c}_p^k, \tag{1.2}$$

$$\hat{\sigma}_p^{2,n} = \frac{1}{N_p^n - 1} \sum_{k=1}^{n} x_p^k (\hat{c}_p^k - \bar{\theta}_p^n)^2. \tag{1.3}$$

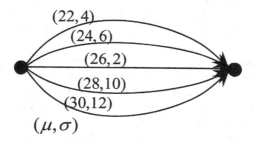

Figure 1.1 A simple shortest path problem, giving the current estimate of the mean and standard deviation (of the estimate) for each path.

Note that $\hat{\sigma}_p^{2,n}$ is our estimate of the variance of $\hat{c}_p$ by iteration n (assuming we have visited path p $N_p^n > 1$ times). The variance of our estimate of the mean, $\bar{\theta}_p^n$, is given by

$$\bar{\sigma}_p^{2,n} = \frac{1}{N_p^n} \hat{\sigma}_p^{2,n}.$$

Now we face the challenge: Which path should we try? Let's start by assuming that you just started a new job and you have been to the Internet to find different paths, but you have not tried any of them. If your job involves commuting from a New Jersey suburb into Manhattan, you have a mixture of options that include driving (various routes) and commuter train, with different transit options once you arrive in Manhattan. But you do have an idea of the length of each path, and you may have heard some stories about delays through the tunnel into Manhattan, as well as a few stories about delayed trains. From this, you construct a rough estimate of the travel time on each path, and we are going to assume that you have at least a rough idea of how far off these estimates may be. We denote these initial estimates using

$\bar{\theta}_p^0$ = initial estimate of the expected travel time on path p,

$\bar{\sigma}_p^0$ = initial estimate of the standard deviation of the difference between $\bar{\theta}_p^0$ and the truth.

If we believe that our estimation errors are normally distributed, then we think that the true mean, μ_p, is in the interval $(\mu_p - z_{\alpha/2}\bar{\sigma}_p^0, \mu_p + z_{\alpha/2}\bar{\sigma}_p^0)$ α percent of the time. If we assume that our errors are normally distributed, we would say that we have an estimate of μ_p that is normally distributed with parameters $(\bar{\theta}_p^0, (\bar{\sigma}_p^0)^2)$.

So which path do you try first? If our priors are as shown in Figure 1.1, presumably we would go with the first path, since it has a mean path time of 22 minutes, which is less than any of the other paths. But our standard deviation around this belief is 4, which means we believe this could possibly be as high as 30. At the same time, there are paths with times of 28 and 30 with standard deviations of 10 and 12. This means that we believe that these paths could have times that are even smaller than 20. Do we always go with the path that we think is the shortest? Or do we try paths that we think are longer, but where we are just not sure, and there is a chance that these paths may actually be better?

If we choose a path we think is best, we say that we are *exploiting* the information we have. If we try a path because it might be better, which would help us make better decisions in the future, we say that we are *exploring*. Exploring a new path, we may find that it is an unexpectedly superior option, but it is also possible that we will simply confirm what we already believed. We may even obtain misleading results – it may be that this one route was experiencing unusual delays on the one day we happened to choose it. Nonetheless, it is often desirable to try something new to avoid becoming stuck on a suboptimal solution just because it "seems" good. Balancing the desire to explore versus exploit is referred to in some communities as the *exploration versus exploitation* problem. Another name is the *learn versus earn* problem. Regardless of the name, the point is the lack of information when we make a decision, along with the value of new information in improving future decisions.

1.2   AREAS OF APPLICATION

The diversity of problems where we have to address information acquisition and learning is tremendous. Below, we try to provide a hint of the diversity.

Transportation

- Responding to disruptions - Imagine that there has been a disruption to a network (such as a bridge failure) forcing people to go through a process of discovering new travel routes. This problem is typically complicated by noisy observations and by travel delays that depend not just on the path but also on the time of departure. People have to evaluate paths by actually traveling them.

- Revenue management - Providers of transportation need to set a price that maximizes revenue (or profit), but since demand functions are unknown, it is often necessary to do a certain amount of trial and error.

- Evaluating airline passengers or cargo for dangerous items - Examining people or cargo to evaluate risk can be time-consuming. There are different policies that can be used to determine who/what should be subjected to varying degrees of examination. Finding the best policy requires testing them in field settings.

- Finding the best heuristic to solve a difficult integer program for routing and scheduling - We may want to find the best set of parameters to use our tabu search heuristic, or perhaps we want to compare tabu search, genetic algorithms, and integer programming for a particular problem. We have to loop over different algorithms (or variations of an algorithm) to find the one that works the best on a particular dataset.

Figure 1.2   The operations center for NetJets®, which manages over 750 aircraft[1]. NetJets® has to test different policies to strike the right balance of costs and service.

- Finding the best business rules - A transportation company needs to determine the best terms for serving customers, the best mix of aircraft, and the right pilots to hire[1] (see Figure 1.2). They may use a computer simulator to evaluate these options, requiring time-consuming simulations to be run to evaluate different strategies.

- Evaluating schedule disruptions - Some customers may unexpectedly ask us to deliver their cargo at a different time, or to a different location than what was originally agreed upon. Such disruptions come at a cost to us, because we may need to make significant changes to our routes and schedules. However, the customers may be willing to pay extra money for the disruption. We have a limited time to find the disruption or combination of disruptions where we can make the most profit.

Energy and the Environment

- Finding locations for wind farms - Wind conditions can depend on micro-geography - a cliff, a local valley, a body of water. It is necessary to send teams with sensors to find the best locations for locating wind turbines in a geographical area. The problem is complicated by variations in wind, making it necessary to visit a location multiple times.

- Finding the best material for a solar panel - It is necessary to test large numbers of molecular compounds to find new materials for converting sunlight to electricity. Testing and evaluating materials is time consuming and very expensive, and there are large numbers of molecular combinations that can be tested.

[1]Includes aircraft under management by Executive Jet® Management.

Figure 1.3   Wind turbines are one form of alternative energy resources (from http://www.nrel.gov/data/pix/searchpix.cgi).

- Tuning parameters for a fuel cell - There are a number of design parameters that have to be chosen to get the best results from a full cell: the power density of the anode or cathode, the conductivity of bipolar plates, and the stability of the seal.

- Finding the best energy-saving technologies for a building - Insulation, tinted windows, motion sensors and automated thermostats interact in a way that is unique to each building. It is necessary to test different combinations to determine the technologies that work the best.

- R&D strategies - There are a vast number of research efforts being devoted to competing technologies (materials for solar panels, biomass fuels, wind turbine designs) which represent projects to collect information about the potential for different designs for solving a particular problem. We have to solve these engineering problems as quickly as possible, but testing different engineering designs is time-consuming and expensive.

- Optimizing the best policy for storing energy in a battery - A policy is defined by one or more parameters that determine how much energy is stored and in what type of storage device. One example might be, "charge the battery when the spot price of energy drops below x." We can collect information in the field or a computer simulation that evaluates the performance of a policy over a period of time.

- Learning how lake pollution due to fertilizer run-off responds to farm policies - We can introduce new policies that encourage or discourage the use of fertilizer, but we do not fully understand the relationship between these policies and lake pollution, and these policies impose different costs on the farmers. We need to test different policies to learn their impact, but each test requires a year to run and there is some uncertainty in evaluating the results.

- On a larger scale, we need to identify the best policies for controlling CO^2 emissions, striking a balance between the cost of these policies (tax incentives on renewables, a carbon tax, research and development costs in new technologies) and the impact on global warming, but we do not know the exact relationship between atmospheric CO^2 and global temperatures.

Homeland Security

- You would like to minimize the time to respond to an emergency over a congested urban network. You can take measurements to improve your understanding of the time to traverse each region of the traffic network, but collecting these observations takes time. How should you structure your observations of links in the network to achieve the best time when you need to find the shortest path?

- You need to manage a group of inspectors to intercept potentially dangerous cargo being smuggled through ports and across borders. Since you do not know the frequency with which smugglers might try to use a port of entry, it is important to allocate inspectors not just to maximize the likelihood of an interception given current beliefs, but to also collect information so that we can improve our understanding of the truth. For example, we may believe that a particular entry point might have a low probability of being used, but we may be wrong.

- Radiation is detected in downtown Manhattan. Inspectors have to be managed around the city to find the source as quickly as possible. Where should we send them to maximize the likelihood of finding the source?

Science and Engineering

- The National Ignition Facility uses large crystals to focus lasers into a very small region to perform nuclear research. The crystals become damaged over time and have to be repaired or replaced, but the process of examining each crystal is time-consuming and reduces the productivity of the facility. NIF has to decide when to examine a crystal to determine its status.

- A company is trying to design an aerosol device whose performance is determined by a number of engineering parameters: the diameter of the tube that pulls liquid from a reservoir, the pressure, the angle of a plate used to direct the spray, and the size of the portal used to project the spray and the angle of

Figure 1.4 Drug discovery requires testing large numbers of molecules.

the departure portal. These have to be varied simultaneously to find the best design.

Health and Medicine

- Drug discovery - Curing a disease often involves first finding a small family of base molecules, and then testing a large number of variations of a base molecule. Each test of a molecular variation can take a day and consumes costly materials, and the performance can be uncertain.

- Drug dosage - Each person responds to medication in a different way. It is often necessary to test different dosages of a medication to find the level that produces the best mix of effectiveness against a condition with minimum side effects.

- How should a doctor test different medications to treat diabetes, given that he will not know in advance how a particular patient might respond to each possible course of treatment?

- What is the best way to test a population for an emerging disease so that we can plan a response strategy?

Sports

- How do you find the best set of five basketball players to use as your starting lineup? Basketball players require complementary skills in defense, passing, and shooting, and it is necessary to try different combinations of players to see which group works the best.

- What is the best combination of rowers for a four person rowing shell? Rowers require a certain synergy to work well together, making it necessary to try different combinations of rowers to see who turns in the best time.

- Who are the best hitters that you should choose for your baseball team? It is necessary to see how a player hits in game situations, and of course these are very noisy observations.

- What plays work the best for your football team? Specific plays draw on different combinations of talents, and a coach has to find out what works best for his team.

Business

- What are the best labor rules or terms in a customer contract to maximize profits? These can be tested in a computer simulation program, but it may require several hours (in some cases, several days) to run. How do we sequence our experiments to find the best rules as quickly as possible?

- What is the best price to charge for a product being sold over the Internet? It is necessary to use a certain amount of trial and error to find the price that maximizes revenue.

- We would like to find the best supplier for a component part. We know the price of the component, but we do not know about the reliability of the service or the quality of the product. We can collect information on service and product quality by placing small orders.

- We need to identify the best set of features to include in a new laptop we are manufacturing. We can estimate consumer response by running market tests, but these are time-consuming and delay the product launch.

- A company needs to identify the best person to lead a division that is selling a new product. The company does not have time to interview all the candidates. How should a company identify a subset of potential candidates?

- Advertising for a new release of a movie - We can choose between TV ads, billboards, trailers on movies already showing, the Internet, and promotions through restaurant chains. What works best? Does it help to do TV ads if you are also doing Internet advertising? How do different outlets interact? You have to try different combinations, evaluate their performance, and use what you learn to guide future advertising strategies.

- Conference call or airline trip? Business people have to decide when to try to land a sale using teleconferencing, or when a personal visit is necessary. For companies that depend on numerous contacts, it is possible to experiment with different methods of landing a sale, but these experiments are potentially expensive, involving (a) the time and expense of a personal trip or (b) the risk of not landing a sale.

E-Commerce

- Which ads will produce the best consumer response when posted on a website? You need to test different ads, and then identify the ads that are the most promising based on the attributes of each ad.

- Netflix can display a small number of movies to you when you log into your account. The challenge is identifying the movies that are likely to be most interesting to a particular user. As new users sign up, Netflix has to learn as quickly as possible which types of movies are most likely to attract the attention of an individual user.

- You need to choose keywords to bid on to get Google to display your ad. What bid should you make for a particular keyword? You measure your performance by the number of clicks that you receive.

- YouTube has to decide which videos to feature on its website to maximize the number of times a video is viewed. The decision is the choice of video, and the information (and reward) is the number of times people click on the video.

- Amazon uses your past history of book purchases to make suggestions for potential new purchases. Which products should be suggested? How can Amazon use your response to past suggestions to guide new suggestions?

The Service Sector

- A university has to make specific offers of admission, after which it then observes which types of students actually matriculate. The university has to actually make an offer of admission to learn whether a student is willing to accept the offer. This information can be used to guide future offers in subsequent years. There is a hard constraint on total admissions.

- A political candidate has to decide in which states to invest his remaining time for campaigning. He decides which states would benefit the most through telephone polls, but has to allocate a fixed budget for polling. How should he allocate his polling budget?

- The Federal government would like to understand the risks associated with issuing small business loans based on the attributes of an applicant. A particular applicant might not look attractive, but it is possible that the government's estimate of risk is inflated. The only way to learn more is to try granting some higher risk loans.

- The Internal Revenue Service has to decide which companies to subject to a tax audit. Should it be smaller companies or larger ones? Are some industries more aggressive than others (for example, due to the presence of lucrative tax write-offs)? The government's estimates of the likelihood of tax cheating may be incorrect, and the only way to improve its estimates is to conduct audits.

Figure 1.5 The Air Force has to design new technologies and determine the best policies for operating them.

The Military

- The military has to collect information on risks faced in a region using UAVs (unmanned aerial vehicles). The UAV collects information about a section of road, and then command determines how to deploy troops and equipment. How should the UAVs be deployed to produce the best deployment strategy?

- A fighter has to decide at what range to launch a missile. After firing a missile, we learn whether the missile hit its target or not, which can be related to factors such as range, weather, altitude and angle-of-attack. With each firing, the fighter learns more about the probability of success.

- The Air Force has to deploy tankers for mid-air refueling. There are different policies for handling the tankers, which include options such as shuttling tankers back and forth between locations, using one tanker to refuel another tanker, and trying different locations for tankers. A deployment policy can be evaluated by measuring (a) how much time fighters spend waiting for refueling and (b) the number of times a fighter has to abort a mission from lack of fuel.

- The military has to decide how to equip a soldier. There is always a tradeoff between cost and the weight of the equipment, versus the likelihood that the soldier will survive. The military can experiment with different combinations of equipment to assess its effectiveness in terms of keeping a soldier alive.

Tuning Models and Algorithms

- There is a large community that models physical problems such as manufacturing systems using Monte Carlo simulation. For example, we may wish to simulate the manufacture of integrated circuits which have to progress through a series of stations. The progression from one station to another may be limited by the size of buffers which hold circuit boards waiting for a particular machine.

We wish to determine the best size of these buffers, but we have to do this by sequential simulations which are time-consuming and noisy.

- There are many problems in discrete optimization where we have to route people and equipment, or scheduling jobs to be served by a machine. These are exceptionally hard optimization problems that are typically solved using heuristic algorithms such as tabu search or genetic algorithms. These algorithms are controlled by a series of parameters which have to be tuned for specific problem classes. One run of an algorithm on a large problem can require several minutes to several hours (or more), and we have to find the best setting for perhaps five or ten parameters.

- Engineering models often have to be calibrated to replicate a physical process such as weather or the spread of a chemical through groundwater. These models can be especially expensive to run, often requiring the use of fast supercomputers to simulate the process in continuous space or time. At the same time, it is necessary to calibrate these models to produce the best possible prediction.

1.3   MAJOR PROBLEM CLASSES

Given the diversity of learning problems, it is useful to organize these problems into major problem classes. A brief summary of some of the major dimensions of learning problems is given below.

- Online versus offline - Online problems involve learning from experiences as they occur. For example, we might observe the time on a path through a network by traveling the path, or adjust the price of a product on the Internet and observe the revenue. We can try a decision that looks bad in the hopes of learning something, but we have to incur the cost of the decision, and balance this cost against future benefits. In offline problems, we might be working in a lab with a budget for making measurements, or we might set aside several weeks to run computer simulations. If we experiment with a chemical or process that does not appear promising, all we care about is the information learned from the experiment; we do not incur any cost from running an unsuccessful experiment. When our budget has been exhausted, we have to use our observations to choose a design or a process that will then be put into production.

- Objectives - Problems differ in terms of what we are trying to achieve. Most of the time we will focus on minimizing the expected cost or maximizing the expected reward from some system. However, we may be simply interested in finding the best design, or ensuring that we find a design that is within five percent of the best.

- The measurement decision - In some settings, we have a small number of choices such as drilling test wells to learn about the potential for oil or natural gas. The number of choices may be small, but each test can cost millions of dollars. Alternatively, we might have to find the best set of 30 proposals out of 100 that have been submitted, which means that we have to choose from 3×10^{25} possible portfolios. Or we may have to choose the best price, temperature, or pressure (a scalar, continuous parameter). We might have to set a combination of 16 parameters to produce the best results for a business simulator. Each of these problems introduce different computational challenges because of the size of the search space.

- The implementation decision - Collecting the best information depends on what you are going to do with the information once you have it. Often, the choices of what to observe (the measurement decision) are the same as what you are going to implement (finding the choice with the best value). But you might measure a link in a graph in order to choose the best path. Or we might want to learn something about a new material to make a decision about new solar panels or batteries. In these problems, the implementation decision (the choice of path or technology) is different from the choice of what to measure.

- What we believe - We may start by knowing nothing about the best system. Typically, we know something (or at least we will know something after we make our first measurement). What assumptions can we reasonably make about different choices? Can we put a normal distribution of belief on an unknown quantity? Are the beliefs correlated (if a laptop with one set of features has higher sales than we expected, does this change our belief about other sets of features)? Are the beliefs stored as a lookup table (that is, a belief for each design), or are the beliefs expressed as some sort of statistical model?

- The nature of a measurement - Closely related to what we believe is what we learn when we make a measurement. Is the observation normally distributed? Is it a binary random variable (success/failure)? Are measurements made with perfect accuracy? If not, do we know the distribution of the error in a measurement?

- Belief states and physical states - All learning problems include a "belief state" (or knowledge state) which captures what we believe about the system. Some problems also include a physical state. For example, to measure the presence of disease at city i, we have to visit city i. After making this measurement, the cost of visiting city j now depends on city i. Our physical location is a physical state.

We are not going to be able to solve all these problems in this book, but we can at least recognize the diversity of problems.

1.4 THE DIFFERENT TYPES OF LEARNING

It is useful to contrast learning problems with other types of optimization problems. Figure 1.1 depicts two optimization problem. The problem in Figure 1.1(a) shows

Table 1.1   (a) A problem involving five known alternatives, and (b) a problem where the value of each alternative is normally distributed with known mean and standard deviation.

Alternative	Value
1	759
2	722
3	698
4	653
5	616

Alternative	Mean	Std. Dev.
1	759	120
2	722	142
3	698	133
4	653	90
5	616	102

(a) The Best of Five Known Alternatives   (b) The Best of Five Uncertain Alternatives

five choices, each of which has a known value. The best choice is obviously the first one, with a value of 759. Of course, deterministic optimization problems can be quite hard, but this happens to be a trivial one.

A harder class of optimization problems arise when there is uncertainty in the parameters. Figure 1.1(b) depicts a problem with five choices where the reward we receive from a choice is normally distributed with known mean and standard deviation. Assume that we have to make a choice before the reward is received, and we want to make a choice that gives us the highest expected return. Again, we would select the first alternative, because it has the highest expected value.

The problems illustrated in Table 1.1 use either known values, or known distributions. This problem is fairly trivial (picking the best out of a list of five), but there are many problems in stochastic optimization that are quite hard. In all of these problems, there are uncertain quantities but we assume that we know the probability distribution describing the likelihood of different outcomes. Since the distributions are assumed known, when we observe an outcome we view it simply as a realization from a known probability distribution. We do not use the observation to update our belief about the probability distribution.

Now consider what happens when you are not only uncertain about the reward, you are uncertain about the probability distribution for the reward. The situation is illustrated in Table 1.2, where after choosing to measure the first alternative, we observe an outcome of 702 and then use this outcome to update our belief about the first alternative. Before our measurement, we thought the reward was normally distributed with mean 759 and standard deviation 102. After the measurement, we now believe the mean is 712 with standard deviation of 92. As a result, alternative 2 now seems to be the best.

Since we are willing to change our belief about an alternative, is it necessarily the case that we should try to evaluate what appears to be the best alternative? Later in this volume, we are going to refer to this as an *exploitation policy*. This means that we exploit our current state of knowledge and choose the alternative that appears to be best. But it might be the case that if we observe an alternative that does not appear

Table 1.2 Learning where we update our beliefs based on observations, which changes our distribution of belief for future measurements.

Alternative	Initial Mean and Std. Dev.		First Obs.	Updated Mean and Std. Dev.		Second Obs.
	Mean	Std. Dev.		Mean	Std. Dev.	
1	759	102	702	712	92	
2	722	133		722	133	734
3	698	78		698	78	
4	653	90		653	90	
5	616	102		616	102	

to be the best to use right now, we may collect information that allows us to make better decisions in the future. The central idea of optimal learning is to incorporate the value of information in the future to make better decisions now.

Now consider another popular optimization problem known as the newsvendor problem. In this problem, we wish to order a quantity (of newspapers, oil, money, energy) x to satisfy a random demand D (that is, D is not known when we have to choose x). We earn p dollars per unit of satisfied demand, which is to say $\min(x, D)$, and we have to pay c dollars per unit of x that we order. The total profit is given by

$$F(x, D) = p \min(x, D) - cx.$$

The optimization problem is to solve

$$\max_x \mathbb{E} F(x, D).$$

There are a number of ways to solve stochastic optimization problems such as this. If the distribution of D is known, we can characterize the optimal solution using

$$P_D[x^* \leq D] = \frac{c}{p},$$

where $P_D()$ is the cumulative distribution function for D. So, as the purchase cost c is decreased, we should increase our order quantity so that the probability that the order quantity is less than demand also decreases.

In many applications, we do not know the distribution of D, but we are able to make observations of D (or we can observe if we have ordered too much or too little). Let x^{n-1} be the order quantity we chose after observing D^{n-1}, which was our best guess of the right order quantity to meet the demand on day n, and let D^n be resulting demand. Now let g^n be the derivative of $F(x, D)$, given that we ordered x^{n-1} and then observed D^n. This derivative is given by

$$g^n = \begin{cases} p - c & \text{if } x \leq D, \\ -c & \text{if } x > D. \end{cases}$$

A simple method for choosing x^n is a stochastic gradient algorithm which looks like

$$x^n = x^{n-1} + \alpha_{n-1} g^n. \tag{1.4}$$

Here, α_{n-1} is a stepsize that has to satisfy certain conditions that are not important here. If the stepsize is chosen appropriately, it is possible to show that in the limit, x^n approaches the optimal solution, even without knowing the distribution of D in advance.

What our algorithm in equation (1.4) ignores is that our choice of x^n allows us to learn something about the distribution of D. For example, it might be that the purchase cost c is fairly high compared to the sales price p, which would encourage us to choose smaller values of x, where we frequently do not satisfy demand. But we might benefit from making some larger orders just to learn more about the rest of the demand distribution. By ignoring our ability to learn, the algorithm may not converge to the right solution, or it may eventually find the right solution, but very slowly. When we use optimal learning, we explicitly capture the value of the information we learn now on future decisions.

1.5   LEARNING FROM DIFFERENT COMMUNITIES

The challenge of efficiently collecting information is one that arises in a number of communities. The result is a lot of parallel discovery, although the questions and computational challenges posed by different communities can be quite different, and this has produced diversity in the strategies proposed for solving these problems. Below we provide a rough list of some of the communities that have become involved in this area.

- Simulation optimization - The simulation community often faces the problem of tuning parameters that influence the performance of a system that we are analyzing using Monte Carlo simulation. These parameters might be the size of a buffer for a manufacturing simulator, the location of ambulances and fire trucks, or the number of advance bookings for a fare class for an airline. Simulations can be time-consuming, so the challenge is deciding how long to analyze a particular configuration or policy before switching to another one.

- The ranking and selection problem - This is a statistical problem that arises in many settings, including the simulation optimization community. It is most often approached using the language of classical frequentist statistics (but not always) and tends to be very practical in its orientation. In ranking and selection, we assume that for each measurement, we can choose equally from a set of alternatives (there is no cost for switching from one alternative to another). Although the ranking and selection framework is widely used in simulation optimization, the simulation community recognizes that it is easier to run the simulation for one configuration a little longer than it is to switch to the simulation of a new configuration.

- The bandit community - There is a subcommunity that has evolved within applied probability and machine learning that studies what has long been referred to as bandit problems. This is the online (pay as you go) version of ranking and selection. A major breakthrough for this problem class was the discovery that a simple index policy (a quantity computed for each alternative that guides which alternative should be tested next) is optimal, producing a line of research (primarily in applied probability) aimed at discovering optimal index policies for more general problems. A separate subcommunity (primarily in computer science) has focused on a simple heuristic known as *upper confidence bounding* which has the property that the number of times we test the wrong alternative is bounded by a logarithmic function, which has then been shown to be the best possible bound. Upper confidence bounding has also been popular in the control theory community.

- Global optimization of expensive functions - The engineering community often finds a need to optimize complex functions of continuous variables. The function is sometimes a complex piece of computer software that takes a long time to run, but the roots of the field come from geospatial applications. The function might be deterministic (but not always), and a single evaluation can take an hour to a week or more.

- Learning in economics - Economists have long studied the value of information in a variety of idealized settings. This community tends to focus on insights into the economic value of information, rather than the derivation of specific procedures for solving information collection problems.

- Active learning in computer science - The machine learning community typically assumes that a dataset is given. When there is an opportunity to choose what to measure, this is known as active learning. This community tends to focus on statistical measures of fit rather than economic measures of performance.

- Statistical design of experiments - A classical problem in statistics is deciding what experiments to run. For certain objective functions, it has long been known that experiments can be designed deterministically, in advance, rather than sequentially. Our focus is primarily on sequential information collection, but there are important problem classes where this is not necessary.

- Frequentist versus Bayesian communities - It is difficult to discuss research in optimal learning without addressing the sometimes contentious differences in styles and attitudes between frequentist and Bayesian statisticians. Frequentists look for the truth using nothing more than the data that we collect, while Bayesians would like to allow us to integrate expert judgment.

- Optimal stopping - There is a special problem class where we have the ability to observe a single stream of information such as the price of an asset. As long as we hold the asset, we get to observe the price. At some point, we have to make a decision whether we should sell the asset or continue to observe prices (a form of learning). Another variant is famously known as the "secretary problem" where we interview candidates for a position (or offers for an asset); after each candidate (or offer) we have to decide if we should accept and stop or reject and continue observing.

- Approximate dynamic programming/reinforcement learning - Approximate dynamic programming, widely known as reinforcement learning in the computer science community, addresses the problem of choosing an action given a state which generates a reward and takes us to a new state. We do not know the exact value of the downstream state, but we might decide to visit a state just to learn more about it. This is generally known as the "exploration versus exploitation" problem, and this setting has motivated a considerable amount of research in optimal learning.

- Psychology - Not surprisingly, the tradeoff between exploration and exploitation is a problem that has to be solved by people (as well as other animals ranging from chimpanzees to ants) for problems ranging from finding foot to finding mates. This has recently attracted attention in the psychology community (Cohen et al. 2007).

Readers who wish to study this field seriously will encounter the contributions of these (and perhaps other) communities. It is not possible to cover all the issues and perspectives of these communities in this volume, but we do provide a foundation that should make it possible for students and researchers to understand the issues and, in some cases, challenge conventional wisdom within specific communities.

1.6 INFORMATION COLLECTION USING DECISION TREES

The simplest types of information collection problems arise when there is a small number of choices to collect information. Should you check the weather report before scheduling a baseball game? Should you purchase an analysis of geologic formulations before drilling for oil? Should you do a statistical analysis of a stock price before investing in the stock?

These are fairly simple problems that can be analyzed using a decision tree, which is a device that works well when the number of decisions, as well as the number of possible outcomes, is small and discrete. We begin by first presenting a small decision tree where collecting information is not an issue.

1.6.1 A Basic Decision Tree

Decision trees are a popular device for solving problems that involve making decisions under uncertainty, because they illustrate the sequencing of decisions and information so clearly. Figure 1.6 illustrates the decision tree that we might construct to help with the decision of whether to hold or sell a stock. In this figure, square nodes are decision

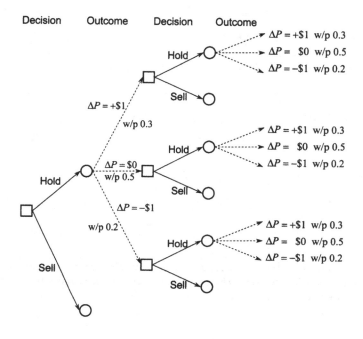

Figure 1.6   Illustration of a decision tree to determine if we should hold or sell a stock, where the stock might go up or down $1 or stay the same in each time period.

nodes, where we get to choose from a set of actions (where the number of actions cannot be too large). Circle nodes are outcome nodes, where something random happens, such as the change in the price of the stock. The solid lines represent decisions, while dashed lines are random outcomes. In our example, there is no cost for holding a stock, and the random outcome represents a change in the price of the stock.

If we hold the stock (which currently can be sold at a price of $50), it might go up or down by a dollar, with probabilities of 0.3 and 0.2, respectively, or hold at the same price with a probability of 0.5. After observing the change in the price, we again have a chance to hold or sell the stock. If we continue holding, the stock might go up or down by a dollar, or stay the same.

We can solve the problem of whether to hold or sell the stock initially by doing what is called "rolling back the decision tree." Figure 1.7(a) shows the tree after the first rollback. Here, we have taken the final random outcome and replaced it with the expected value, which gives us the result that we expect the price to go up by $0.10 if we hold the stock. We now have the option of holding or selling, which is a decision that we control. Since the price is likely to go up if we hold, we make this choice.

In Figure 1.7(b), we now use the expected value of the second decision to give us what we will earn at the end of each random outcome resulting from the first decision.

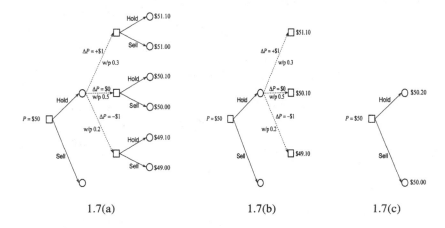

1.7(a)   1.7(b)   1.7(c)

Figure 1.7   (a) Decision tree with second outcome replaced by expected value. (b) Decision tree with second decision replaced by best expected value. (c) Decision tree with first outcome replaced by expected value, producing a deterministic decision.

We have the same three outcomes (up or down $1, or stay the same), but each outcome produces an expected return of $51.10, $50.10, or $49.10. Now we again have to find the expectation over these returns, which gives us the expected value of $50.20. Finally, we have to evaluate our original decision to hold or sell, and of course we are willing to hold for $50.20 rather than sell now for $50.

1.6.2 Decision Tree for Offline Learning

The previous example provided a quick illustration of a basic decision tree. A common decision problem is whether or not we should collect information to make a decision. For example, consider a bank that is considering whether it should grant a short-term credit loan of $100,000. The bank expects to make $10,000 if the loan is paid off on time. If the loan is defaulted, the bank loses the amount of the loan.

From history, the bank knows that 95 percent of loans are repaid in full, while 5 percent default. If the bank purchases the credit report, this information will allow the bank to classify the customer into one of three groups: 52 percent fall into the top A rating, 30 percent fall into the middle B rating, while 18 percent fall into the lower C rating with the highest risk of default. The company selling the credit report provides the joint distribution $P(\text{Credit, Default})$ that a customer will receive each credit rating, and whether it defaulted or not. This data are summarized in Table 1.3.

We need to understand how the information from the credit report changes our belief about the probability of a default. For this, we use a simple application of

Table 1.3   The marginal probability of each credit rating, the joint probability of a credit rating and whether someone defaults on a loan, and the conditional probability of a default given a credit rating.

Credit Rating	P(Credit)	P(Credit,Default) No	Yes	P(Default | Credit) No	Yes
A	0.52	0.51	0.01	0.981	0.019
B	0.30	0.28	0.02	0.933	0.067
C	0.18	0.16	0.02	0.889	0.111
	P(Default) =	0.95	0.05		

Bayes' theorem, which states

$$P(\text{Default} \mid \text{Credit}) = \frac{P(\text{Credit} \mid \text{Default})P(\text{Default})}{P(\text{Credit})}$$
$$= \frac{P(\text{Credit}, \text{Default})}{P(\text{Credit})}.$$

Bayes' theorem allows us to start with our initial estimate of the probability of a default, $P(\text{Default})$, and then use the information "Credit" from the credit history and turn it into a *posterior* distribution $P(\text{Default} \mid \text{Credit})$. The results of this calculation are shown in the final two columns of Table 1.3.

Using this information, we can construct a new decision tree, shown in Figure 1.8. Unlike our first decision tree in Figure 1.7, we now see that the decision to collect information changes the downstream probabilities.

We repeat the exercise of rolling back the decision tree in Figure 1.9. Figure 1.9(a) shows the expected value of the decision to grant the loan given the information about the credit history. We see that if the grantee has an A or B credit rating, it makes sense to grant the loan, but not if the rating is C. Thus, the information from the credit report has the effect of changing the decision of whether or not to grant the loan. After we roll the tree back to the original decision of whether to purchase the credit report, we find that the credit report produces an expected value of $4,900, compared to $4,500 that we would expect to receive without the credit report. This means that we would be willing to pay up to $400 for the credit report.

1.6.3   Decision Tree for Online Learning

Now consider a problem where we learn as we go. We use the setting of trying to identify the best hitter on a baseball team. The only way to collect information is to put the hitter into the lineup. Assume that we are evaluating hitters for the fourth position in the lineup, typically reserved for power hitters. Part of what we are trying to learn is how a hitter actually performs in game situations.

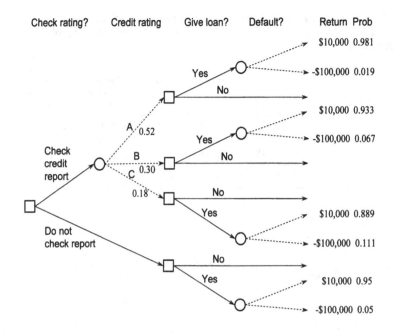

Figure 1.8   The decision tree for the question of whether we should purchase a credit risk report.

Assume that we have three candidates for the position. The information we have on each hitter from previous games is given in Table 1.4. If we choose player A, we have to balance the likelihood of getting a hit, and the value of the information we gain about his true hitting ability, since we will use the event of whether or not he gets a hit to update our assessment of his probability of getting a hit. We are going to again use Bayes' theorem to update our belief about the probability of getting a hit. To do this, we have to make some probabilistic assumptions that are not relevant to our discussion here; we defer until Chapter 2, Section 2.3.4, the discussion of the model that we use to calculate the updated probabilities. Fortunately, this model produces

Table 1.4   History of hitting performance for three candidates.

Player	No. Hits	No. At-Bats	Average
A	36	100	0.360
B	1	3	0.333
C	7	22	0.318

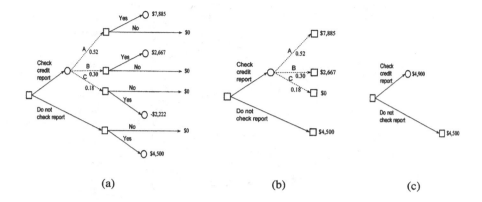

Figure 1.9 (a) Decision tree with final default replaced with expected value. (b) Decision tree with second decision replaced by best expected value. (c) Decision tree with the uncertainty of the credit risk report replaced with its expected value.

some very intuitive updating equations. Let H^n be number of hits a player has made in n at-bats. Let $\hat{H}^{n+1} = 1$ if a hitter gets a hit in his $(n+1)$st at-bat. Our prior probability of getting a hit after n at-bats is

$$\mathbb{P}[\hat{H}^{n+1} = 1 | H^n, n] = \frac{H^n}{n}.$$

Once we observe $\hat{H}^{n+1}$, it is possible to show that the posterior probability is

$$\mathbb{P}[\hat{H}^{n+2} = 1 | H^n, n, \hat{H}^{n+1}] = \frac{H^n + \hat{H}^{n+1}}{n+1}.$$

In other words, all we are doing is computing the batting average (hits over at-bats). In Chapter 2 we are going to put a lot more rigor behind this, but for now, we are going to take advantage of the simple, intuitive updating equations that this theory provides.

Our challenge is to determine whether we should try player A, B, or C right now. At the moment, A has the best batting average of .360, based on a history of 36 hits out of 100 at-bats. Why would we try player B, whose average is only .333? We easily see that this statistic is based on only three at-bats, which would suggest that we have a lot of uncertainty in this average.

We can study this formally by setting up the decision tree shown in Figure 1.10. For practical reasons, we can only study a problem that spans two at-bats. We show the current prior probability of a hit, or no hit, in the first at-bat. For the second at-bat, we show only the probability of getting a hit, to keep the figure from becoming too cluttered.

Figure 1.11 shows the calculations as we roll back the tree. Figure 1.11(c) shows the expected value of playing each hitter for exactly one more at-bat using the information

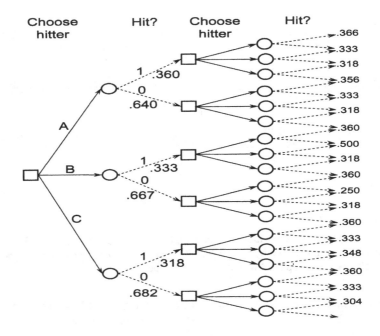

Figure 1.10 The decision tree for finding the best hitter.

obtained from our first decision. It is important to emphasize that after the first decision, only one hitter has had an at-bat, so the batting averages only change for that hitter. Figure 1.11(b) reflects our ability to choose what we think is the best hitter, and Figure 1.11(a) shows the expected value of each hitter before any at-bats have occurred. We use as our reward function the expected number of total hits over the two at-bats. So, if we choose batter A, the expected value is

$$.720 = .360(1 + .366) + .640(0 + .356),$$

where .360 is our prior belief about his probability of getting a hit; .366 is the expected number of hits in his second at-bat (the same as the probability of getting a hit) given that he got a hit in his first at-bat. If player A did not get a hit in his first at-bat, his updated probability of getting a hit, .356, is still higher than any other player. This means that if we have only one more at-bat, we would still pick player A even if he did not get a hit in his first at-bat.

Although player A initially has the highest batting average, our analysis says that we should try player B for the first at-bat. Why is this? On further examination, we realize that it has a lot to do with the fact that player B has had only three at-bats. If this player gets a hit, our estimate of his probability of getting a hit jumps to 0.500, although it drops to .250 if he does not get a hit. If player A gets a hit, his batting average moves from .360 to .366, reflecting the weight of his much longer record.

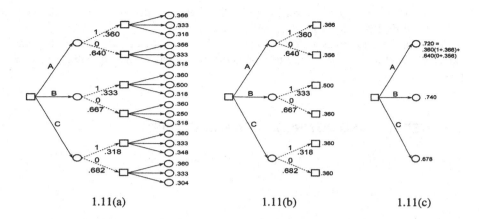

1.11(a)   1.11(b)   1.11(c)

Figure 1.11   (a) Expected value of a hit in the second at-bat. (b) Value of best hitter after one at-bat. (c) Expected value of each hitter before first at-bat.

This is our first hint that it can be useful to collect information about choices where there is the greatest uncertainty.

1.6.4 Discussion

These simple examples illustrate some of the issues that are fundamental to information collection. First, we see that collecting information changes our beliefs about uncertain quantities. Second, when we change our beliefs, we change our decisions which produces an economic impact. From this analysis, we can compute the expected value of the information.

In the credit risk example, the *measurement decision* (purchasing the credit risk report) was completely separate from the *implementation decision* (whether or not to give a loan). There are many problems where we learn from our actions. For example, we might grant a loan and then learn from this experience. This means giving a loan allows us to observe whether or not someone defaults on the loan, which might then change our behavior in the future. We can ignore the value of this information when making a decision, but one of the goals of this book is to use this value of information in our decisions.

Most of the problems that we address in this book can be visualized as a decision tree. Decisions trees, however, grow very quickly with the number of decisions being made, the number of random outcomes that might occur, and the number of time periods. For example, the very simple decision tree that we used in Section 1.6.3 to analyze which baseball player we should use grows quickly if we look more than two at-bats into the future. With three players to choose from, and only two possible outcomes, the tree grows by a factor of six for every at-bat we add to our planning horizon. If we want to plan over the next 10 at-bats, our decision tree would have $6^{10} = 60,466,176$ end points. And this is a tiny problem.

Although decision trees are often impractical as a computational device, they are extremely useful as a conceptual mechanism for understanding sequential decision problems.

1.7 WEBSITE AND DOWNLOADABLE SOFTWARE

The book is supported by additional material at the website

http://optimallearning.princeton.edu/

The website will provide additional readings, chapter notes and comments, sample applications (many drawn from the projects generated by students in the course *Optimal Learning* taught at Princeton University), and downloadable software. Most of the software is in the form of MATLAB modules that offer implementations of some of the algorithms. One module, the Optimal Learning Calculator, uses a spreadsheet front-end which talks to a series of Java-based modules which implement some of the algorithms. The spreadsheet offers an interactive environment which makes it possible to experiment with a variety of learning policies. You can solve a learning problem manually, or simulate any of a series of policies on problems. These problems can be randomly generated, or entered manually by hand.

1.8 GOALS OF THIS BOOK

There are a number of communities that address the general problem of collecting information. These communities span economics, computer science, statistics, and operations research (which in turn includes subcommunities such as decision analysis, applied probability, and simulation). One of our goals is to bring these communities together under a common vocabulary and notation.

Given the diversity of problems where learning arises, it is not surprising that a multitude of techniques have evolved to address these problems. Our presentation reviews a number of these techniques, but considerable emphasis is put on a concept called the *knowledge gradient* which guides measurements based on the marginal value of a single measurement. The knowledge gradient is comparable to the gradient in classical optimization, but focuses on the value of information. The power of the knowledge gradient is that it is a simple concept that can be applied to a wide range of problems. In fact, this technique opens up new problems, and it allows us to consider learning in settings where it has previously not been considered. Our empirical work to date suggests that it is a surprisingly robust strategy, in that it is usually competitive with competing techniques, while often outperforming other methods without the need for tunable parameters.

The book is aimed at students and professionals with a basic course in statistics and a full course in probability. The presentation emphasizes concepts and practical tools over heavy mathematical development.

PROBLEMS

1.1 Pick a problem that involves sequential information and decision processes. Ideally, the problem is one of special interest to you. Give a short description of the problem in plain English. Then describe the following components listed below. In each case, a candidate answer is provided using the setting of finding a new compound for storing energy in a battery (this is a fairly complicated example - there are many settings which are much simpler).

a) What decision are you making that determines the information you collect? [Example: Testing a particular molecule.]

b) Precisely what information is being observed as a result of your choice? [Example: The amount of energy stored per pound.]

c) What decision are you going to make with this information? [Example: We will use this information to decide which type of battery is the most economical, which in turn will impact our decision of whether to use batteries, flywheels, or hydroelectric reservoirs as our major form of storage.]

d) What is the economic impact of the decision? [Example: The information will allow the Department of Energy to determine if it should invest in rooftop solar panels, where batteries are needed to smooth out variations in solar energy, or more distant wind turbines, which can be used in conjunction with water reservoirs.]

1.2 For each of the situations below, identify whether it is an online or offline learning problem. Then, identify the measurement decision (what is being measured), and how you evaluate the quality of the measurements (that is, what is the value of the information you are collecting).

a) The adventurer Steve Fossett was lost somewhere in Nevada. The problem is to design a search process that might identify the location of the plane he was flying.

b) You would like to find the best price for a downloadable song by adjusting the price.

c) A bank evaluates loans based on a set of attributes determined from a loan application. Some loans are turned down, others are approved. For the loans that are approved, the bank can observe if the loan is repaid or defaults. The bank can then later use this data to correlate the default rate to the attributes of the loan. Since the bank is not able to observe defaults on loans which are not approved, it occasionally may decide to grant a loan which its policy suggests should be turned down, just to observe the information.

d) A doctor administers drugs to control the blood pressure in a patient. The doctor will adjust both the type of medication and the dosage, observing the effect on the patient's blood pressure.

1.3   In Section 1.6.2, we addressed the problem of whether a bank should request a credit report. Repeat the exercise of finding out whether the bank should purchase the credit report, and determine how much the bank would be willing to pay for the report. As before, assume a loan is for $100,000. If the loan is paid off, the bank makes $10,000. If the loan defaults, assume that the bank loses on average $30,000 (since a portion of the loan would have been repaid). Assume that 85 percent of the loans are repaid. The bank has collected past statistics on credit reports which are expressed as the conditional probabilities in Table 1.5 below. So, 72 percent of the loans that did not default had a credit rating of A.

Table 1.5   Data for exercise 1.3.

Default	P(Credit)	P(Credit rating | Default)		
		A	B	C
No	0.85	0.72	0.24	0.04
Yes	0.15	0.25	0.57	0.18

1.4   In Table 1.3 for the credit risk problem, the probability $P(Credit = C, Default = Yes) = 0.16$ and $P(Credit = C, Default = No) = 0.01$. Change these probabilities to 0.17 and 0.01, respectively, and solve the decision tree in Figure 1.9 again. How did this change in the data change the behavior of the system after reviewing the credit risk report? What is the new value of the information in the credit risk report? Given an intuitive explanation.

1.5   Return to the decision tree in Section 1.6.3 where we are trying to decide which hitter to use. This decision tree has been implemented in the spreadsheet available on the book website at

http://optimallearning.princeton.edu/exercises/BaseballTree3levels.xlsx

Note that the decision tree considers three successive at-bats.

- The decision tree takes into account the information we gain from observing the outcome of the first at-bat (whether it be player A, B, or C). How would your answer have changed if we formulated the decision tree without taking advantage of the information gained from the first at-bat?

- Use the spreadsheet to compute the value of using each of the three batters, while changing the batting statistics from player B from 1 for 3 to 2 for 6 to 3 for 9 to 4 for 12 and finally 5 for 15. Draw a graph giving the value of choosing each of the three hitters (in the first at-bat) for all five sets of batting statistics. Explain intuitively why your choice of who to use for the first at-bat changes.

CHAPTER 2

ADAPTIVE LEARNING

At the heart of any learning problem is a probabilistic statement of what we believe about parameters that characterize the behavior of our system. Most of our attention will focus on uncertainty about the performance of a choice (such as a person or technology), a design (such as policies for running a business) or a policy (such as rules for testing cargo for explosives or when to sell an asset). Our beliefs are influenced by observations that we make.

There are two perspectives that we can take when forming a belief, known as the *frequentist view* and the *Bayesian view*. In the frequentist view, we begin with no knowledge at all about our parameters, and our beliefs are formed entirely by measurements that we make. Since measurements are inherently noisy, we can repeat a series of measurements and obtain different estimates of the parameters. If we repeat the measurements often enough, we can form a frequency distribution of any parameter that we estimate using the measurements.

In the Bayesian view, we start with initial beliefs about parameters, known as the *prior distribution*, which is formed before we make any observations. After a measurement, we combine the prior distribution with a measurement to form a posterior distribution. This then becomes the next prior distribution. This distribution is our *distribution of belief* about the true values of a set of parameters, formed from

Optimal Learning. By Warren B. Powell and Ilya O. Ryzhov
Copyright © 2012 John Wiley & Sons, Inc.

a combination of our initial belief and subsequent observations. By contrast, in the frequentist view, we form a probability distribution of estimates of parameters that reflect the variation in the observations.

Below, we provide a brief illustration of how to form estimates using the frequentist view. In the remainder of the volume, there is a bias toward the Bayesian view of knowledge since this tends to be the most natural for many learning situations. However, both views are important, and the choice of Bayesian or frequentist perspectives is generally determined by the characteristics of an individual problem. We do return to the frequentist view in Chapter 14.

2.1 THE FREQUENTIST VIEW

The frequentist view is arguably the approach that is most familiar to people with an introductory course in statistics. Assume we are trying to estimate the mean μ of a random variable W which might be the performance of a device or policy. Let W^n be the nth sample observation. Also let $\bar{\theta}^n$ be our estimate of μ, and $\hat{\sigma}^{2,n}$ be our estimate of the variance of W. We know from elementary statistics that we can write $\bar{\theta}^n$ and $\hat{\sigma}^{2,n}$ using

$$\bar{\theta}^n = \frac{1}{n}\sum_{m=1}^{n} W^m, \tag{2.1}$$

$$\hat{\sigma}^{2,n} = \frac{1}{n-1}\sum_{m=1}^{n}(W^m - \bar{\theta}^n)^2. \tag{2.2}$$

The estimate $\bar{\theta}^n$ is a random variable (in the frequentist view) because it is computed from other random variables, namely $W^1, W^2, \ldots, W^n$. Imagine if we had 100 people each choose a sample of n observations of W. We would obtain 100 different estimates of $\bar{\theta}^n$, reflecting the variation in our observations of W. The best estimate of the variance of the estimator $\bar{\theta}^n$ is easily found to be

$$\bar{\sigma}^{2,n} = \frac{1}{n}\hat{\sigma}^{2,n}.$$

Note that as $n \to \infty$, $\bar{\sigma}^{2,n} \to 0$, but $\hat{\sigma}^{2,n} \to \sigma^2$, where σ^2 is the true variance of W. If σ^2 is known, there would be no need to compute $\hat{\sigma}^{2,n}$ and $\bar{\sigma}^{2,n}$ would be given as above with $\hat{\sigma}^{2,n} = \sigma^2$.

We can write these expressions recursively using

$$\bar{\theta}^n = \left(1 - \frac{1}{n}\right)\bar{\theta}^{n-1} + \frac{1}{n}W^n, \tag{2.3}$$

$$\hat{\sigma}^{2,n} = \begin{cases} \frac{1}{n}(W^n - \bar{\theta}^{n-1})^2, & n = 2, \\ \frac{n-2}{n-1}\hat{\sigma}^{2,n-1} + \frac{1}{n}(W^n - \bar{\theta}^{n-1})^2, & n > 2. \end{cases} \tag{2.4}$$

We will often speak of our "state of knowledge," which captures what we know about the parameters we are trying to estimate. Given our observations, we would write our state of knowledge as

$$K^n_{freq} = (\bar{\theta}^n, \hat{\sigma}^{2,n}, n).$$

Equations (2.3) and (2.4) describe how our state of knowledge evolves over time.

The state of knowledge (or belief state) is supposed to communicate a probability distribution as opposed to statistics such as mean and variance. When we are forming an average, we can apply the law of large numbers and assume that our estimate $\bar{\theta}^n$ is approximately normally distributed. This is true exactly if W is normally distributed, but it is generally a very good approximation even when W is described by other distributions.

2.2 THE BAYESIAN VIEW

The Bayesian perspective casts a different interpretation on the statistics we compute, which is particularly useful in the context of optimal learning. In the frequentist perspective, we do not start with any knowledge about the system before we have collected any data. It is easy to verify from equations (2.3) and (2.4) that we never use $\bar{\theta}^0$ or $\hat{\sigma}^{2,0}$.

By contrast, in the Bayesian perspective we assume that we begin with a prior distribution of belief about the unknown parameter μ. In other words, any number whose value we do not know is interpreted as a random variable, and the distribution of this random variable represents our belief about how likely μ is to take on certain values. So if μ is the true but unknown mean of W, we might say that while we do not know what this mean is, we think it is normally distributed around θ^0 with standard deviation σ^0. Thus, the true mean μ is treated as a random variable with a known mean and variance, but we are willing to adjust our estimates of the mean and variance as we collect additional information. If we add a distributional assumption such as the normal distribution, we would say that this is our initial distribution of belief, known generally as the Bayesian prior.

The Bayesian perspective is well suited to information collection since it is often the case that we approach these problems with some sort of prior knowledge. For example, when we set a price for a product, we can draw on our past experience and sales figures from previous years. However, even more important is the conceptual framework that there exists a truth that we are trying to discover. Optimal learning can be viewed as the problem of trying to learn μ as efficiently as possible.

We note a subtle change in notation from the frequentist perspective, where $\bar{\theta}^n$ was our statistic giving our estimate of μ. In the Bayesian view, we let θ^n be our estimate of the mean of the random variable μ after we have made n observations. It is important to remember that μ is a random variable whose distribution reflects our prior belief about μ. The parameter θ^0 is not a random variable. This is our initial estimate of the mean of our prior distribution. By contrast, θ^n, for $n \geq 1$, is a random variable for the same reason that $\bar{\theta}^n$ is random in the frequentist view: θ^n is computed from a series of random observations $W^1, W^2, \ldots, W^n$, and therefore the distribution of θ^n reflects the distribution of all of our measurements. However, in the Bayesian perspective we are primarily interested in the mean and variance of μ.

Below we first use some simple expressions from probability to illustrate the effect of collecting information. We then give the Bayesian version of (2.3) and (2.4) for the case of independent beliefs, where observations of one choice do not influence our beliefs about other choices. We follow this discussion by giving the updating equations for correlated beliefs, where an observation of μ_x for alternative x tells us something about $\mu_{x'}$. We round out our presentation by touching on other important types of distributions.

2.2.1 The Updating Equations for Independent Beliefs

We begin by assuming (as we do through most of our presentation) that our random variable W is normally distributed. Let σ_W^2 be the variance of W, which captures the noise in our ability to observe the true value. To simplify the algebra, we define the *precision* of W as

$$\beta^W = \frac{1}{\sigma_W^2}.$$

Precision has an intuitive meaning: Smaller variance means that the observations will be closer to the unknown mean, that is, they will be more precise. Now let θ^n be our estimate of the true mean μ after n observations, and let β^n be the precision of this estimate. That is, having already seen the values $W^1, W^2, ..., W^n$, we believe that the mean of μ is θ^n, and the variance of μ is $\frac{1}{\beta^n}$. We say that we are "at time n" when this happens; note that all quantities that become known at time n are indexed by the superscript n, so the observation W^{n+1} is not known until time $n + 1$. Higher precision means that we allow for less variation in the unknown quantity, that is, we are more sure that μ is equal to θ^n. After observing W^{n+1}, the updated mean and precision of our estimate of μ is given by

$$\theta^{n+1} = \frac{\beta^n \theta^n + \beta^W W^{n+1}}{\beta^n + \beta^W}, \tag{2.5}$$

$$\beta^{n+1} = \beta^n + \beta^W. \tag{2.6}$$

Equation (2.5) can be written more compactly as

$$\theta^{n+1} = (\beta^{n+1})^{-1}(\beta^n \theta^n + \beta^W W^{n+1}). \tag{2.7}$$

There is another way of expressing the updating which provides insight into the structure of the flow of information. First define

$$\tilde{\sigma}^{2,n} = Var^n[\theta^{n+1}] \tag{2.8}$$

$$= Var^n[\theta^{n+1} - \theta^n], \tag{2.9}$$

where $Var^n[\cdot] = Var[\cdot \,|\, W^1, ..., W^n]$ denotes the variance of the argument given the information we have through n observations. For example,

$$Var^n[\theta^n] = 0$$

since, given the information after n observations, θ^n is a number that we can compute deterministically from the prior history of observations.

The parameter $\tilde{\sigma}^{2,n}$ can be described as the variance of θ^{n+1}, given the information we have collected through iteration n, which means the only random variable is W^{n+1}. Equivalently, $\tilde{\sigma}^{2,n}$ can be thought of as the *change* in the variance of θ^n as a result of the observation of W^{n+1}. Equation (2.9) is an equivalent statement since, given the information collected up through iteration n, θ^n is deterministic and is therefore a constant. We use equation (2.9) to offer the interpretation that $\tilde{\sigma}^{2,n}$ is the *change* in the variance of our estimate of the mean of μ.

It is possible to write $\tilde{\sigma}^{2,n}$ in different ways. For example, we can show that

$$\tilde{\sigma}^{2,n} = \sigma^{2,n} - \sigma^{2,n+1} \tag{2.10}$$

$$= \frac{(\sigma^{2,n})}{1 + \sigma_W^2/\sigma^{2,n}} \tag{2.11}$$

$$= (\beta^n)^{-1} - (\beta^n + \beta^W)^{-1}. \tag{2.12}$$

The proof of (2.10) is given in Section 2.5.1. Equations (2.11) and (2.12) come directly from (2.10) and (2.6), using either variances or precisions.

Just as we let $Var^n[\cdot]$ be the variance given what we know after n measurements, let $\mathbb{E}^n$ be the expectation given what we know after n measurements. That is, if $W^1,\ldots,W^n$ are the first n measurements, we can write

$$\mathbb{E}^n\theta^{n+1} \equiv \mathbb{E}(\theta^{n+1}|W^1,\ldots,W^n) = \theta^n.$$

We note in passing that $\mathbb{E}\theta^{n+1}$ refers to the expectation before we have made any measurements, which means $W^1,\ldots,W^n$ are all random, as is W^{n+1}. By contrast, when we compute $\mathbb{E}^n\theta^{n+1}$, $W^1,\ldots,W^n$ are assumed fixed, and only W^{n+1} is random. By the same token, $\mathbb{E}^{n+1}\theta^{n+1} = \theta^{n+1}$, where θ^{n+1} is some number which is fixed because we assume that we already know $W^1,\ldots,W^{n+1}$. It is important to realize that when we write an expectation, we have to be explicit about what we are conditioning on. In practice, conditioning on history occurs implicitly when we collect a sequence of observations. In particular, if we have made n observations and have not yet made the $(n+1)$st observation, then $W^1,\ldots,W^n$ is known and W^{n+1} is random.

Using this property, we can write θ^{n+1} in a different way that brings out the role of $\tilde{\sigma}^n$. Assume that we have made n observations and let

$$Z = \frac{\theta^{n+1} - \theta^n}{\tilde{\sigma}^n}.$$

We note that Z is a random variable only because we have not yet observed W^{n+1}. Normally we would index $Z = Z^{n+1}$ since it is a random variable that depends on W^{n+1}, but we are going to leave this indexing implicit. It is easy to see that $\mathbb{E}^n Z = 0$ and $Var^n Z = 1$. Also, since W^{n+1} is normally distributed, then Z is normally distributed, which means $Z \sim N(0,1)$. This means that we can write

$$\theta^{n+1} = \theta^n + \tilde{\sigma}^n Z. \tag{2.13}$$

Equation (2.13) makes it clear how θ^n evolves over the observations. It also reinforces the idea that $\tilde{\sigma}^n$ is the change in the variance due to a single observation.

Equations (2.5) and (2.6) are the Bayesian counterparts of (2.3) and (2.4), although we have simplified the problem a bit by assuming that the variance of W is known. The state of knowledge in the Bayesian view (with normally distributed beliefs) is given by

$$K^n_{Bayes} = (\theta^n, \beta^n).$$

As we show below in Section 2.5.2, if our prior belief about μ is normally distributed with mean θ^n and precision β^n, and if W is normally distributed, then our posterior belief after $n + 1$ observations is also normally distributed with mean θ^{n+1} and precision β^{n+1}. We often use the term *Gaussian prior*, when we want to say that our prior is normally distributed. We also allow ourselves a slight abuse of notation: we use $\mathcal{N}\left(\mu, \sigma^2\right)$ to mean a normal distribution with mean μ and variance σ^2, but we also use the notation $\mathcal{N}\left(\mu, \beta\right)$ when we want to emphasize the precision instead of the variance.

Needless to say, it is especially convenient if the prior distribution and the posterior distribution are of the same basic type. When this is the case, we say that the prior and posterior are *conjugate*. This happens in a few special cases when the prior distribution and the distribution of W are chosen in a specific way. When this is the case, we say that the prior distribution is a *conjugate family*. The property that the posterior distribution is in the same family as the prior distribution is called *conjugacy*. The normal distribution is unusual in that the conjugate family is the same as the sampling family (the distribution of the measurement W is also normal). For this reason, this class of models is sometimes referred to as the "normal–normal" model (this phraseology becomes clearer below when we discuss other combinations).

In some cases, we may impose conjugacy as an approximation. For example, it might be the case that the prior distribution on μ is normal, but the distribution of the observation W is not normal (for example, it might be nonnegative). In this case, the posterior may not even have a convenient analytical form. But we might feel comfortable approximating the posterior as a normal distribution, in which case we would simply use (2.5)-(2.6) to update the mean and variance and then assume that the posterior distribution is normal.

2.2.2 The Expected Value of Information

The previous section described how we update the mean and variance (or precision) for a particular observation of W. It is useful to see what happens when we look at the *expected* change in the mean and variance when we average over all possible realizations. Let μ be the unknown true value of some quantity. Then, μ is a random variable with (assumed) mean θ^0 and variance $\sigma^{2,0}$. Let W be a random observation of whatever we are measuring, from which we might update our estimate of the mean and variance. Note that

$$\mathbb{E}W = \mathbb{E}\mathbb{E}\left(W \mid \mu\right) = \mathbb{E}\mu = \theta^0.$$

We observe that

$$\mathbb{E}\mu = \theta^0,$$
$$Var(\mu) = \sigma^{2,0}.$$

Assume that we observe $W = w$. For example, we might assume that the travel time on a link has mean $\theta^0 = 22$ minutes, but we observe an actual travel time of $W = w = 27$ minutes. Then, $\mathbb{E}(\mu|W = w)$ would be the updated mean, and $Var(\mu|W = w)$ would be the updated variance. But now let's consider what happens *on average* to our estimates of the mean and variance when we consider all possible outcomes of our observation of W.

Let $\mathbb{E}_\mu \mu$ be the expected value of μ (over our density function for μ), which is equal to θ^0. Now let $\mathbb{E}_\mu(\mu|W)$ be the expected value of μ given a particular observation of W. $\mathbb{E}_\mu(\mu|W)$ is a random variable (since it depends on W), but we are interested in its expectation over all possible values of W, which we can write as $\mathbb{E}_W \mathbb{E}_\mu(\mu|W)$. We can compute this by taking expectations of equation (2.7), given what we know at iteration n (which means that only W^{n+1} is random). We start by observing that

$$\mathbb{E}_\mu(\mu|W) = \theta^1 = (\beta^1)^{-1}(\beta^0\theta^n + \beta^W W),$$

where $\beta^1 = \beta^0 + \beta^W$. We then take expectations of both sides over the random observation W to obtain

$$\begin{aligned}\mathbb{E}_W \theta^1 &= \mathbb{E}_W((\beta^1)^{-1}(\beta^0\theta^0 + \beta^W W)) \\ &= (\beta^1)^{-1}(\beta^0\theta^0 + \beta^W \mathbb{E}W) \\ &= (\beta^1)^{-1}(\beta^0\theta^0 + \beta^W \theta^0) \\ &= (\beta^1)^{-1}(\beta^0 + \beta^W)\theta^0 \\ &= \theta^0.\end{aligned}$$

This result seems to be saying that collecting an observation of W does not change our belief of the true mean μ. This is not the case. As we saw in the previous section, a particular observation of W will, in fact, change our belief about the mean of μ. But if we look at all possible realizations of W, before the observation occurs, *on average* our estimate of the mean does not change.

This simple equation provides an insight into priors and learning. Imagine if

$$\mathbb{E}\mu = \mathbb{E}(\mathbb{E}(\mu|W)) = \theta^0 + a.$$

That is, observing W will, on average, shift our belief about the true mean from θ^0 to $\theta^0 + a$, where a is a constant which would have to be known before we do our measurement. If this were true (for a other than zero), then this would mean that our initial estimate θ^0 is not a true prior. That is, we could shift our prior by a so that it becomes an unbiased estimate of the mean.

Now consider what happens to the variance after a measurement. We use some basic relationships from probability to obtain

$$Var(\mu) = \mathbb{E}(\mu^2) - (\mathbb{E}(\mu))^2 \qquad (2.14)$$
$$= \mathbb{E}(\mu^2) - \mathbb{E}[(\mathbb{E}(\mu|W))^2] + \mathbb{E}[(\mathbb{E}(\mu|W))^2] - (\mathbb{E}(\mu))^2 \qquad (2.15)$$
$$= \mathbb{E}[(\mathbb{E}(\mu^2|W)) - (\mathbb{E}(\mu|W))^2] + \mathbb{E}[(\mathbb{E}(\mu|W))^2] - (\mathbb{E}[\mathbb{E}(\mu|W)])^2 \ (2.16)$$
$$= \mathbb{E}[Var(\mu|W)] + Var[\mathbb{E}(\mu|W)]. \qquad (2.17)$$

Equation (2.14) is the definition of the variance. In (2.15), we add and subtract $\mathbb{E}[(\mathbb{E}(\mu|W))]^2$. In (2.16), we turn two expectations into conditional expectations, setting up the final (and classic) result given by equation (2.17). Our interest is primarily in equation (2.17). Above, we pointed out that $\mathbb{E}(\mathbb{E}(\mu|W)) = \mathbb{E}\mu$. This is not the case with the variance, where equation (2.17) tells us that

$$\mathbb{E}[Var(\mu|W)] = Var(\mu) - Var[\mathbb{E}(\mu|W)]. \qquad (2.18)$$

This means that the variance after a measurement will, on average, always be smaller than the original variance. Of course, it might be the case that $Var[\mathbb{E}(\mu|W)] = 0$. This would happen if W were an irrelevant measurement. For example, assume that μ is our estimate of the travel time on a path in a network, and W is an observation of the change in the S&P stock index yesterday. The S&P stock index does not tell us anything about the travel time on the path, which means that $\mathbb{E}(\mu|W)$ is a constant (in our example, it would be θ^0). Clearly, $Var(\theta^0) = 0$, since θ^0 is not a random variable (it is just a number).

We collect observations one at a time. So, the above discussion continues to apply after we observe $W^1, W^2, ..., W^n$. Since the posterior mean θ^n is a known, fixed quantity at time n, we can simply view it as a new prior. Our problem essentially restarts after each observation, just with different parameters for our distribution of belief. The advantage of recursive updating is that it allows us to turn a problem with a long-time horizon into a sequence of small problems – a concept that will also inform our solution techniques in later chapters.

2.2.3 Updating for Correlated Normal Priors

A particularly important problem class in optimal learning involves problems where there are multiple choices, where our beliefs about the choices are correlated. Some examples of correlated beliefs are as follows:

- We are interested in finding the price of a product that maximizes total revenue. We believe that the function $R(p)$ that relates revenue to price is continuous. Assume that we set a price p^n and observe revenue R^{n+1} that is higher than we had expected. If we raise our estimate of the function $R(p)$ at the price p^n, our beliefs about the revenue at nearby prices should be higher.

- We choose five people for the starting lineup of our basketball team and observe total scoring for one period. We are trying to decide if this group of five people is better than another lineup that includes three from the same group with two different people. If the scoring of these five people is higher than we had expected, we would probably raise our belief about the other group, since there are three people in common.

- A physician is trying to treat diabetes using a treatment of three drugs, where she observes the drop in blood sugar from a course of a particular treatment. If one treatment produces a better-than-expected response, this would also increase our belief of the response from other treatments that have one or two drugs in common.

- We are trying to find the highest concentration of a virus in the population. If the concentration of one group of people is higher than expected, our belief about other groups that are close (either geographically, or due to other relationships) would also be higher.

Correlated beliefs are a particularly powerful device in optimal learning, allowing us to generalize the results of a single observation to other alternatives that we have not directly measured.

Let θ_x^n be our belief about alternative x after n measurements. Now let

$$Cov^n(\mu_x, \mu_y) \quad = \quad \text{the covariance in our belief about } \mu_x \text{ and } \mu_y.$$

We let Σ^n be the covariance matrix, with element $\Sigma_{xy}^n = Cov^n(\mu_x, \mu_y)$. Just as we defined the precision β_x^n to be the reciprocal of the variance, we are going to define the precision matrix B^n to be

$$B^n = (\Sigma^n)^{-1}.$$

Let e_x be a column vector of zeroes with a 1 for element x, and as before we let W^{n+1} be the (scalar) observation when we decide to measure alternative x. We could label W^{n+1} as W_x^{n+1} to make the dependence on the alternative more explicit. For this discussion, we are going to use the notation that we choose to measure x^n and the resulting observation is W^{n+1}. If we choose to measure x^n, we can also interpret the observation as a column vector given by $W^{n+1}e_{x^n}$. Keeping in mind that θ^n is a column vector of our beliefs about the expectation of μ, the Bayesian equation for updating this vector in the presence of correlated beliefs is given by

$$\theta^{n+1} \quad = \quad (B^{n+1})^{-1}(B^n\theta^n + \beta^W W^{n+1}e_{x^n}), \qquad (2.19)$$

where B^{n+1} is given by

$$B^{n+1} \quad = \quad (B^n + \beta^W e_{x^n}(e_{x^n})^T). \qquad (2.20)$$

Note that $e_x(e_x)^T$ is a matrix of zeroes with a one in row x, column x, whereas β^W is a scalar giving the precision of our measurement W.

It is possible to perform these updates without having to deal with the inverse of the covariance matrix. This is done using a result known as the Sherman–Morrison formula. If A is an invertible matrix (such as Σ^n) and u is a column vector (such as e_x), the Sherman–Morrison formula is

$$[A + uu^T]^{-1} = A^{-1} - \frac{A^{-1}uu^T A^{-1}}{1 + u^T A^{-1} u}. \tag{2.21}$$

Let $\lambda^W = \sigma_W^2 = 1/\beta^W$ be the variance of our measurement W^{n+1}. We are going to simplify our notation by assuming that our measurement variance is the same across all alternatives x, but if this is not the case, we can replace λ^W with λ_x^W throughout. Using the Sherman–Morrison formula, and letting $x = x^n$, we can rewrite the updating equations as

$$\theta^{n+1}(x) = \theta^n + \frac{W^{n+1} - \theta_x^n}{\lambda^W + \Sigma_{xx}^n} \Sigma^n e_x, \tag{2.22}$$

$$\Sigma^{n+1}(x) = \Sigma^n - \frac{\Sigma^n e_x (e_x)^T \Sigma^n}{\lambda^W + \Sigma_{xx}^n}. \tag{2.23}$$

where we express the dependence of $\theta^{n+1}(x)$ and $\Sigma^{n+1}(x)$ on the alternative x which we have chosen to measure.

To illustrate, assume that we have three alternatives with mean vector

$$\theta^n = \begin{bmatrix} 20 \\ 16 \\ 22 \end{bmatrix}.$$

Assume that $\lambda^W = 9$ and that our covariance matrix Σ^n is given by

$$\Sigma^n = \begin{bmatrix} 12 & 6 & 3 \\ 6 & 7 & 4 \\ 3 & 4 & 15 \end{bmatrix}.$$

Assume that we choose to measure $x = 3$ and observe $W^{n+1} = W_3^{n+1} = 19$. Applying equation (2.22), we update the means of our beliefs using

$$\begin{aligned} \theta^{n+1}(3) &= \begin{bmatrix} 20 \\ 16 \\ 22 \end{bmatrix} + \frac{19 - 22}{9 + 15} \begin{bmatrix} 12 & 6 & 3 \\ 6 & 7 & 4 \\ 3 & 4 & 15 \end{bmatrix} \begin{bmatrix} 0 \\ 0 \\ 1 \end{bmatrix} \\ &= \begin{bmatrix} 20 \\ 16 \\ 22 \end{bmatrix} + \frac{-3}{24} \begin{bmatrix} 3 \\ 4 \\ 15 \end{bmatrix} \\ &= \begin{bmatrix} 19.625 \\ 15.500 \\ 20.125 \end{bmatrix}. \end{aligned}$$

The update of the covariance matrix is computed using

$$
\Sigma^{n+1}(3) = \begin{bmatrix} 12 & 6 & 3 \\ 6 & 7 & 4 \\ 3 & 4 & 15 \end{bmatrix} - \cfrac{\begin{bmatrix} 12 & 6 & 3 \\ 6 & 7 & 4 \\ 3 & 4 & 15 \end{bmatrix} \begin{bmatrix} 0 \\ 0 \\ 1 \end{bmatrix} [0\ 0\ 1] \begin{bmatrix} 12 & 6 & 3 \\ 6 & 7 & 4 \\ 3 & 4 & 15 \end{bmatrix}}{9 + 15}
$$

$$
= \begin{bmatrix} 12 & 6 & 3 \\ 6 & 7 & 4 \\ 3 & 4 & 15 \end{bmatrix} - \frac{1}{24} \begin{bmatrix} 3 \\ 4 \\ 15 \end{bmatrix} [3\ 4\ 15]
$$

$$
= \begin{bmatrix} 12 & 6 & 3 \\ 6 & 7 & 4 \\ 3 & 4 & 15 \end{bmatrix} - \frac{1}{24} \begin{bmatrix} 9 & 12 & 45 \\ 12 & 16 & 60 \\ 45 & 60 & 225 \end{bmatrix}
$$

$$
= \begin{bmatrix} 12 & 6 & 3 \\ 6 & 7 & 4 \\ 3 & 4 & 15 \end{bmatrix} - \begin{bmatrix} 0.375 & 0.500 & 1.875 \\ 0.500 & 0.667 & 2.500 \\ 1.875 & 2.500 & 9.375 \end{bmatrix}
$$

$$
= \begin{bmatrix} 11.625 & 5.500 & 1.125 \\ 5.500 & 6.333 & 1.500 \\ 1.125 & 1.500 & 5.625 \end{bmatrix}.
$$

These calculations are fairly easy, which means we can execute them even if we have thousands of alternatives. But we will run up against the limits of computer memory if the number of alternatives is in the 10^5 range or more, which arises when we consider problems where an alternative x is itself a multidimensional vector.

2.2.4 Bayesian Updating with an Uninformative Prior

What if we truly have no prior information about a parameter before we start collecting information? We can use what is known in the Bayesian statistics literature as an uninformative prior, which is equivalent to a normal density with infinite variance (or zero precision). We note that it is not necessary for the prior to be a true density (which integrates to 1.0). For example, our prior on a random variable x can be $f(x) = .01$ for all $-\infty < x < \infty$, which of course integrates to infinity. This is known as an *improper prior*.

When we look at the Bayesian updating equations in (2.5) and (2.6), we see that if we use $\beta^0 = 0$, then it simply means that we put no weight on the initial estimates. It is easy to see that if $\beta^0 = 0$, then $\theta^1 = W^1$ (the first observation) and $\beta^1 = \beta^W$ (the precision of our measurement).

The problem with uninformative priors is that we have no guidance at all regarding our first measurement x^0. Fortunately, in most applications of information collection we start with some prior knowledge, but this is not always the case.

Another strategy we can use if we have no prior information is known as *empirical Bayes*. Put simply, empirical Bayes requires that we collect a small initial sample and then use the results of this sample to form a "prior" (obviously, this "prior" is formed *after* we calculate our small sample, but before we do any guided learning).

Empirical Bayes sounds like a different name for frequentist (essentially our prior belief is created using a frequentist procedure), but the interpretation of what is random is different than with a frequentist model. The main distinguishing feature of the Bayesian approach is that it puts a number on the likelihood that the unknown value takes on a certain value or falls within a particular interval.

2.3   UPDATING FOR NON-GAUSSIAN PRIORS

So far, we have considered a setting where the random observations W^n are assumed to come from a normal distribution, whose mean is the true value that we wish to estimate. We have also used a normal distribution to describe our prior belief about the unknown value. These are two very different normal distributions, and it is important to distinguish between them. We use the term "sampling distribution" to refer to the distribution of the observations W^n. This distribution depends on certain unknown parameters, like the value μ in the preceding section. Although we do not know this distribution, we have a way of obtaining samples from it.

By contrast, the term "prior distribution" describes the distribution of our own beliefs about the unknown parameters of the sampling distribution. Unlike the sampling distribution, which is determined by nature and unknown to us, the prior distribution is something that we construct to encode our own uncertainty about the truth. The prior distribution changes as we accumulate observations, reflecting the changes in our beliefs and the reduction in our uncertainty that result from collecting new information.

In the preceding section, both the sampling distribution and the prior distribution are assumed to be normal. This model is particularly intuitive and versatile because the unknown parameter in the sampling distribution is precisely the mean of the sampled observations. The mean θ^0 of the prior distribution can easily be interpreted as our "best guess" as to the unknown value, with σ^0 representing our uncertainty. Thus, whenever we have a rough idea of what the unknown mean might be, a normal distribution provides a very intuitive and understandable way to encode that idea in a mathematical model.

Unfortunately, the normal–normal model (normal prior, normal samples) is not always suitable. For one thing, a normal sampling distribution assumes that our random observations can take on any real value. In reality, this might not be the case: For instance, we might be observing the waiting times of customers in a service center, where we are trying to estimate the service rate. Waiting times are always positive, so an exponential distribution would seem to be a more appropriate choice than a normal distribution. Even more troubling is a situation where our observations are obviously discrete. For instance, we might be observing the success or failure of a medical test, where the outcome is 0 or 1. A normal distribution is certainly a poor choice for a sampling model in this setting.

The normal distribution is not always the best choice for a prior, either. For example, if we are observing exponentially distributed service times, the service rate is necessarily a strictly positive number. If we were to put a normal prior on the service rate, then even with a positive prior mean, we would essentially be allowing the possibility that our exponential sampling distribution has a negative rate, which is impossible. Our uncertainty about the service rate should be encoded using a different kind of distribution that accounts for the fact that the rate must be positive. In the case of 0/1 observations (success or failure of a test), the sample mean is the probability of success. We know that this number must be between 0 and 1, so a normal distribution is again not the best choice to represent our beliefs.

In this section, we discuss several other possible learning models where the sampling and prior distributions are not normal. However, all the distributions that we consider will retain the conjugacy property, which means that the posterior distribution is of the same type as the prior distribution.

2.3.1 The Gamma–Exponential Model

The gamma–exponential model is one possible choice for a situation where the observations are continuous and positive. Suppose that we are trying to estimate the service time distribution at a car repair shop by combining our prior belief with observations of actual service times. We feel comfortable assuming that the sampling distribution governing the service times is exponential with parameter λ. The service rate is the unknown value that we wish to estimate.

Since λ is itself unknown, we view it as a random variable. Clearly it is not appropriate to assume that it follows a normal distribution, since this would mean that we believe that λ might be negative. Assume instead that λ comes from a gamma distribution with parameters a and b. This distribution is given by

$$f(x|a,b) = b(bx)^{a-1} \frac{e^{-bx}}{\Gamma(a)},$$

where a is typically an integer (as it will be for our applications) and $\Gamma(a) = (a-1)!$. Figure 2.1 illustrates several examples of the gamma density. If $a = 1$, the gamma is an exponential distribution. For $a > 1$, it takes on a skewed shape which approaches a normal distribution for larger values of a.

The mean of this distribution is given by

$$\mathbb{E}(\lambda) = \frac{a^0}{b^0}. \tag{2.24}$$

The quantities a^0 and b^0 should be chosen by us in such a way so that (2.24) represents our initial beliefs about the service rate.

We can let a^n/b^n be our estimate of λ after n observations, as before. After observing W^{n+1}, we update our beliefs using the equations

$$a^{n+1} = a^n + 1, \tag{2.25}$$
$$b^{n+1} = b^n + W^{n+1}. \tag{2.26}$$

Our belief about λ is that it follows a gamma distribution with parameters a^{n+1} and b^{n+1}. Despite the complexity of the gamma density, the updating equations governing the way in which we learn about λ are actually quite simple.

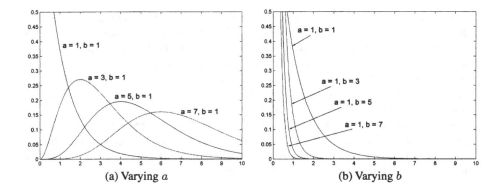

(a) Varying a (b) Varying b

Figure 2.1   Illustration of a family of gamma distributions (a) varying a and (b) varying b.

Equations (2.25) and (2.26) give a simple explanation of the gamma prior that makes the parameters a^n and b^n seem a bit less mysterious. Essentially, b^n is the sum of the first n service times (plus some prior constant b^0), whereas a^n is roughly equal to n (again, plus a prior constant). Thus, after n observations, our estimate of the service rate is given by

$$\mathbb{E}\left(\lambda \mid W^1, ..., W^n\right) = \frac{a^n}{b^n}.$$

This estimate is roughly the number of customers that were served per single unit of time. This is precisely the meaning of a service rate.

While the gamma–exponential model (gamma prior, exponential sampling distribution) is useful for modeling problems with continuous, positive observations, it is incapable of handling correlated beliefs. There is no easy multivariate analog for the gamma distribution, the way there is with the normal distribution, and thus no analog of the correlated normal updating equations (2.22) and (2.23). In a setting where there are multiple unknown values with heavy correlations, it is important to consider the tradeoff between (a) using a multivariate normal model to capture the correlations, and (b) using a different type of model to more accurately represent the individual distributions of the alternatives.

2.3.2 The Gamma–Poisson Model

The gamma–Poisson model is similar to the gamma–exponential model, but the observations are now assumed to be discrete. For example, we may now be interested in the arrival rate of customers to the car repair shop, rather than the service time. Suppose that the total number of customers N that visit the shop in a single day follows a Poisson distribution with rate λ customers per day. Our observations are now the actual numbers of customers that arrive on different days. If the arrival rate is λ, the distribution of N follows the Poisson distribution given by

$$\mathbb{P}[N = x] = \frac{\lambda^x e^{-\lambda}}{x!},$$

where $x = 0, 1, \ldots$. The problem is that we do not know λ, and we wish to estimate it from observations N^n where N^n is the observed number of arrivals on the nth day.

Once again, we assume that λ comes from a gamma distribution with parameters a^0 and b^0. The prior distribution changes after each observation according to the equations

$$a^{n+1} = a^n + N^{n+1}, \tag{2.27}$$
$$b^{n+1} = b^n + 1. \tag{2.28}$$

After n observations, our estimate of the Poisson rate,

$$\mathbb{E}\left(\lambda \mid W^1, \ldots, W^n\right) = \frac{a^n}{b^n},$$

is roughly equal to the average number of customers that arrived per day. This is in line with the meaning of the Poisson rate.

The gamma–Poisson case highlights the distinction between the sampling distribution and the prior. While the individual Poisson observations are discrete, the Poisson rate itself can be any positive real number and thus can be modeled using the gamma distribution.

2.3.3 The Pareto-Uniform Model

Suppose that W is uniform on the interval $[0, B]$, where B is unknown. Our problem is thus to estimate the maximum of a uniform distribution. This problem is the continuous version of a production estimation problem, in which we can observe a sequence of serial numbers, and the goal is to guess the highest serial number produced. We can also use this model to estimate the maximum possible demand for a product or other extreme values.

We assume that B comes from a Pareto distribution with parameters $b > 0$ and $\alpha > 1$. The density of this distribution is given by

$$f(x|\alpha, b) = \begin{cases} \frac{\alpha b^\alpha}{x^{\alpha+1}} & \text{if } x > b, \\ 0 & \text{otherwise.} \end{cases}$$

Thus, our prior estimate of B using priors $\alpha = \alpha^0$ and $b = b^0$ is given by

$$\mathbb{E}(B) = \frac{\alpha^0 b^0}{\alpha^0 - 1}.$$

The parameter b^0 estimates the $\frac{\alpha^0-1}{\alpha^0}$-quantile of the uniform distribution, and α^0 gives us the multiplier used to obtain the estimate of the maximum.

Although this model looks somewhat peculiar, it also has the conjugacy property. Our beliefs continue to have a Pareto distribution as we make observations, and the parameters of the distribution evolve according to the equations

$$b^{n+1} = \max(b^n, W^{n+1}), \qquad (2.29)$$
$$\alpha^{n+1} = \alpha^n + 1. \qquad (2.30)$$

Thus, b^n is roughly the maximum of the first n uniform observations. Our beliefs tell us that the true maximum of the distribution must be larger than b^n. However, if we have made many observations, it is likely that b^n is fairly close to the maximum. The degree of this "closeness" is represented by α^n.

2.3.4   Models for Learning Probabilities*

In many problems, our objective is to learn the probability that a certain event will occur, rather than the economic value of the event. For example, in a medical setting, our observations might simply be whether or not a certain medical treatment is successful. Such an observation can be modeled as a Bernoulli random variable, which is equal to 1 (success) with probability ρ, and 0 (failure) with probability $1 - \rho$. The success probability ρ is the unknown true value in this case.

We assume that ρ comes from a beta distribution with parameters α and β. Recall that the beta density is given by

$$f(x|\alpha, \beta) = \begin{cases} \frac{\Gamma(\alpha+\beta)}{\Gamma(\alpha)\Gamma(\beta)} x^{\alpha-1} (1-x)^{\beta-1} & \text{if } 0 < x < 1, \\ 0 & \text{otherwise.} \end{cases}$$

As before, $\Gamma(y) = y!$ when y is integer. In this setting, α and β are always integer. Figure 2.2 illustrates the beta distribution for different values of α and β.

Our prior estimate of ρ using $\alpha = \alpha^0$ and $\beta = \beta^0$ is given by

$$\mathbb{E}(\rho) = \frac{\alpha^0}{\alpha^0 + \beta^0}. \qquad (2.31)$$

Thus, α^0 and β^0 are weights that, when normalized, give us the probabilities of success and failure, respectively. If α^0 is large relative to β^0, this means that we believe success to be more likely than failure.

A common trait of all the learning models we have discussed thus far is that, while the prior or sampling distributions can have fairly complicated densities, the resulting updating equations are simple and often have an intuitive meaning. This is also the case for the beta–Bernoulli model. The conjugacy property holds, and the parameters evolve according to the equations

$$\alpha^{n+1} = \alpha^n + W^{n+1}, \qquad (2.32)$$
$$\beta^{n+1} = \beta^n + (1 - W^{n+1}), \qquad (2.33)$$

where the observations W^n are 1 or 0, indicating a success or a failure. We see that the parameters α^n and β^n roughly keep track of the number of successes and failures

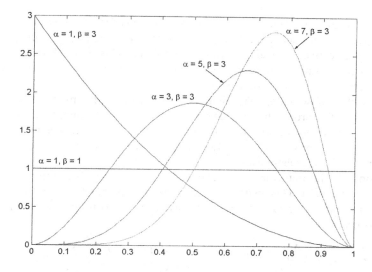

Figure 2.2 Illustration of a family of beta distributions.

in n observations. For instance, the parameter α^n is the number of successes in n observations, plus a prior constant α^0. This prior constant serves as a scaling factor of sorts. To see the importance of this scaling, consider the following three scenarios:

$$\begin{array}{lr} \text{Scenario 1:} & \alpha^0 = \beta^0 = 0.1 \\ \text{Scenario 2:} & \alpha^0 = \beta^0 = 1 \\ \text{Scenario 3:} & \alpha^0 = \beta^0 = 10 \end{array}$$

In each scenario, our estimate of ρ is $1/2$, because the prior constants α^0 and β^0 are equal. However, suppose now that $W^1 = 1$ in all three cases, and consider how our beliefs change:

$$\begin{array}{llr} \text{Scenario 1:} & \alpha^1 = 1.1, \beta^1 = 0.1, & \mathbb{E}\left(\rho \,|\, W^1\right) = 0.916, \\ \text{Scenario 2:} & \alpha^1 = 2, \beta^1 = 1, & \mathbb{E}\left(\rho \,|\, W^1\right) = 0.666, \\ \text{Scenario 3:} & \alpha^0 = 11, \beta^0 = 10, & \mathbb{E}\left(\rho \,|\, W^1\right) = 0.524. \end{array}$$

In the first scenario, observing a single success leads us to greatly increase our estimate of the success probability. In the other two scenarios, we also increase our estimate of ρ, but by a much smaller amount. Thus, the prior values of α^0 and β^0 can be viewed as a measure of our confidence in our estimate of ρ. High values of α^0 and β^0 show that we are very confident in our prior estimate, and hence this estimate is not likely to change by very much. Low values of α^0 and β^0 show that we have little prior knowledge about ρ, and our prior estimate can easily be changed by only a few observations.

The beta–Bernoulli model can be easily generalized to a multivariate setting. Suppose that, instead of a simple 0/1 value, each observation can be classified as belonging to one of K different categories. For example, instead of merely measuring the success or failure of a medical treatment, we also consider cases where the treatment is generally effective, but causes certain side effects. This result should be viewed differently from either total failure or unqualified success. Consequently, we now have more than two possible outcomes.

We model our observations as individual trials from a multinomial distribution with K categories. The probability that an observation belongs to category $k = 1, ..., K$ is $P(W^n = k) = \rho_k$, with each $\rho_k \in [0, 1]$ and $\sum_{k=1}^{K} \rho_k = 1$. The unknown true values are now the probabilities ρ_k. Let us use $\rho = (\rho_1, ..., \rho_K)$ to denote the vector containing all these probabilities.

Our prior is the multivariate generalization of the beta distribution, called the *Dirichlet distribution*. This distribution has a vector of parameters $\alpha = (\alpha_1, ..., \alpha_K)$, with one parameter for each category, satisfying $\alpha_k \geq 0$ for all k. The Dirichlet density is given by

$$f(x) = \begin{cases} \frac{\Gamma(\alpha_1 + ... + \alpha_K)}{\prod_{k=1}^{K} \Gamma(\alpha_k)} \prod_{k=1}^{K} x_k^{\alpha_k - 1} & \text{if } x_k \geq 0 \text{ for all } k \text{ and } \sum_{k=1}^{K} x_k = 1, \\ 0 & \text{otherwise.} \end{cases} \quad (2.34)$$

Essentially, the Dirichlet density is a probability distribution for the probability that an observation belongs in a particular category. The density can only be nonzero on the set of points x (the probabilities) in a k-dimensional space such that every component of x is positive, and the sum of the components is 1 (since the sum of the probabilities of being in each category has to sum to 1). For example, in two dimensions, this set is the part of the line $x_1 + x_2 = 1$ that lies in the non-negative quadrant.

By computing the marginal densities of (2.34), it can be shown that our estimate of the probability of observing an outcome in category k given a prior $\alpha = \alpha^0$ is

$$\mathbb{E}(\rho_k) = \frac{\alpha_k^0}{\alpha_1^0 + ... + \alpha_K^0},$$

a straightforward generalization of (2.31). Just as with the beta–Bernoulli model, the prior values α_k^0 represent the weight that we want to assign to the kth category. Large values of α_k^0 relative to $\alpha_{k'}^0$ indicate that we are more likely to observe category k than category k'. Our earlier discussion of scaling applies here as well.

To update our beliefs, we apply the equation

$$\alpha_k^{n+1} = \begin{cases} \alpha_k^n + 1 & \text{if } W^{n+1} \text{ belongs to category } k, \\ \alpha^n & \text{otherwise.} \end{cases} \quad (2.35)$$

We can write this more concisely if we model W as taking values of the form e_k, where e_k is a K-vector of zeroes, with only the kth component equal to 1. Then, the probability mass function of W is given by $P(W = e_k) = \rho_k$, and (2.35) can be rewritten as

$$\alpha^{n+1} = \alpha^n + W^{n+1}. \quad (2.36)$$

It is important to remember that (2.36) is a vector equation, where only one component of W^{n+1} is equal to 1, and the other components are equal to zero. Simply put, if observation $n + 1$ belongs to category k, we increment α_k^n by 1 and leave the other components of α^n unchanged.

2.3.5 Learning an Unknown Variance*

Our last learning model takes us back to the basic setting of one-dimensional Gaussian priors from Section 2.2.1. As before, we assume that the observation $W \sim \mathcal{N}(\mu, \beta^W)$, where $\beta^W = 1/\sigma_W^2$ is the precision. However, we will now suppose that both the true mean μ and the precision β^W are unknown. We will have to learn both of these quantities at the same time.

It is easy to imagine applications where β^W is unknown. In fact, the precision of an observation is often more difficult to estimate than the mean. For example, in finance, the return of a stock can be directly observed from market data, but the volatility has to be indirectly inferred from the returns. We often assume that β^W is known because it makes our model cleaner, but even then, in practice the value that we plug in for this quantity will be some sort of statistical estimate.

Because the mean and precision are both estimated using the same data, our beliefs about these two quantities are correlated. We create a joint prior distribution on $\left(\mu, \beta^W\right)$ in the following way. First, the marginal distribution of β^W is $Gamma(a, b)$, where $a, b > 0$ are prior parameters of our choosing. Next, given that $\beta^W = r$, the conditional distribution of μ is $\mathcal{N}(\theta, \tau r)$, where $-\infty < \theta < \infty$ and $\tau > 0$ are also prior parameters. Note that τr denotes the conditional precision of μ, not the conditional variance. We can write the joint density of $\left(\mu, \beta^W\right)$ as

$$f\left(x, r \mid \theta, \tau, a, b\right) = \frac{1}{\sqrt{2\pi\tau^{-1}r^{-1}}} \frac{b\left(br\right)^{a-1} e^{-br}}{\Gamma\left(a\right)} e^{-\frac{(x-\theta)^2}{2\tau^{-1}r^{-1}}}.$$

This is widely known as a "normal–gamma" distribution. It is closely related to Student's t-distribution (often simply called the t-distribution), used in statistics to estimate the mean of a sample under unknown variance. In fact, if $\left(\mu, \beta^W\right)$ is normal–gamma with parameters θ, τ, a and b, the marginal distribution of μ can be connected back to the t distribution. The random variable $\sqrt{\frac{\tau a}{b}}\left(\mu - \theta\right)$ follows the standard t distribution with $2a$ degrees of freedom, analogous to expressing a Gaussian random variable in terms of the standard Gaussian distribution.

The estimates of the unknown quantities that we obtain from the normal–gamma distribution are given by

$$\mathbb{E}\mu = \theta, \qquad \mathbb{E}\beta^W = \frac{a}{b}.$$

The parameter τ only affects the amount of uncertainty in our beliefs, with

$$Var\left(\mu\right) = \frac{b}{\tau(a-1)}, \qquad Var(\beta^W) = \frac{a}{b^2}.$$

Recall that τ affects the precision, so lower τ leads to more uncertainty and higher prior variance.

Like the other distributions considered in this chapter, the normal–gamma prior is conjugate when combined with normal observations. Suppose that $(\theta^n, \tau^n, a^n, b^n)$

represent our beliefs after n observations, and we make an observation $W^{n+1} \sim \mathcal{N}\left(\mu, \beta^W\right)$. Then, the posterior distribution of $\left(\mu, \beta^W\right)$ is normal–gamma with parameters

$$\theta^{n+1} = \frac{\tau^n \theta^n + W^{n+1}}{\tau^n + 1}, \tag{2.37}$$

$$\tau^{n+1} = \tau^n + 1, \tag{2.38}$$

$$a^{n+1} = a^n + \frac{1}{2}, \tag{2.39}$$

$$b^{n+1} = b^n + \frac{\tau^n \left(W^{n+1} - \theta^n\right)^2}{2\left(\tau^n + 1\right)}. \tag{2.40}$$

The equations are more complicated than their analogs in Section 2.2.1. However, (2.37) is actually a straightforward generalization of (2.5), replacing the precisions β^n and β^W from the known-variance model by scale factors, τ^n for the prior precision and 1 for the observation. In this way, we can see that the parameter τ^n is roughly equal to n, plus a prior constant.

Later on, we will use the normal–gamma model in a setting where, instead of collecting one observation at a time, we can obtain a batch of k observations simultaneously. In this case, we can easily update our beliefs by calculating (2.37)–(2.40) k times for the individual observations $W^{n+1}, ..., W^{n+k}$. It is instructive, however, to look at the equivalent "batch version" of the updating equations. Let $\bar{W}^{n,k} = \frac{1}{k}\sum_{i=1}^{k} W^{n+i}$ be the average of our k observations. Then, the posterior distribution of $\left(\mu, \beta^W\right)$ is normal–gamma with parameters

$$\theta^{n+k} = \frac{\tau^n \theta^n + k\bar{W}^{n,k}}{\tau^n + k}, \tag{2.41}$$

$$\tau^{n+k} = \tau^n + k, \tag{2.42}$$

$$a^{n+k} = a^n + \frac{k}{2}, \tag{2.43}$$

$$b^{n+k} = b^n + \frac{1}{2}\sum_{i=1}^{k}\left(W^{n+i} - \bar{W}^{n,k}\right)^2 + \frac{\tau^n k \left(\bar{W}^{n,k} - \theta^n\right)^2}{2\left(\tau^n + k\right)}. \tag{2.44}$$

The differences between (2.37)–(2.40) and (2.41)–(2.44) are mostly straightforward. For example, in (2.41), the scale factor of k observations is simply k times the scale factor of a single observation. Notice, however, that (2.44) now involves a sum of squared errors $\left(W^{n+i} - \bar{W}^{n,k}\right)^2$ for $i = 1, ..., k$.

This sum of squares will be automatically computed if we apply (2.40) k times in a row, much like the recursive expression in (2.4) computes the sum of squares in (2.2). In this way, we see that the frequentist and Bayesian models are estimating the variance in roughly the same way. The Bayesian model simply adds a correction to the frequentist estimate in the form of the last term in (2.40), which uses the prior information θ^n and τ^n. Larger values of τ^n correspond to a more precise prior and will place more weight on this term.

The Bayesian model allows us to learn the unknown variance from a single observation, whereas the frequentist model requires at least two. Essentially, the prior stands in for the missing "first" observation. The difference is mostly cosmetic: In practice, the prior is frequently constructed using old or preliminary observations. The true difference between the frequentist and Bayesian philosophies is not in the updating equations, but in the philosophical interpretation of μ as an "unknown, but fixed, number" (frequentist) or a "random variable" (Bayesian).

2.4 MONTE CARLO SIMULATION

There are many settings where we need to take a sample realization of a random variable, a process known widely as Monte Carlo simulation. There are many good books on this subject. This section provides only a brief introduction to some elementary methods for generating samples of a random variable.

All computer languages include a utility for generating a random number that is uniformly distributed between 0 and 1. For example, in Microsoft Excel, this function is called "RAND()", and we can generate a random number from this function by entering "=RAND()" in a cell.

We can use this simple function to generate random variables with virtually any distribution. Let U be a random variable that is uniformly distributed between 0 and 1. If we want a random variable that is uniformly distributed between a and b, we first compute U and then calculate

$$X = a + (b - a)U.$$

It is fairly easy to generate random variables with general distributions if we can compute the inverse cumulative distribution function. Let $F(x) = P[X \leq x]$ be the cdf of a random variable X. If we have a way of generating a sample realization of X with the cumulative distribution $F(x)$, then if we let $Y = F(X)$ (where X is a realization of our random variable), then it is a simple probability exercise to show that Y is uniformly distributed between 0 and 1. Now assume that we can find the inverse of the cumulative distribution function, which we represent as $F^{-1}(u)$ for a value $0 \leq u \leq 1$. If U is a random variable that is uniformly distributed between 0 and 1, then if we compute X using

$$X = F_X^{-1}(U),$$

we can show that X has the cumulative distribution $F(x)$.

For example, consider the case of an exponential density function $\lambda e^{-\lambda x}$ with cumulative distribution function $1 - e^{-\lambda x}$. Setting $U = 1 - e^{-\lambda x}$ and solving for x gives

$$X = -\frac{1}{\lambda} \ln(1 - U).$$

Since $1 - U$ is also uniformly distributed between 0 and 1, we can use

$$X = -\frac{1}{\lambda} \ln(U).$$

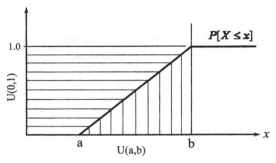

(a) Generating uniform random variables.

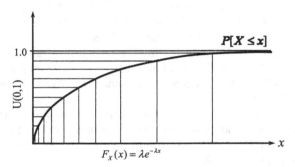

(b) Generating exponentially-distributed random variables.

Figure 2.3   Generating (a) uniformly and (b) exponentially distributed random variables using the inverse cumulative distribution method.

Figure 2.3 demonstrates how we can use a random variable that is uniformly distributed between 0 and 1 to create a random variable that is uniformly distributed between a and b (in Figure 2.3a), and a random variable that has an exponential distribution (in Figure 2.3b).

We can use this method to compute random variables with a normal distribution. Excel and MATLAB, for example, have a function called $\text{NORMINV}(p, \mu, \sigma)$, where p is a probability and μ is the mean of a normally distributed random variable with standard deviation σ. Writing

$$x = \text{NORMINV}(p, \mu, \sigma)$$

in MATLAB (in Excel, you would enter "$= \text{NORMINV}(p, \mu, \sigma)$" in a cell) generates the value of a normally distributed random variable X such that $P(X \leq x) = p$. To generate a random variable that is normally distributed with mean μ and standard deviation σ, simply generate a uniform random variable U and them compute

$$x = \text{NORMINV}(U, \mu, \sigma).$$

Now imagine that we want to generate a Monte Carlo sample for a vector of correlated random variables. Let μ be a M-dimensional vector of means, and let Σ be a $M \times M$ covariance matrix. We would like to find a sample realization $u \sim N(\mu, \Sigma)$. This can be done very simply. Let C be the "square root" of Σ which is computed using Cholesky decomposition. In MATLAB, this is done using

$$C = \mathtt{chol}(\Sigma).$$

The result is an upper triangular matrix C, which is sometimes called the square root of Σ because

$$\Sigma = CC^T.$$

Let Z be an M-dimensional vector of independent, standard normal variables generated as we just described above. Given Z, we can compute a sample realization of μ using

$$u = \mu + CZ.$$

The vector u satisfies $\mathbb{E}u = 0$. To find the variance, we use

$$Cov(u) = Var(u + CZ) = CCov(Z)C^T.$$

Since the elements of the vector Z are independent with variance 1, $Cov(Z) = I$ (the identity matrix), which means $Cov(u) = CC^T = \Sigma$.

To illustrate, assume our vector of means is given by

$$\mu = \begin{bmatrix} 10 \\ 3 \\ 7 \end{bmatrix}.$$

our covariance matrix is given by

$$\Sigma = \begin{bmatrix} 9 & 3.31 & 0.1648 \\ 3.31 & 9 & 3.3109 \\ 0.1648 & 3.3109 & 9 \end{bmatrix}.$$

The Cholesky decomposition computed by MATLAB using $C = \mathtt{chol}(\Sigma)$ is

$$C = \begin{bmatrix} 3 & 1.1033 & 0.0549 \\ 0 & 3 & 1.1651 \\ 0 & 0 & 3 \end{bmatrix}.$$

Imagine that we generate a vector Z of independent standard normal deviates

$$Z = \begin{bmatrix} 1.1 \\ -0.57 \\ 0.98 \end{bmatrix}.$$

Using this set of sample realizations of Z, a sample realization u would be

$$u = \begin{bmatrix} 10.7249 \\ 2.4318 \\ 7.9400 \end{bmatrix}.$$

Using computers to generate random numbers has proven to be an exceptionally powerful tool in the analysis of stochastic systems. Not surprisingly, then, the field has matured into a rich and deep area of study. This presentation is little more than a hint at the many tools available to help with the process of generating random numbers.

2.5   WHY DOES IT WORK?*

2.5.1   Derivation of $\tilde{\sigma}$

An important quantity in optimal learning is the variance $\tilde{\sigma}_x^n$ of θ_x^{n+1} given what we know after n measurements.

Proposition 2.5.1 *The variance of θ^{n+1}, defined as*

$$\begin{aligned} \tilde{\sigma}^n &= Var^n[\theta^{n+1}] \\ &= Var^n[\theta^{n+1} - \theta^n], \end{aligned}$$

is given by $(\tilde{\sigma}_x^n)^2 = (\sigma_x^n)^2 - (\sigma_x^{n+1})^2$.

Proof: Keep in mind that after n measurements, θ_x^n is deterministic. We are dealing with two random variables: (a) the truth μ_x and (b) the estimate θ_x^{n+1} after we have made n measurements (the $(n+1)$st measurement, W^{n+1}, is unknown). We begin with the relation

$$(\theta_x^{n+1} - \mu_x) = (\theta_x^{n+1} - \theta_x^n) + (\theta_x^n - \mu_x). \tag{2.45}$$

Recall that $(\sigma_x^{n+1})^2 = \mathbb{E}^{n+1}\left[(\theta_x^{n+1} - \mu_x)^2\right]$. Squaring both sides of (2.45) and taking the conditional expectation $\mathbb{E}^{n+1}(\cdot) = \mathbb{E}(\cdot|W^1, \ldots, W^{n+1})$ gives

$$\begin{aligned} (\sigma_x^{n+1})^2 &= \mathbb{E}^{n+1}\left[(\theta_x^n - \mu_x)^2\right] + 2\mathbb{E}^{n+1}\left[(\theta_x^n - \mu_x)(\theta_x^{n+1} - \theta_x^n)\right] \\ &\quad + \mathbb{E}^{n+1}\left[(\theta_x^{n+1} - \theta_x^n)^2\right] \\ &= \mathbb{E}^{n+1}\left[(\theta_x^n - \mu_x)^2\right] + 2(\theta_x^n - \theta_x^{n+1})(\theta_x^{n+1} - \theta_x^n) + (\theta_x^{n+1} - \theta_x^n)^2 \\ &= \mathbb{E}^{n+1}\left[(\theta_x^n - \mu_x)^2\right] - (\theta_x^{n+1} - \theta_x^n)^2. \end{aligned}$$

Keep in mind that $\mathbb{E}^{n+1}\theta_x^{n+1} = \theta_x^{n+1}$ and $\mathbb{E}^{n+1}\mu_x = \theta_x^{n+1}$. We then observe that while θ_x^{n+1} is random given the first n observations, σ_x^{n+1} is deterministic, because σ_x^{n+1} does not depend on W^{n+1} – it only depends on our decision of what to measure x^n. Using this property, we can take the expectation given $W^1, \ldots, W^n$ to obtain

$$\begin{aligned} \mathbb{E}^n(\sigma_x^{n+1})^2 = (\sigma_x^{n+1})^2 &= \mathbb{E}^n\left[\mathbb{E}^{n+1}\left[(\theta_x^n - \mu_x)^2\right]\right] - \mathbb{E}^n\left[(\theta_x^{n+1} - \theta_x^n)^2\right] \\ &= \mathbb{E}^n\left[(\theta_x^n - \mu_x)^2\right] - \mathbb{E}^n\left[(\theta_x^{n+1} - \theta_x^n)^2\right] \\ &= (\sigma_x^n)^2 - (\tilde{\sigma}_x^n)^2. \end{aligned}$$

□

2.5.2 Derivation of Bayesian Updating Equations for Independent Beliefs

Bayesian analysis begins with a simple formula that everyone learns in their first probability course. Given the events A and B, the basic properties of conditional probability imply

$$P(A, B) = P(A|B)P(B) = P(B|A)P(A),$$

which implies

$$P(B|A) = \frac{P(A|B)P(B)}{P(A)}.$$

This expression is famously known as Bayes' theorem. In a learning setting, the event A refers to a measurement (or some type of new information), while B refers to the event that a parameter (say, the mean of a distribution) takes on a particular value. $P(B)$ refers to our initial (or prior) distribution of belief about the unknown parameter before we make a measurement, and $P(B|A)$ is the distribution of belief about the parameter after the measurement. For this reason, $P(B|A)$ is known as the *posterior distribution*.

We can apply the same idea for continuous variables. We replace B with the event that $\mu = u$ (to be more precise, we replace B with the event that $u \leq \mu \leq u + du$), and A with the event that we observed $W = w$. Let $g(u)$ be our prior distribution of belief about the mean μ, and let $g(u|w)$ be the posterior distribution of belief about μ given that we observed $W = w$. We then let $f(w|u)$ be the distribution of the random variable W if $\mu = u$. We can now write our posterior $g(u|w)$, which is the density of μ given that we observe $W = w$, as

$$g(u|w) = \frac{f(w|u)g(u)}{f(w)},$$

where $f(w)$ is the unconditional density of the random variable W, which we compute using

$$f(w) = \int_u f(w|u)g(u).$$

Equation (2.46) gives us the density of μ given that we have observed $W = w$.

We illustrate these calculations by assuming that our prior $g(u)$ follows the normal distribution with mean θ^0 and variance $\sigma^{2,0}$, given by

$$g(u) = \frac{1}{\sqrt{2\pi}\sigma^0} \exp\left(-\frac{1}{2}\frac{(u - \theta^0)^2}{\sigma^{2,0}}\right).$$

We further assume that the observation W is also normally distributed with mean μ and variance σ_ϵ^2, which is sometimes referred to as the measurement or observation error. The conditional distribution $f(w|u)$ is

$$f(w|u) = \frac{1}{\sqrt{2\pi}\sigma} \exp\left(-\frac{1}{2}\frac{(w-u)^2}{\sigma_\epsilon^2}\right).$$

We can compute $f(w)$ from $f(w|u)$ and $g(u)$, but it is only really necessary to find the density $g(u|w)$ up to a normalization constant ($f(w)$ is part of this normalization constant). For this reason, we can write

$$g(u|w) \propto f(w|u)g(u). \tag{2.46}$$

Using this reasoning, we can drop coefficients such as $\frac{1}{\sqrt{2\pi}\sigma^0}$ and write

$$
\begin{aligned}
g(u|w) &\propto \left[\exp\left(-\frac{1}{2}\frac{(w-u)^2}{\sigma^2}\right)\right] \cdot \left[\exp\left(-\frac{1}{2}\frac{(u-\theta^0)^2}{\sigma^{2,0}}\right)\right], \\
&\propto \exp\left[-\frac{1}{2}\left(\frac{(w-u)^2}{\sigma^2} + \frac{(u-\theta^0)^2}{\sigma^{2,0}}\right)\right].
\end{aligned}
\tag{2.47}
$$

After some algebra, we find that

$$g(u|w) \propto \exp\left(-\frac{1}{2}\beta^1(u-\theta^1)^2\right), \tag{2.48}$$

where

$$\theta^1 = \frac{(\beta^W w + \beta^0\theta^0)}{\alpha + \beta^0}, \tag{2.49}$$

$$\beta^1 = \beta^W + \beta^0. \tag{2.50}$$

The next step is to find the normalization constant (call it K), which we do by solving

$$K \int_u g(u|w)\,du = 1.$$

We could find the normalization constant by solving the integral and picking K so that $g(u|w)$ integrates to 1, but there is an easier way. What we are going to do is look around for a known probability density function with the same structure as (2.48) and then simply use its normalization constant. It is fairly easy to see that (2.48) corresponds to a normal distribution, which means that the normalization constant is $K = \sqrt{\frac{\beta^1}{2\pi}}$. This means that our posterior density is given by

$$g(u|w) = \sqrt{\frac{\beta^1}{2\pi}} \exp\left(-\frac{1}{2}\beta^1(u-\theta^1)^2\right). \tag{2.51}$$

From equation (2.51), we see that the posterior density $g(u|w)$ is also normally distributed with mean θ^1 given by (2.49), and precision β^1 given by (2.50) (it is only now that we see that our choice of notation θ^1 and β^1 in equations (2.49) and (2.50) was not an accident). This means that as long as we are willing to stay with our assumption of normality, then we need only to carry the mean and variance (or precision). The implication is that we can write our knowledge state as $K^n = (\theta^n, \beta^n)$ (or $K^n = (\theta^n, \sigma^{2,n})$) and that (2.49)–(2.50) is our knowledge transition function.

Our derivation above was conducted in the context of the normal distribution, but we followed certain steps that can be applied to other distributions. These include:

1) We have to be given the prior $g(u)$ and the conditional measurement density $f(w|u)$.

2) We use equation (2.46) to find the posterior up to the constant of proportionality, as we did in (2.47) for the normal distribution.

3) We then manipulate the result in the hope of finding a posterior distribution, recognizing that we can discard terms that do not depend on u (these are absorbed into the normalization constant). If we are lucky, we will find that we have a conjugate family, and that we end up with the same class of distribution we started with for the prior. Otherwise, we are looking for a familiar distribution so that we do not have to compute the normalization constant ourselves.

4) We identify the transition equations that relate the parameters of the posterior to the parameters of the prior and the measurement distribution, as we did with equations (2.49) and (2.50).

2.6 BIBLIOGRAPHIC NOTES

Sections 2.1-2.2 - There are numerous books on both frequentist and Bayesian statistics. An excellent reference for frequentist statistics is Hastie et al. (2009), which is available as of this writing as a free download in PDF form. An excellent reference for Bayesian statistics is Gelman et al. (2004). Another classical reference, with a focus on using Bayesian statistics to solve decision problems, is DeGroot (1970).

Section 2.3 - See Gelman et al. (2004) for a thorough treatment of Bayesian models.

Section 2.4 - There are a number of outstanding references on Monte Carlo simulation, including Banks et al. (1996), Roberts & Casella (2004), Glasserman (2004) and Rubinstein & Kroese (2008), to name a few.

PROBLEMS

2.1 In a spreadsheet, implement both the batch and recursive formulas for the frequentist estimates mean and variance of a set of random numbers (just use RAND() to produce random numbers between 0 and 1). Use a sequence of 20 random numbers. Note that Excel has functions to produce the batch estimates (AVERAGE and VAR) of the mean and variance. Compare your results to Bayesian estimates of the mean and variance, assuming that your prior is a mean of .5 and a variance of .2 (the prior estimate of the mean is correct, while the variance is incorrect).

2.2   Download the spreadsheet from the book website:

http://optimallearning.princeton.edu/exercises/FiveAlternative.xls

On day 0, the spreadsheet shows your initial estimate of the travel time on each path. In the column painted yellow, enter the path that you wish to follow the *next* day. You will see an observed time, and in the column to the right, we record the time you experience the next day when you follow your chosen path. To the right there is a work area where you can code your own calculations. Assume all random variables are normally distributed.

a) In the work area, use the Bayesian updating formulas to compute updated estimates of the mean and variance. Assume that the standard deviation of the observed travel time for any path is 5 minutes.

b) Using your estimates from part a, simulate a policy where you always choose the path that you thought was fastest. Record your total travel time, along with the path that you thought was best.

c) You should find that you are getting stuck on one path and that you do not "discover" the best path (you can quickly find that this is path 1). Suggest a policy that could be applied to any dataset (there cannot be any hint that you are using your knowledge of the best path). Report your total travel time and the final path you choose.

2.3   You are trying to determine the distribution of how much people weigh in a population. Your prior distribution of belief about the mean μ is that it is normally distributed with mean 180 and standard deviation 40. You then observe the weights of n students drawn from this population. The average weight in the sample is $\bar{y} = 150$ pounds. Assume that the weights are normally distributed with unknown mean μ and a known standard deviation of 20 pounds.

a) Give your posterior distribution for μ given the sample you have observed. Note that your answer will be a function of n.

b) A new observation is made and has a weight of $\tilde{y}$ pounds. Give the posterior distribution for $\tilde{y}$ (again, your answer will be a function of n).

c) Give a 95 percent posterior confidence intervals for μ and $\tilde{y}$ if $n = 10$.

d) Repeat part c with $n = 100$.

2.4   Show that, for a fixed $k > 1$, equation (2.44) is equivalent to applying (2.40) k times in a row. Use the equivalence of (2.2) and (2.4). Now implement both the recursive and batch formulas in a spreadsheet and verify that they produce the same numbers.

2.5   In this problem, you will derive the Bayesian updating equations (2.25) and (2.26) for the gamma–exponential model. Suppose that W is a continuous random variable that follows an exponential distribution with parameter λ. The parameter λ is also random, reflecting our distribution of belief. We say that λ has a gamma prior by writing $\lambda \sim Gamma(a, b)$, meaning that our distribution of belief is gamma. Each time we observe W, we use this observation to update our belief about the true distribution of W, reflecting our uncertainty about λ.

a) Write down the prior density $f(u)$. What does u refer to?

b) Write down the conditional density $g(w|u)$ of the observation.

c) Write down the unconditional density $g(w)$ of the observation. (Hint: Start with $g(w|u)$ and take an expectation over the prior. Remember that the gamma function has the property that $\Gamma(a) = (a - 1)\Gamma(a - 1)$.)

d) Apply Bayes' rule to get $f(u|w)$, the posterior density after the observation. What distribution does this density correspond to? How does this verify equations (2.25) and (2.26)?

2.6   In this problem, you will see how the updating equations you derived in exercise 2.5 works in practice. Suppose that the customer service time at a certain store is exponentially distributed, and we are using the gamma–exponential model to learn the service rate as we observe the service times of individual customers.

a) Let U be a random variable that is uniformly distributed between 0 and 1. Let $R = -\frac{1}{\lambda}\log U$. Show that R follows an exponential distribution with parameter λ. This gives us a way to create samples from any exponential distribution by transforming samples from a uniform distribution (see section 2.4).

b) In a spreadsheet, use the above method to simulate 10 observations from an exponential distribution with parameter $\lambda = 3$. Now suppose that we do not know that λ has this value, and model our beliefs using a gamma prior with parameters $\alpha^0 = 1$ and $\beta^0 = 0.2$. What is our best initial estimate of λ? In your spreadsheet, record the values of α^n and β^n, as well as the resulting estimate of λ, for each n. Copy the results into a table and hand in a paper copy.

CHAPTER 3

THE ECONOMICS OF INFORMATION

Underlying the general area of optimal learning is the notion of the economics of information. While the cost of information is highly problem–dependent, the benefits of information can often be captured using models that combine the issues of uncertainty in the context of simple decision problems.

In this chapter, we are going to approach information collection as an economist would. We will use simple models to derive insights into the value of information in a decision problem. In the rest of the book, we will focus on deriving algorithms for information collection, but some of the basic concepts underlying these algorithms can already be seen in the simple models discussed in this chapter.

3.1 AN ELEMENTARY INFORMATION PROBLEM

We begin by considering an elementary game where we have to decide whether to first acquire a signal that provides information into the probability of winning a game. The problem is illustrated in the decision tree shown in Figure 3.1. The game has two elementary outcomes. If we win ("W"), we receive a reward R, while if we lose ("L"), we lose -1. Without any new information (an information state we designate by "N" for "None"), the probability of winning the game is p, with an expected

Optimal Learning. By Warren B. Powell and Ilya O. Ryzhov
Copyright © 2012 John Wiley & Sons, Inc.

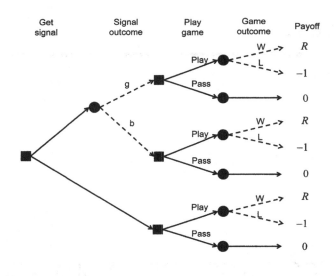

Figure 3.1 Decision tree for an elementary learning game, where we have to choose whether to use a signal that changes our distribution of belief about the probability of winning.

value V of

$$\mathbb{E}[V|N] = \max\{0, pR - (1 - p)\},\tag{3.1}$$

where we assume that we will not play the game at all if its expected value is less than zero.

Before we play the game, we have the option of acquiring an information signal S, which might represent purchasing a report, checking the Internet or making a phone call. The signal may be good ("g") or bad ("b"). We assume that the signal will correctly predict the outcome of the game with probability q. This means that $P[S = g|W] = P[S = b|L] = q$.

We would like to understand the value of purchasing the signal (an elementary information acquisition problem) and the value of the quality of the signal, represented by the probability q. To begin, we need to understand how the signal changes the expected payoff from the game. We can write the conditional value of the game given the signal as

$$\mathbb{E}[V|S = g] = \max\{0, RP[W|S = g] - P[L|S = g]\}.\tag{3.2}$$

Equation (3.2) captures our ability to observe the signal, and then decide whether we want to play the game or not. If the signal is bad, the expected winnings are

$$\mathbb{E}[V|S = b] = \max\{0, RP[W|S = b] - P[L|S = b]\}.\tag{3.3}$$

We now want to find the value of the game given that we have decided to acquire the signal, but before we know its value. This is given by

$$\mathbb{E}[V|S] = \mathbb{E}[V|S = g]P[S = g] + \mathbb{E}[V|S = b]P[S = b].$$

For this calculation, we need to compute the unconditional probabilities $P[S = g]$ and $P[S = b]$. Up to now, we have only been given the conditional probabilities $P[S = g|W] = P[S = b|L] = q$. We can find the unconditional probability of a good signal using

$$
\begin{aligned}
P[S = g] &= P[S = g|W]P[W] + P[S = g|L]P[L] \\
&= qp + (1 - q)(1 - p).
\end{aligned}
\tag{3.4}
$$

Similarly, the unconditional probability of a bad signal is found using

$$
\begin{aligned}
P[S = b] &= P[S = b|W]P[W] + P[S = b|L]P[L] \\
&= (1 - q)p + q(1 - p).
\end{aligned}
\tag{3.5}
$$

We are now ready to find the conditional probability of winning or losing given the outcome of the signal. We use Bayes' theorem to write the probability of winning given a good signal as

$$
\begin{aligned}
P[W|S = g] &= \frac{P[W]P[S = g|W]}{P[S = g]} \\
&= \frac{pq}{qp + (1 - q)(1 - p)}.
\end{aligned}
$$

Similarly, the probability of winning given a bad signal is given by

$$
\begin{aligned}
P[W|S = b] &= \frac{P[W]P[S = b|W]}{P[S = b]} \\
&= \frac{p(1 - q)}{(1 - q)p + q(1 - p)}.
\end{aligned}
$$

Of course, $P[L|S = g] = 1 - P[W|S = g]$.

We are now ready to calculate the value of the signal. We let S represent the decision to acquire the signal before we know the outcome of the signal. Combining (3.2), (3.3), (3.4), and (3.5), we can find the expected value of the game given that

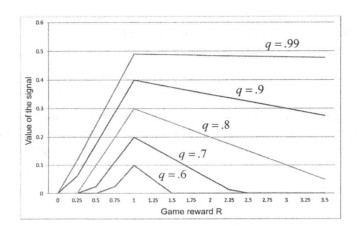

Figure 3.2 The value of the signal as a function of R for different values of q.

we have chosen to acquire the signal is

$$
\begin{aligned}
\mathbb{E}[V|S] &= \mathbb{E}[V|S=g]P[S=g] + \mathbb{E}[V|S=b]P[S=b] \\
&= \max\{0, RP[W|S=g] - P[L|S=g]\}(qp + (1-q)(1-p)) \\
&\quad + \max\{0, RP[W|S=b] - P[L|S=b]\}((1-q)p + q(1-p)) \\
&= \max\left\{0, R\left(\frac{pq}{qp + (1-q)(1-p)}\right)\right\}(qp + (1-q)(1-p)) \\
&\quad + \max\left\{0, R\frac{p(1-q)}{(1-q)p + q(1-p)} - \left(1 - \frac{p(1-q)}{(1-q)p + q(1-p)}\right)\right\} \\
&\quad \times \big((1-q)p + q(1-p)\big) \\
&= \max\left\{0, R\left(\frac{pq}{qp + (1-q)(1-p)}\right)\right\}(qp + (1-q)(1-p)) \\
&\quad + \max\left\{0, R\frac{p(1-q)}{(1-q)p + q(1-p)} - \left(\frac{q(1-p)}{(1-q)p + q(1-p)}\right)\right\} \\
&\quad \times \big((1-q)p + q(1-p)\big).
\end{aligned}
$$

The value of the signal, which depends on the game reward R, the probability of winning p, and the quality of the signal q is given by

$$
V^s(R, p, q) = \mathbb{E}[V|S] - \mathbb{E}[V|N].
$$

The behavior of $V^s(R, p, q)$ is illustrated in Figure 3.2, which gives the value of the signal as a function of the reward R, for different values of the signal quality q, with $p = .5$. Keep in mind that if we lose the game, we lose \$1. The quality of the signal is highest when $R = 1$ (a byproduct of the choice of p, as we see below). When $R = 1$ and $p = .5$, we are most ambivalent about playing the game, and as a result

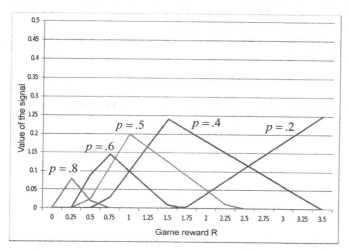

Figure 3.3 The value of the signal as a function of R for different values of q.

the additional information provided by the signal is most important. For sufficiently small values of R, we are less inclined to play the game regardless of the information provided by the signal, just as we are more inclined to play the game for larger values of R. Thus, we see that the value of the signal is greatest when it has the greatest potential for changing a decision.

Figure 3.3 shows how the value of the signal changes as we vary p while holding q constant at .7. As p increases, the point at which the value of the signal is highest moves down and to the left, reflecting both the reduced value of additional information as the probability of winning the game increases (which lowers the maximum height of the curve) and the lowering of the break even point (which moves the maximum to the left).

3.2 THE MARGINAL VALUE OF INFORMATION

Our elementary game in Section 3.1 allowed us to understand the value of a discrete piece of information in a stylized setting. We now turn to a variant of this problem where we are allowed to take multiple measurements to increase the precision of the information we are gaining.

Imagine that we have a choice between doing nothing (and receiving a reward of 0) and choosing a random reward with unknown mean μ. Assume that our prior belief about μ is normally distributed with mean θ^0 and precision $\beta^0 = 1/\sigma^{2,0}$. Before playing the game, we are allowed to collect information in the form of a series of measurements $W^1, \ldots, W^n$, presumably at a cost which we are going to ignore in order to focus on the value of information. We assume that W has the unknown mean

μ (of course, because it is coming from the true distribution) and a precision β^W, which, for convenience, we assume is known.

If we choose to make n measurements, the precision of our estimate of the reward would be given by

$$\beta^n = \beta^0 + n\beta^W.$$

The updated estimate of our reward (using our Bayesian model) would be

$$\theta^n = \frac{\beta^0\theta^0 + n\beta^W \bar{W}^n}{\beta^0 + n\beta^W},$$

where

$$\bar{W}^n = \frac{1}{n}\sum_{k=1}^{n} W^k.$$

We are going to create a random variable that captures our belief about the reward, which we are then going to use to make a decision about whether or not to play the game. We start by using the well-known identity from probability given by

$$Var(\mu) = \mathbb{E}[Var(\mu\,|\,W^1, ..., W^n)] + Var[\mathbb{E}(\mu\,|\,W^1, ..., W^n)],$$

where $Var(\mu\,|\,W^1, ..., W^n) = 1/\beta^n = (\beta^0 + n\beta^W)^{-1}$ and $\mathbb{E}\left(\mu\,|\,W^1, ..., W^n\right) = \theta^n$. Let $\tilde{\sigma}^2(n)$ be the variance of θ^n given what we knew before we took our n measurements (keep in mind that θ^0 is deterministic). Equivalently, $\tilde{\sigma}^2(n)$ is the *change* in the variance due to the n measurements. We can calculate

$$
\begin{aligned}
\tilde{\sigma}^2(n) &= Var(\theta^n) \\
&= Var(\mu) - \mathbb{E}[1/\beta^n] \\
&= 1/\beta^0 - 1/\beta^n \\
&= 1/\beta^0 - 1/(\beta^0 + n\beta^W).
\end{aligned}
$$

Let Z be a standard normal random variable with mean 0 and variance 1. We can write our random variable θ^n as

$$\theta^n = \theta^0 + \tilde{\sigma}^2(n)Z.$$

We can quickly verify that $\mathbb{E}\theta^n = \theta^0$, and $Var[\theta^n] = \tilde{\sigma}^2(n)$.

After our n measurements, we are going to choose to play the game if we believe that its value is greater than zero, otherwise we will pass. Thus, the value of the game V^n after n measurements is

$$V^n = \mathbb{E}\max\{0, \theta^n\}.$$

This expectation can be calculated simply using

$$V^n = \tilde{\sigma}_x(n_x)f\left(\frac{\theta^n}{\tilde{\sigma}(n)}\right),$$

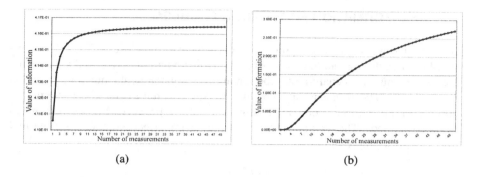

Figure 3.4   Illustrations of when the value of information is (a) concave and (b) nonconcave.

where $f(\zeta)$ is given by

$$f(\zeta) = \zeta\Phi(\zeta) + \phi(\zeta),$$

and where $\Phi(\zeta)$ and $\phi(\zeta)$ are, respectively, the cumulative standard normal distribution and the standard normal density. That is,

$$\phi(\zeta) = \frac{1}{\sqrt{2\pi}}e^{-\zeta^2},$$

and

$$\Phi(\zeta) = \int_{-\infty}^{\zeta} \phi(x)\,dx.$$

This simple model allows us to learn about the marginal value of information. It seems reasonable to expect that the value of information would be concave, as illustrated in Figure 3.4(a). This curve was generated assuming that $\beta^W = 4$ (which means the standard deviation of a measurement σ_W is .5) and $\sigma^0 = 5$. Note that the standard deviation of a measurement is smaller than the standard deviation of our belief, so each measurement significantly reduces our uncertainty about the value of the game.

It turns out, however, that the value of information may be nonconcave. Figure 3.4(b) demonstrates an example of this, which was created assuming that $\sigma_W = 20$, or $\beta^W = 1/\sigma_W^2 = .0025$. Now, the noise in a measurement is much larger than the uncertainty in our belief. A single measurement does little to change our belief about the reward. However, the accumulation of knowledge from multiple measurements has the same effect as a single, much more precise measurement. So, as n grows, we eventually move back into a region where the marginal value of information decreases.

We will revisit this discussion in Chapter 5, where the decision problem will become more complicated. We will calculate the value of information for multiple decisions and use this quantity to distinguish between choices. Our discussion here introduces the basic concept for a simple problem with two decisions.

3.3 AN INFORMATION ACQUISITION PROBLEM

Imagine that we are selling a product where we have to choose which price p_t we should charge during month t. Let $p*$ be the unknown optimal price, and assume that the lost revenue from charging a suboptimal price is given by the quadratic function $\beta(p_t - p*)^2$. The company may conduct market research in any given month at a cost $\$c$ per unit sold (the company continues to sell the product while the research is being conducted). The market research provides an imperfect estimate of the optimal price which we represent as

$$\hat{p}_t = p^* + \epsilon_t,$$

where the sampling errors ϵ_t are independent and identically distributed with mean zero and variance σ^2.

Let

$$x_t = \begin{cases} 1 & \text{if the company conducts a market research study during time period } t, \\ 0 & \text{otherwise.} \end{cases}$$

We assume that our market research study does not affect market behavior or our pricing strategy. Our goal is to design a policy for deciding when we should perform market research that minimizes expected costs over a finite horizon $t = 1, 2, \ldots, T$.

Since each market research study gives us an unbiased estimate of the true optimal price, it makes sense for us to set our price to be the average of the prices $\hat{p}_t$ obtained from each of the market research studies. Let

$$n_t = \sum_{t'=1}^{t} x_{t'}$$

be the number of market research studies we have performed up to (and including) time t. Thus

$$\bar{p}_t = \begin{cases} \frac{n_t - 1}{n_t} \bar{p}_{t-1} + \frac{1}{n_t} \hat{p}_t & \text{if } x_t = 1, \\ \bar{p}_{t-1} & \text{otherwise.} \end{cases}$$

$\bar{p}_t$ is an unbiased estimate of p^* with variance

$$\bar{\sigma}_t^2 = \frac{\sigma^2}{n_t},$$

where we are assuming that σ^2 is known.

We have chosen the structure of the problem to make it especially easy to solve. We start by observing that our choice of loss function allows us to write the expected lost revenue as

$$\mathbb{E}\beta(\bar{p}_t - p^*)^2 = Var(p^* \mid \bar{p}_t) = \beta\bar{\sigma}_t^2.$$

Using this nice structure, we can write the optimization problem for deciding the vector $x_1, \ldots, x_T$, which determines when we will perform market research studies, as follows:

$$\min_x F(x) = \mathbb{E} \sum_{t=1}^{T} \left(\beta(\bar{p}_t - p^*)^2 + cx_t \right)$$

$$= \sum_{t=1}^{T} \left(\beta \bar{\sigma}_t^2 + cx_t \right).$$

It seems intuitively obvious that we should perform market research studies for the first τ time periods and then stop (the reader has a chance to prove this in exercise 3.2). This means that $x_t = 1$, $t = 1, 2, \ldots, \tau$ with $x_t = 0$, $t > \tau$, which also means that $n_t = t$ for $t \leq \tau$. Using this structure, we may rewrite $F(x)$ as

$$F(\tau) = \sum_{t=1}^{\tau} \left(\beta \frac{\sigma^2}{t} + c \right) + \sum_{t=\tau+1}^{T} \beta \frac{\sigma^2}{\tau}.$$

An analytically convenient way of solving this problem is to treat time as continuous, which gives us

$$F(\tau) = \int_{t=1}^{\tau} \left(\beta \frac{\sigma^2}{t} + c \right) dt + \int_{t=\tau}^{T} \beta \frac{\sigma^2}{\tau} dt$$

$$= \left(\beta \sigma^2 \ln t + ct \right) \big|_1^\tau + \beta \frac{\sigma^2}{\tau}(t) \Big|_\tau^T$$

$$= \beta \sigma^2 \ln \tau + c(\tau - 1) + \beta \frac{\sigma^2}{\tau}(T - \tau).$$

Differentiating with respect to τ and setting the result equal to zero gives

$$\frac{\partial F(\tau)}{\partial \tau} = \beta \sigma^2 \frac{1}{\tau} + c - \beta \sigma^2 \frac{T}{\tau^2} = 0.$$

Finding the optimal point τ^* to stop collecting information requires solving

$$c\tau^2 + \beta \sigma^2 \tau - \beta \sigma^2 T = 0.$$

Applying the familiar solution to quadratic equations and recognizing that we are interested in a positive solution, we obtain

$$\tau = \frac{-\beta \sigma^2 + \sqrt{(\beta \sigma^2)^2 + 4c\beta \sigma^2 T}}{2c}.$$

This result allows us to understand the how the solution to this problem behaves as a function of the parameters of the problem. For example, we see that the amount of time we should be collecting information increases with σ^2, β, and T and decreases with c, as we would expect. If there is no noise ($\sigma^2 = 0$), then we should not collect any information. Most importantly, it highlights the concept that there is an optimal strategy for collecting information, and that we should collect more information when our level of uncertainty is higher.

3.4 BIBLIOGRAPHIC NOTES

Section 3.1 The material in this section is based on Chapters 4 and 5 in Birchler & Butler (2007), which is a very nice introduction to the economics of information from the perspective of the field of economics.

Section 3.2 The marginal value of information has been studied by a number of authors. Howard (1966) appears to be the first to have presented a problem (in an auction setting) where the value of information is superadditive (two sources of information combined contribute more than each of them individually). Radner & Stiglitz (1984) present the first thorough investigation of the nonconcavity of information, providing general conditions under which the value of information is nonconcave. Chade & Schlee (2002) and Delara & Gilotte (2007) further generalize these results. Weibull et al. (2007) and Frazier & Powell (2010) investigate the implications of nonconcavity in the value of information in terms of allocating resources for collecting information.

Section 3.3 This section is based on material from Chapter 10 in Powell (2007).

PROBLEMS

3.1   Program a spreadsheet to recreate the graph in Figure 3.2, which assumes that the signal precision $q = .7$.

  a) Create the graph for $q = .6$.

  b) Create the graph for $q = .8$.

  c) Compare the two graphs and describe the effect of q on the value of information as a function of the reward. Why is there little value for both high and low values of the reward?

  d) What happens when $q = .5$? Explain.

3.2   Prove our intuitive claim in Section 3.3 that it is optimal to first collect information and then, at some time τ, stop collecting information and use it. This means that if $x_t = 0$ for some time t, then $x_{t+1} = 0$ (and so on). [Hint: This can be proved with an interchange argument. Assume that x^1 is one policy with, for some t, $x_t = 0$ and $x_{t+1} = 1$, and let x^2 be the same policy except that $x_t = 1$ and $x_{t+1} = 0$ (we have simply exchanged the decision to collect information at time $t + 1$ to time t). Show that x^2 is better than x^1. Use this to prove your result.]

CHAPTER 4

RANKING AND SELECTION

Ranking and selection problems arise in many settings. We may have to choose a type of cholesterol drug, the parameters of a chemical process (e.g., temperature, relative mix of certain inputs), or the design of a manufacturing process (choice of equipment, size of buffers, routing of jobs). Testing an alternative might involve running a time-consuming computer simulation, or it might require a physical experiment. We assume that in either case, the measurement involves some noise, which means we may need to repeat the experiment several times. We assume that we have a budget that determines how many times we can perform these experiments, after which we have to take the best design and live with it.

We assume that we have a finite set of alternatives $\mathcal{X} = (1, 2, \ldots, M)$ where M is "not too large." We then assume that we have a budget for evaluating these alternatives, and when our budget is exhausted, we have to choose the alternative that appears to be best. For this reason, this is called offline learning, since the cost of a measurement is distinct from the cost of using the resulting design in production.

The assumption of a "finite" set of alternatives is fairly important, and will prove to be somewhat restrictive. It will generally be the case that we plan on testing all the alternatives at least once, although this is not essential. We may use our prior knowledge about the performance of an alternative to decide that it is so poor that

Optimal Learning. By Warren B. Powell and Ilya O. Ryzhov
Copyright © 2012 John Wiley & Sons, Inc.

we do not even have to try it. But there are many applications where the number of alternatives is much larger than what the measurement budget will allow us to evaluate. In these cases, we will introduce extensions to the basic model.

4.1 THE MODEL

Recall the Bayesian philosophy that any unknown number is a random variable. We start by assuming that we have a prior distribution of belief about the performance μ_x of each alternative $x \in \mathcal{X}$ which is normally distributed with mean θ_x^0 and precision β_x^0. Throughout this book, it is important to recognize that when we talk about the true mean μ (or μ_x), that μ is a random variable. Whenever we solve a problem, there is an underlying truth which is unknown to us, which can be represented as a realization of the random variable μ.

At the nth iteration (starting with $n = 0$), we choose x^n and observe $W_{x^n}^{n+1}$. We then use this new information to update our belief about μ_{x^n} (remember that μ_{x^n} is a random variable). We do not update priors that are not observed, so our updating equations now look like

$$\theta_x^{n+1} = \begin{cases} \frac{\beta_x^n \theta_x^n + \beta_x^W W_x^{n+1}}{\beta_x^n + \beta_x^W} & \text{if } x^n = x, \\ \theta_x^n & \text{otherwise,} \end{cases} \quad (4.1)$$

$$\beta_x^{n+1} = \begin{cases} \beta_x^n + \beta_x^W & \text{if } x^n = x, \\ \beta_x^n & \text{otherwise.} \end{cases} \quad (4.2)$$

We are going to make a series of observations $W_{x^0}^1, W_{x^1}^2, \ldots, W_{x^{N-1}}^N$. These observations depend in part on our decisions of what to measure, and then in part on the random outcome of a measurement. In addition, the probability distribution that describes a measurement is based on the underlying truth, which is also unknown to us. It is useful in our presentation to think of a set Ψ of potential outcomes of μ. The element $\psi \in \Psi$ might be thought of as a state of the world, where $\mu(\psi)$ is the truth when the state of the world is ψ. This is easiest to visualize if you assume that there are finitely many states of the world, but everything we are going to do works fine if the set of all possible true values is continuous (for example, if our distribution of belief is normal).

For now, we assume that we have some rule or decision function $X^\pi(K^n)$ which depends on our state of knowledge $K^n = (\theta_x^n, \beta_x^n)_{x \in \mathcal{X}}$ and returns the alternative to be tested in iteration $n + 1$. We refer to $X^\pi(K)$ as a policy indexed by $\pi \in \Pi$. We will discuss the precise meaning of π and Π later. We refer to the decision function X^π and the policy π interchangeably.

We use the notation ω to denote a sample realization of all the random quantities in our problem. Think of a matrix $W(\omega)$ of numbers, such that the number in the nth row and xth column represents a sample realization $W_x^n(\omega)$ of the measurement W_x^n. Figure 4.1 illustrates this idea with three sample realizations of W_x^n (three different values of ω) for $n = 1, \ldots, 10$ and $x = 1, 2, 3$. It is useful to think of generating

Table 4.1 Three sample realizations of three alternatives over 10 observations.

n	$\omega = \omega_1$			$\omega = \omega_2$			$\omega = \omega_3$		
	W_1^n	W_2^n	W_3^n	W_1^n	W_2^n	W_3^n	W_1^n	W_2^n	W_3^n
1	37.4	64.1	59.2	38.1	66.9	55.7	33.4	66.5	40.1
2	24.3	65.2	56.0	28.0	57.6	59.3	25.4	60.1	59.5
3	30.5	64.5	56.5	38.9	59.3	50.2	22.7	59.5	48.7
4	20.5	63.2	55.9	34.1	57.7	57.1	24.1	59.8	58.8
5	29.1	64.5	44.8	36.3	60.1	56.4	37.3	65.1	57.1
6	36.7	62.1	59.6	37.0	53.6	42.9	20.7	51.8	42.9
7	27.3	53.5	44.8	27.7	60.9	48.9	32.0	53.4	53.3
8	26.8	57.7	48.1	28.8	68.6	54.4	34.3	53.0	48.6
9	40.0	58.6	42.8	39.6	52.8	53.5	31.1	64.0	59.9
10	21.4	51.6	56.4	33.1	54.5	42.7	33.4	56.7	42.7

all these realizations, but then we are going to choose a measurement policy π that determines which of these realizations that we are actually going to see. That is, we only get to see W_x^n if we choose to observe $x = X^\pi(K^n)$.

Since the true values μ_x are also random variables, we can let ψ be a sample realization of the truth. That is, $\mu(\psi)$ is a particular set of truth values $\mu_1(\psi), ..., \mu_M(\psi)$. Our Bayesian model makes the fundamental assumption that $\mu_x \sim \mathcal{N}\left(\theta_x^0, \beta_x^0\right)$, that is, our prior distribution is assumed to be accurate *on average*. Therefore, the sample path ψ is generated from the prior distribution.

With this, we finally can compute a sample realization of the value of a policy using

$$F^\pi(\mu(\psi), W(\omega)) = \max_{x \in \mathcal{X}} \theta_x^N(\psi, \omega, \pi). \tag{4.3}$$

In other words, we take our estimate of θ_x^N for each x after N measurements and then choose the alternative with the highest estimate. The estimates θ_x^N depend on both types of sample realizations, ω and ψ. Our beliefs change over time in a way that depends on the exact numbers $W^1(\omega), ..., W^N(\omega)$. At the same time, the probability distribution of the observations (the likelihood of observing, say, ω_1 instead of ω_3) is determined by the truth $\mu(\psi)$.

Moreover, the measurement policy π determines which observations we actually see: recall again that $W^n = (W_x^n)_{x \in \mathcal{X}}$ is the vector of outcomes for all alternatives, but that we observe only one component of this vector. Thus, the precise distribution of F^π depends on π. We did not observe this distinction in Chapter 2, where we were given an observation. Now, we have the ability to choose what we want to observe.

If we want to evaluate a policy, we need to compute $F^\pi(\mu(\psi), W(\omega))$ for different samples $W(\omega)$ and different truths $\mu(\psi)$, and take an average. Let ω_k be the kth sample of measurements where $k = 1, 2, ..., K$, and let ψ_ℓ be a particular truth

Table 4.2 Three sample realizations of both a truth μ and a set of sample realizations W drawn from this truth.

	$\omega = \omega_1$			$\omega = \omega_2$			$\omega = \omega_3$	
$\mu_1(\omega_1)$	$\mu_2(\omega_1)$	$\mu_3(\omega_1)$	$\mu_1(\omega_2)$	$\mu_2(\omega_2)$	$\mu_3(\omega_2)$	$\mu_1(\omega_3)$	$\mu_2(\omega_3)$	$\mu_3(\omega_3)$
32.1	57.8	55.8	29.5	62.3	60.1	34.4	59.2	59.7
n	W_1^n	W_2^n	W_3^n	W_1^n	W_2^n	W_3^n	W_1^n	W_2^n	W_3^n
1	33.8	57.7	53.3	32.6	62.7	60.5	29.6	55.7	62.6
2	37.0	55.0	59.7	26.0	60.8	55.3	36.8	62.1	59.1
3	36.0	54.9	51.9	34.3	65.0	58.3	38.0	60.3	60.4
4	30.8	53.3	58.6	30.0	57.7	57.8	36.1	59.9	63.2
5	30.9	58.6	58.3	30.7	59.4	61.0	38.6	58.3	63.2
6	29.7	56.3	55.6	26.3	62.4	60.9	33.9	54.6	64.5
7	32.1	57.0	57.9	31.6	57.8	59.9	30.8	63.3	55.3
8	31.1	53.9	54.4	29.1	59.6	59.0	31.6	58.9	60.8
9	35.9	55.5	50.8	31.5	65.9	60.8	36.0	62.3	62.9
10	35.4	60.5	54.7	33.7	62.4	55.9	38.9	54.9	64.2

where $\ell = 1, 2, \ldots, L$ is a set of potential truths. We can compute an average using

$$\bar{F}^\pi = \frac{1}{L} \sum_{\ell=1}^{L} \left(\frac{1}{K} \sum_{k=1}^{K} F^\pi(\mu(\psi_\ell), W(\omega_k)) \right). \qquad (4.4)$$

This hints at how we would like to state our objective in a formal way. Let $\mathbb{E}_W$ be the expectation over all possible measurements. The measurements we observe depend on our measurement policy π, since this determines which elements of the vector W that we actually observe. For this reason, we write the expectation over measurements as $\mathbb{E}_W$. Let $\mathbb{E}_\mu$ be the expectation over all possible truths. We can write our objective as

$$\max_{\pi \in \Pi} \mathbb{E}_\mu \mathbb{E}_W F^\pi(\mu, W), \qquad (4.5)$$

where $F^\pi(\mu, W) = \max_{x \in \mathcal{X}} \theta_x^N$ is the sample estimate of the objective function, computed using the truth $\mu(\psi)$, the measurements $W^1(\omega), \ldots, W^N(\omega)$ and the policy π that determines which measurements we actually observe.

Mathematically, we can lump the uncertainty about the truth and the uncertainty about measurements into a single space of outcomes. Table 4.2 illustrates how we might represent 3 sample realizations. Here, ω represents both a realization of the truth μ and a realization of the measurements W_x^n drawn from this truth. We would calculate an estimate of $\bar{F}^\pi$ using

$$\bar{F}^\pi = \frac{1}{K} \sum_{k=1}^{K} F^\pi(\mu(\omega_k), W(\omega_k)). \tag{4.6}$$

Note that calculating $\bar{F}^\pi$ using (4.6) or (4.4) is conceptually the same, although the actual numbers may be slightly different when we code up the calculation. Using (4.6), we calculate $F^\pi(\mu(\omega_k), W(\omega_k))$ K times, while if we use (4.4), we calculate $F^\pi(\mu(\omega_\ell), W(\omega_k))$ $L \times K$ times.

Now let $\mathbb{E}$ be the expectation over the outcomes $\omega \in \Omega$ where now both μ and W are changing with ω. In this case, we would write our objective function as

$$\max_{\pi \in \Pi} \mathbb{E} F^\pi(\mu, W). \tag{4.7}$$

This is a perfectly legitimate way of writing our objective function. But it is important to realize that the expectation is summing (or integrating) over our uncertainty in our belief about the truth as well as the randomness in our measurements.

The preceding discussion becomes relevant when we simulate the performance of a policy. In (4.3), we first generate L truth values, and then generate K sets of observations for each one. We thus obtain a more precise estimate of the average performance of the policy on each truth. In (4.6), we generate a new truth for *each* set of observations. We can thus simulate a larger number of truth values, but we will sacrifice some precision in our estimate of average performance for any given truth.

4.2 MEASUREMENT POLICIES

Central to the concept of optimal learning is a measurement policy. This is a rule that tells us which action x we should take next in order to observe something new. In addition, we may also be receiving rewards or incurring costs, which have to be balanced against the value of the information being gained.

In this section, we contrast deterministic versus sequential policies and then provide a mathematical framework for finding optimal sequential policies. Unfortunately, this framework does not provide us with practical policies that we can compute. The section closes with a presentation of a number of the more popular heuristic policies that have been used on this problem class.

4.2.1 Deterministic Versus Sequential Policies

Before we begin our presentation, it is important to make a distinction between what we call *deterministic policies* and *sequential policies*. In a deterministic policy, we decide what we are going to measure before we begin making any measurements.

For example, a business may decide to perform four market research studies in different parts of the country before finalizing the pricing and advertising strategy in a full roll-out to the entire country. The decision to do four studies (and their locations) is made before we have any information from any of the studies.

There are problem classes where deterministic policies are optimal. For example, we might be interested in making measurements that minimize some function of the variance of the quantities that we are trying to estimate. If you take a close look at our formula for updating the variance (or equivalently the precision) in equation (4.2), we see that our estimate of the variance is a deterministic function of what we choose to measure. This means that any rule that depends purely on the variance can be solved deterministically.

Our interest is primarily in sequential policies, where the decision of what to measure next may depend on past measurements. For example, when we are trying to find the shortest path, we may decide to continue sampling a path if it remains competitive, or give up on a path if the observed travel times are simply too long. Our decisions of what to measure in this case depend on the outcomes of prior measurements.

4.2.2 Optimal Sequential Policies

It is possible to provide a mathematical characterization of an optimal measurement policy. Imagine that we are moving over a graph, where we gain information about the graph (including other links) every time we make a transition from one node to the next. For this system, let S be the state variable which captures both our physical state (which might be the node we are sitting at) and the knowledge state (what we know about all the links in the graph as a result of our previous measurements). Further, let S^n be the state after n measurements. We note in passing that the ranking and selection problem does not have a physical state; S^n consists purely of the state of knowledge. Whenever we discuss this problem, we can use S^n interchangeably with our earlier notation K^n.

Assume that we are in state S^n and we make a decision x^n which might change both our physical state (we move from i to j) and our knowledge state (which might be updated using equations (4.1) and (4.2)). Let W^{n+1} be the next observation we observe which is used to determine S^{n+1}. When we do not want to get into the details of how the state changes, we introduce a transition function, often referred to as a state model (or sometimes simply "model") $S^M(\cdot)$ which updates the state using

$$S^{n+1} = S^M(S^n, x^n, W^{n+1}).$$

When we have normally distributed beliefs, our transition function (or knowledge transition function) is given by equations (4.1) and (4.2).

Next let $V(S^n)$ be the value of being in state S^n. The quantity $C(S^n, x)$ captures our total contribution (or reward), minus any measurement costs, from being in state S^n and taking action x. Assume we wish to maximize the total discounted reward, with discount factor γ. Bellman's equation characterizes the optimal decision using

$$V(S^n) \quad = \quad \max_{x} \big(C(S^n, x) + \gamma \mathbb{E}\{ V(S^{n+1}) | S^n \} \big). \qquad (4.8)$$

We let x^n represent the optimal solution to (4.8). We let $X^*(S)$ be the complete mapping of states $S \in \mathcal{S}$ to actions $x \in \mathcal{X}$, where $\mathcal{X}$ describes the set of feasible

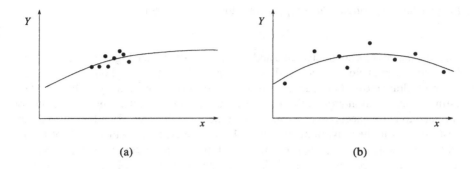

Figure 4.1 Estimating a function where we sample closest to the point that might be best (a), versus sampling a wide range of points so that we get a better estimate of the function (b).

actions. We refer to the function $X^*(S)$ as the optimal policy if it describes the solution to (4.8) for all states $S^n \in \mathcal{S}$.

It may be mathematically comforting to characterize the optimal policy, but equation (4.8) is virtually impossible to solve, even for very small problems. The simplest knowledge state uses at least one continuous variable for each action x. Calculating a value function with as few as two continuous dimensions can, in practice, be quite a challenge. Needless to say, we do not have very many problems of practical significance that meets this modest criterion. Not surprisingly, the field of optimal learning consists primarily of finding shortcuts or, failing this, good heuristics.

4.2.3 Heuristic Policies

Our goal, ultimately, is to find the best possible policies for learning. The reality, however, is that most of the time we are happy to find good policies. Below are some popular methods that have been suggested for problems that are typically associated with discrete selection problems, which is to say that the set of measurement decisions is discrete and "not too large."

Pure Exploration A pure exploration strategy might sample a decision $x^n = x$ with probability $1/M$ (the probabilities do not have to be the same - they just have to be strictly positive). We would only use a pure exploration policy if we were focusing purely on estimating the value of each choice, as opposed to making a good economic decision. If we really are trying to find the best value of μ_x, a pure exploration strategy means that we would spend a lot of time measuring suboptimal choices.

Pure exploration can be effective for offline learning problems, especially when the number of choices is extremely large (and especially if a measurement x is multidimensional). This is often what has been used when we are given a dataset of observations from which we have to fit a model so that we can find the best choice or design. With offline learning, it does not hurt us to observe a poor choice, and extreme choices can give us the best estimates of a function. For example, consider the problem of fitting a linear regression model of the form

$$Y = \theta_0 + \theta_1 x + \theta_2 x^2 + \epsilon.$$

Imagine that x is a scalar between 0 and 10, and we believe the highest values of Y are likely to be found close to the middle of the range. Figure 4.1(a) shows what happens when we draw most of our samples from a narrow range. We may get a lot of data to estimate the function around those points, but we do not get the kind of information we need to get an accurate estimate of the function, which would allow us to do the best job finding the maximum. In Figure 4.1(b), we explore a wider range of points, which allows us to do a better job of estimating the entire curve. For instance, we discover that Y decreases once x is large enough, whereas 4.1(a) leads us to believe that Y always increases with x.

Pure Exploitation   Exploitation means making the best decision given our current set of estimates (we are "exploiting" our knowledge). So, after iteration n, we would next measure

$$x^n = \arg\max_{x \in \mathcal{X}} \theta_x^n.$$

This strategy would seem to focus our energy on the options that appear to be the best. However, it is very easy to get stuck measuring choices that seem to be the best, especially when we simply had some bad luck measuring the better choices.

Pure exploitation is a common strategy in online problems, where we have to live with the results of each measurement. With pure exploitation, we can always defend our choice because we are doing what we believe is the best, but we may be ignoring errors in our own beliefs.

Epsilon-Greedy Exploration   A simple strategy that avoids the limitations of pure exploration and pure exploitation is to use a mixed strategy, where we explore with probability ϵ (known as the exploration rate) and we exploit with probability $1 - \epsilon$. The value of ϵ has to be tuned for each application.

Mixing exploration and exploitation is appealing because it allows you to spend more time evaluating the choices that appear to be best (to make sure this is the case) while still doing a certain amount of exploration. As our measurement budget goes to infinity, we can still provide guarantees that we will find the best alternative because of the exploration component. But this policy still suffers from the significant limitation that when we do choose to explore, we sample from the entire population of alternatives, which may be extremely large, including choices that are clearly suboptimal.

The problem with a mixed exploration/exploitation strategy with fixed ϵ is that the correct balancing of exploration and exploitation changes with the number of iterations. In the beginning, it is better to explore. As we build confidence in our estimates, we would prefer to exploit more. We can do this by using an exploration probability ϵ^n at iteration n that declines with n. We have to make sure it does not decline too quickly. We do this by setting

$$\epsilon^n = c/n$$

for $0 < c < 1$. If we explore, we would choose measurement x with probability $1/|\mathcal{X}|$. This means that in the limit, the number of times we will measure x is given by

$$\sum_{n=1}^{\infty} \frac{c}{n|\mathcal{X}|} = \infty.$$

This assures us that we will estimate each measurement x perfectly, but as the measurements progress, we will spend more time measuring what we think are the best choices.

Boltzmann Exploration A different strategy for balancing exploration and exploitation is known as Boltzmann exploration. With this strategy, we sample measurement x with probability p_x^n given by

$$p_x^n = \frac{e^{\rho \theta_x^n}}{\sum_{x' \in \mathcal{X}} e^{\rho \theta_{x'}^n}}. \tag{4.9}$$

This policy is also known as the *soft max* policy. If $\rho = 0$, we are going to sample each measurement with equal probability (pure exploration). As $\rho \to \infty$, we will sample the measurement with the highest value of θ^n with probability 1 (pure exploitation). In between, we explore the better options with higher probability. Furthermore, we can make ρ increase with n (which is typical), so that we explore more in the beginning, converging to a pure exploitation strategy.

Care should be used when computing the probabilities using a Boltzmann distribution, especially if you are increasing ρ as you progress to focus attention on the best alternatives. The problem is that the exponent $\rho \theta_x^n$ can become so large as to make it impossible to evaluate $e^{(\rho \theta_x^n)}$. A better way is to first compute

$$\bar{\mu}^n = \max_{x \in \mathcal{X}} \theta_x^n,$$

and then compute the probabilities using

$$p_x^n = \frac{e^{\rho(\theta_x^n - \bar{\mu}^n)}}{\sum_{x' \in \mathcal{X}} e^{\rho(\theta_{x'}^n - \bar{\mu}^n)}}.$$

This calculation can be further streamlined by excluding any choices x where θ_x^n is sufficiently far from $\bar{\mu}^n$ (for example, where $\rho|\bar{\mu}^n - \theta_x^n| > 10$).

A limitation of Boltzmann exploration is that computing these probabilities can be fairly expensive when there are large numbers of potential measurements. It is a popular policy in computer science where problems typically have fewer than 100 alternatives, but x might be a vector, producing an exponentially large number of choices. However, Boltzmann exploration has the distinct advantage that our exploration is based on our belief about the quality of each alternative. As a result, we do not spend much time evaluating truly bad alternatives.

Interval Estimation   Imagine that instead of choosing a measurement that we think is best, we will choose a measurement that we think might eventually be best if we were to take enough measurements. With this idea, we might construct a confidence interval and then value an option based on the upper limit of, say, a 95% confidence interval. Letting α be our confidence level and denoting by z_α the standard normal deviate leaving α in the upper tail, our upper limit would be

$$\nu_x^{IE,n} = \theta_x^n + z_\alpha \sigma_x^n, \tag{4.10}$$

where $\sigma_x^n = \sqrt{\frac{1}{\beta_x^n}}$ is the standard deviation of the distribution of our belief at time n. When we use an interval exploration policy, we choose the measurement x^n with the highest value of $\nu_x^{IE,n}$.

Although the interpretation as the upper limit of a confidence interval is appealing, the confidence level α carries no particular meaning. Instead, z_α is simply a tunable parameter. It has been reported in the literature that values around 2 or 3 work best for many applications, but it is possible to construct problems where the best value may be anywhere from 0.1 to 100 (the high values arise when the priors are really poor). Furthermore, it has been found that the algorithm can be very sensitive to the choice of z_α. However, if properly tuned, this policy can work extremely well in many settings.

Interval estimation introduces two important dimensions relative to policies based on exploration and exploitation. First, like Boltzmann estimation, the likelihood of measuring a choice depends on how well we think the choice may work. This means that we will avoid exploring options that seem to be genuinely bad. Second, we are going to be more willing to explore options which we are more uncertain about. The term $z_\alpha \sigma_x^n$ has been called the "uncertainty bonus." As we explore an option with a high value of $\nu_x^{IE,n}$, σ_x^n will decrease, which will often push us to try other options.

IE is not guaranteed to find the best option, even with an infinitely large measurement budget. It is possible for us to get some poor initial estimates of what might be the best option. If θ_x^n is low enough for some choice x, we may never revisit this choice again. Our experimental work suggests that this can happen, but rarely.

Chernoff Interval Estimation   A variant of interval estimation is a method where the confidence interval is derived using Chernoff's inequality. This is computed using

$$\nu_x^{Ch,n} = \theta_x^n + \frac{\alpha + \sqrt{2n\theta_x^n \alpha + \alpha^2}}{n}, \tag{4.11}$$

where

$$\alpha = \ln \frac{2nM}{\delta}.$$

Here, M is the number of alternatives and $0 < \delta < 2$ is a tolerance parameter that controls the likelihood that $\nu_x^{Ch,n}$ is greater than the true value μ_x. The parameter δ comes from the result that if (4.11) is calculated n times for all M alternatives, then $\nu_x^{Ch,n} \geq \mu_x$ with probability $1 - \frac{\delta}{2}$ for all n and x. Alternatively, it is possible to show that the probability that $\nu_x^{Ch,n} < \mu_x$ for a particular iteration n and alternative x is less than $\frac{\delta}{2nM}$.

4.3 EVALUATING POLICIES

Now that we have an initial sense of some possible measurement policies, we have to start thinking about how to compare them so we can choose the best one. It turns out that comparing one measurement policy to another is a relatively subtle exercise.

To start, consider a simple deterministic optimization problem, where we are trying to find the best value of a function $f(x)$. If we have two solutions x^1 and x^2, all we have to do is compare $f(x^1)$ and $f(x^2)$ to see which is bigger. Although finding a good solution can be hard for certain classes of deterministic problems, deciding when one answer is better than another is relatively simple.

When we have a stochastic problem, the comparison is a bit harder. Imagine that we have a function $F(x, W)$ that involves a random variable W. We would like to find the value x that maximizes $\mathbb{E}F(x, W)$. This could be a shortest path problem, where the random variable W is the vector of costs on all the links. It could be a problem involving the allocation of vaccines to different parts of the country; x could be a vector determining how many vaccine doses are sent to each region, after which we observe the random demands W for the vaccine. Assume now that we cannot compute the expectation. Instead, we might use Monte Carlo methods to observe n sample realizations $\omega_1, \omega_2, \ldots, \omega_n$ (see Section 2.4 for a brief review of Monte Carlo simulation methods). We would then evaluate a solution x^1 using

$$\bar{F}(x^1) = \frac{1}{n} \sum_{k=1}^{n} F(x^1, W(\omega_k)).$$

This is an estimate of how well x^1 performs. We would similarly calculate $\bar{F}(x^2)$, and we could compare $\bar{F}(x^1)$ and $\bar{F}(x^2)$. Of course, even if one is better than the other, we should compute confidence intervals around each to see if the difference is statistically different. So, this is a bit harder, but conceptually it is the same comparison that we would undertake with the deterministic case.

Now think about what happens when we compare two measurement policies. Consider our ranking and selection problem where we use a measurement policy π to determine how often we sample the value of a decision x. Perhaps we want to compare Boltzmann exploration to interval estimation, or we want to compare Boltzmann exploration with parameter $\rho = .1$ to the results when we use $\rho = .2$.

After N measurements, θ_x^N is our belief about the value of alternative x. We can use these beliefs to find the best choice, which we can write using

$$x^\pi = \arg \max_{x \in \mathcal{X}} \theta_x^N. \tag{4.12}$$

How well did we do with our measurements? We might evaluate our solution using

$$F^\pi = \max_{x \in \mathcal{X}} \theta_x^N = \theta_{x^\pi}^N.$$

Of course, $\theta_{x^\pi}^N$ is random. The crucial point to remember is the same issue that we raised in Section 4.1: The value of $\theta_{x^\pi}^N$ that we observe in our simulations depends on the generated outcomes of the observations, as well as the generated value of the truth. In Section 4.1, we distinguished these two types of generated outcomes by using ω to denote a sample realization of an observation and using ψ to denote a sample realization of the truth.

We proceed as before. For $\ell = 1, ..., L$, let $\mu(\psi_\ell)$ be sampled truth values, while $W(\omega_k)$ for $k = 1, ..., K$ are sampled observation values. We let $x^\pi(\mu(\psi_\ell), W(\omega_k))$ be our choice of the best alternative, which depends on the truth $\mu(\psi)$, the sampled observations $W(\omega)$, and the measurement policy π. The estimates after N observations $\theta_x^N(\mu, W)$ also depend on the truth $\mu(\psi)$, the sampled observations $W(\omega)$ and the policy π (which we suppress for notational compactness). We can create an estimate

$$\bar{F}^\pi = \frac{1}{L} \sum_{\ell=1}^{L} \left(\frac{1}{K} \sum_{k=1}^{K} \theta_{x^\pi(\mu(\psi_\ell), W(\omega_k))}^N (\mu(\psi_\ell), W(\omega_k)) \right), \tag{4.13}$$

representing average performance across L truths and K sets of observations per truth. We can then compare policies π_1 and π_2 by computing $\bar{F}^{\pi_1}$ and $\bar{F}^{\pi_2}$. If $\bar{F}^{\pi_1} > \bar{F}^{\pi_2}$, we might conclude that π_1 is better. Of course, we should again find confidence intervals for each to see if the difference is significant.

There is a neat trick that can make policy evaluation much more efficient. Since we are generating the truth values $\mu(\psi_\ell)$, they are now known to us, meaning that we can use them directly to evaluate performance. Define

$$F^\pi(\mu(\psi), W(\omega)) = \mu_{x^\pi(\mu(\psi), W(\omega))}(\psi),$$

to be the *true value* of the alternative selected by policy π. The policy does not get to see the true values $\mu(\psi)$. The alternative $x^\pi = \arg\max_x \theta_x^N$ is chosen purely based on the time-N beliefs. However, since we have generated a sample realization of μ, we can use the true value of this alternative to evaluate the policy. Thus, (4.13) becomes

$$\bar{F}^\pi = \frac{1}{L} \sum_{\ell=1}^{L} \left(\frac{1}{K} \sum_{k=1}^{K} \mu_{x^\pi(\mu(\psi_\ell), W(\omega_k))}(\psi) \right). \tag{4.14}$$

On average over many sample realizations, θ_x^N is just a noisy estimate of μ_x, so these two approaches are equivalent. However, (4.14) provides a more precise estimate with less variance. Essentially we are eliminating the noise due to the observations. Alternatively, we could define the regret $\bar{R}^\pi$ as

$$\bar{R}^\pi = \frac{1}{L} \sum_{\ell=1}^{L} \left(\frac{1}{K} \sum_{k=1}^{K} \left(\mu_{x*}(\psi) - \mu_{x^\pi(\mu(\psi_\ell), W(\omega_k))}(\psi) \right) \right), \tag{4.15}$$

where $\mu_{x*}(\psi) = \max_x \mu_x(\psi)$ is the best we can do for a particular truth. The regret has a lower bound of zero, which provides a nice reference point.

We can use either equation (4.13) (which uses estimates of the value of each alternative) or (4.14) (which uses the simulated truths) to evaluate the value of a policy. In Section 4.4.2, we show that (in expectation) $\bar{F}^\pi$, given by equation (4.13) using the estimates θ_x^N, provides an unbiased estimate of $\bar{F}^\pi$ as given by equation (4.14), where we use the assumed truth $\mu(\psi)$. If you are running simulations to find the best policy, it is better to use equation (4.14), since (4.13) introduces additional statistical errors.

We suggest that the best environment for identifying good learning policies is inside the computer, where you can assume a truth and then try to discover the truth. Once you have decided on a good learning policy, you can go to the field where the truth is unknown, and you only have access to the estimates θ_x^N.

4.4 MORE ADVANCED TOPICS*

Ranking and selection is a relatively simple problem, as are the basic policies that are introduced in this chapter. But this simple problem hides some fairly advanced probabilistic concepts. In this section we introduce readers with an interest in a more advanced treatment to some of these concepts.

4.4.1 An Alternative Representation of the Probability Space

In Section 4.1, we represented a sample realization of observations W in Table 4.1 (or Table 4.2) as if specifying ω had the effect of telling us what all possible realizations of W *might* be. That is, if we fix ω, we create a table of what W would be for every alternative x and every measurement n. Let $P(\omega)$ be the probability of outcome ω (assuming these are discrete). We write our objective function $F^\pi(\mu, \Omega)$ as a function of π because the policy for choosing which alternative to measure is embedded in the objection function. If we wanted to take its expectation, we would write

$$F^\pi = \mathbb{E}F^\pi(\mu, W) = \sum_{\omega \in \Omega} P(\omega)F^\pi(\mu(\omega), W(\omega)).$$

For some (and possibly many), this is not the most natural way of thinking about a sample realization. Another way of thinking about a sample path is to assume that for iteration n, we first choose x^n and then observe $W_{x^n}(\omega_n^\pi)$. We essentially "separate" the distributions of our observations from one another, so ω_n^π now refers to the outcome of the nth observation only. That is, we only get to see the outcome after we choose the alternative we are going to observe according to the policy π. With this construction, a sample realization would consist of

$$(x^0, W_{x^0}^1(\omega_1^\pi), x^1(\omega_1^\pi), W_{x^1}^2(\omega_2^\pi), \ldots, x^n(\omega_n^\pi), \ldots, x^{N-1}(\omega_{N-1}^\pi), W_{x^{N-1}}^N(\omega_N^\pi)).$$

That is, a sample path is now a sequence of decisions and observations. In this interpretation, our decisions are also random variables. For instance, our decision at time n is based on the knowledge state K^n, which has just been updated with the most recent observation $W^n(\omega_n^\pi)$. Thus, x_n implicitly depends on ω_n^π.

Let $\omega \in \Omega$ be a sample realization using our original construction from Section 4.1, as illustrated in Table 4.1. For each $\omega \in \Omega$, and given a measurement policy π

that determines the decision, we can construct a new sample realization of decisions $x^n(\omega_n^\pi)$ and observations $W_{x^n}^{n+1}(\omega_{n+1}^\pi)$. Let $\omega^\pi = (\omega_1^\pi, \omega_2^\pi, ..., \omega_N^\pi)$ be a sample realization of both decisions and the measurements that go with the decision, as depicted in the sample path above. Given a policy π, we can construct a set of outcomes Ω^π where every outcome $\omega \in \Omega$ can be mapped to an outcome $\omega^\pi \in \Omega^\pi$. We would then write the probability of ω^π as $P^\pi(\omega^\pi)$.

If we want to find our objective function, we no longer have to embed the measurement policy π in the objective function, since ω^π already contains the decisions that we want to make. In this case, we can write $F(\mu(\omega^\pi), W(\omega^\pi))$ since we no longer need to index $F(\cdot)$ by the policy π. Now if we wanted to take an expectation of $F(\mu, W)$ we would write

$$F^\pi = E^\pi F(\mu, W) = \sum_{\omega^\pi \in \Omega^\pi} P^\pi(\omega) F(\mu(\omega^\pi), \Omega(\omega^\pi)).$$

In the research literature, it is possible to see authors writing $\mathbb{E} F^\pi(\mu, W)$ and $\mathbb{E}^\pi F(\mu, W)$. The ultimate meaning is the same, but these two ways of writing the expectation implicitly represent different constructions of the underlying probability space.

4.4.2   Equivalence of Using True Means and Sample Estimates

We formally state the equivalence of the two approaches to policy evaluation from Section 4.3, and present an argument as to why it holds.

Theorem 4.4.1 *Let π be a policy, and let x^π be the alternative selected by the policy. Then,*

$$\mathbb{E}\mu_{x^\pi} = \mathbb{E} \max_x \theta_x^N.$$

Proof: Recall that $\mathbb{E}^N \mu_x = \theta_x^N$ for any fixed x. By the tower property of conditional expectations,

$$\mathbb{E}\mu_{x^\pi} = \mathbb{E}\mathbb{E}^N \mu_{x^\pi} = \mathbb{E}\theta_{x^\pi}^N,$$

because $x^\pi = \arg\max_x \theta_x^N$ is known at time N (that is, it is fixed from the point of view of our time-N beliefs). However, $\theta_{x^\pi}^N = \max_x \theta_x^N$ by definition of x^π.   □

Theorem 4.4.1 has an interesting corollary. Denote by χ an "implementation policy" for selecting an alternative at time N. We can think of χ as a function mapping the knowledge state K^N to an alternative $\chi(K^n) \in \{1, ..., M\}$. Then,

$$\max_\chi \mathbb{E}\mu_{\chi(K^N)} = \max_x \theta_x^N.$$

In other words, the optimal decision at time N is always to select θ_x^N. If we have no more opportunities to learn, the best possible decision we can make is to go with our final set of beliefs. This result addresses the issue of why most of our policies seek to maximize $\max_x \theta_x^N$ in some way, even though what we really want to learn is the unknown true value $\max_x \mu_x$.

4.5 BIBLIOGRAPHIC NOTES

Section 4.1 - We present here a standard Bayesian framework for ranking and selection; see, e.g., Gupta & Miescke (1996) or Chick (2006). Our presentation
is based on the measure-theoretic idea of random variables as functions on a
space of outcomes or sample paths; although measure theory is far outside the
scope of this book, interested readers are directed to Cinlar (2011), a definitive
rigorous exposition of measure-theoretic probability.

Section 4.2 - The design of policies for taking observations (or measurements) of
noisy functions has its roots in the 1950s and 1960s. It evolved originally
under the umbrella of stochastic optimization over a continuous domain from
the seminal paper of Robbins and Monro (Robbins & Monro 1951), but this
literature did not focus on the issue of maximizing the information gained from
each observation (there was more attention on asymptotic convergence than
rate of convergence). The ranking and selection community evolved with a
focus on the problem of finding the best out of a set of discrete alternatives; see
Barr & Rizvi (1966) for an early review, and see Fu (2002) for a more current
review. The challenge of collecting information in an optimal way has its roots
in DeGroot (1970), which appears to give the first presentation of optimal
learning as a dynamic program (but without an algorithm). Interval estimation
was introduced by Kaelbling (1993). The adaptation of interval estimation
using Chernoff bounds was done by Streeter & Smith (2006). Epsilon-greedy is
described in Sutton & Barto (1998), with an analysis of convergence properties
given in Singh et al. (2000). There is an emerging area of research which
uses the concept of upper confidence bounding (UCB), originally developed
for online problems, in an offline setting. We introduce the idea of upper
confidence bounding in Chapter 6 for online ("bandit") problems. Drawing
on this framework, Audibert et al. (2010) presents a frequentist approach to
ranking and selection with finite alternatives, with provable guarantees.

PROBLEMS

4.1 We wish to find a good learning policy to solve the problem in Table 6.6.

a) Briefly describe the epsilon-greedy policy, the Boltzmann policy, and the interval-
estimation policy. Evaluate each policy in terms of its ability to capture important characteristics of a good learning policy.

b) Define the expected opportunity cost (EOC). Describe in words how you would
approximate the EOC (since computing it exactly is impossible) using Monte
Carlo simulation.

c) Let ω^n be the index for the nth sample realization of the random observations.
Give an expression for the EOC for some policy π and give a precise formula
for the confidence interval for the true performance of a policy π.

4.2   This exercise requires that you test an exploration policy in MATLAB. You will need to download two files from the course website:

http://optimallearning.princeton.edu/exercises/exploration.m
http://optimallearning.princeton.edu/exercises/explorationRun.m

The MATLAB file `exploration.m` executes a pure exploration policy for a general ranking and selection problem. The file `explorationRun.m` creates the data for a problem with 10 alternatives, where it simulates 1000 truths and 50 samples per truth. The program `exploration.m` computes the average opportunity cost "o_cost" and the standard deviation "se_result."

   a) Write out the meaning of "o_cost" mathematically using the notation we have been using in the course.

   b) The standard deviation "se_result" is the standard deviation of what variable?

   c) Construct a 95 percent confidence interval for the value of the exploration policy, and modify the code to produce this confidence interval.

4.3   In this exercise you have to implement the Boltzmann learning policy, which you should model after the `exploration.m` routine in the previous exercise.

   a) Create a new file `boltzmann.m` which implements the Boltzmann learning policy. Keep in mind that you will have to introduce a tunable parameter ρ. The Boltzmann policy gives you the probability that you should sample an alternative. Imagine you have three alternatives, and the Boltzmann distribution gives you sampling probabilities of .2, .5 and .3. To sample from this distribution, generate a random number between 0 and 1. If this number is less than or equal to .2, you choose the first alternative; if it is between .2 and .7, choose the second alternative; otherwise choose the third. You will need to generalize this for an arbitrary number of alternatives.

   b) Using $N = 5000$, vary ρ over .001, .01, .1, 1.0, until it appears that you have found the range which bounds the best value of ρ. For each value of ρ that you try, record it in a table and report the value of the policy and the standard error. Narrow your range until you begin getting results that are indistinguishable from each other.

   c) Compare the performance of the best Boltzmann policy that you can find to a pure exploration policy that was developed in exercise 4.2.

4.4   Implement the interval estimation policy using the `exploration.m` routine provided in exercise 4.2.

   a) Create a file called `ie.m` which implements the interval estimation policy. Again, you will need a tunable parameter z_α which you might call *zalpha*.

b) The optimal value of z_α will be in the range from 0 to 5. Search over this range, first in increments of 1.0, and then in smaller increments, until you again are unable to distinguish between values (continue using $N = 5000$).

c) Compare the performance of the best Boltzmann policy that you can find to a pure exploration policy that was developed in exercise 4.2.

4.5   This exercise builds on the exploration policy, Boltzmann policy and interval estimation policy that was implemented in exercises 4.2, 4.3, and 4.4.

a) Run each policy using $N = 5000$, $M = 50$ and report the confidence intervals, using your best estimate of the tunable parameters. Can you conclude that one policy is better than the other two? If so, which one? If not, use the fact that the standard deviation declines inversely with $\sqrt{N}$ (it also decreases inversely with $\sqrt{M}$, but we are going to hold M constant for our study). Use this to determine how small you need the standard error to be, and then how large N needs to be to produce a confidence interval that is small enough to conclude that one policy is best. You have may to repeat this exercise a few times for the numbers to settle down.

b) Now set $N = 1$, $M = 1$ and run each policy 10 times, reporting the actual outcome (noticed that you will not get a standard error in this case). This simulates the application of a learning policy in a specific situation, where you put it into practice (but we are going to pretend that you can replicate this 10 times). Record how often each policy discovers the best alternative (o_cost = 0). Comment on the likelihood that your best policy would outperform the other two policies on a single sample path.

4.6   You have to choose the best of three medications to treat a disease. The performance of each medication depends on the genetic characteristics of the individual. From the perspective of this medication, people can be divided into five genetic subgroups. If a doctor knew a patient's subgroup, he would know the medication he should use, but this information is not available. Lacking this information, the doctor has to resort to trial and error. Complicating the process is that measuring the performance of a medication involves a certain level of noise.

Table 4.3 gives the average performance of each type of medication on the five patient types, based on extensive prior records. The five patient types occur with equal probability in the population. The doctor will typically test a medication on a patient for one month, after which a blood test provides a measure of the performance of the medication. But these measurements are not precise; the error between the measurement and the actual impact of the medication is normally distributed with a standard deviation of 2.2 (we assume this is constant for all medications and patient types).

a) What is the prior vector θ^0 that you would use for this problem, given the data in the table?

Table 4.3 The performance of each type of medication for each type of patient.

	Patient Type				
Medication	A	B	C	D	E
M1	6.2	7.3	5.4	7.2	5.4
M2	8.4	6.9	6.8	6.6	4.2
M3	5.2	5.8	6.3	5.5	3.7

b) What is your prior probability distribution for the performance of medication M1? Note: It is *not* normally distributed.

c) Test each of the following policies below. You may use the MATLAB routines developed in the previous exercises, or perform the exercise manually with a budget of $N = 10$ observations. If you are using the MATLAB routines, use a budget of $N = 50$ observations.

   1) Pure exploitation.

   2) Boltzmann exploration, using scaling factor $\rho = 1$.

   3) Epsilon-greedy exploration, where the exploration probability is given by $1/n$.

   4) Interval estimation. Test the performance of $z_\alpha = 1.0, 2.0, 3.0$ and 4.0 and select the one that performs the best.

For each policy, report the average performance based on 100 trials, and compute a 95 percent confidence interval.

CHAPTER 5

THE KNOWLEDGE GRADIENT

The knowledge gradient is the name we apply to the simple idea of measuring the
alternative that produces the greatest value, in expectation, from a single observation.
This idea is also sometimes known as the "economic approach" to ranking and selection, indicating a connection to the ideas we sketched out in Chapter 3. In this
chapter, we develop the concept of the knowledge gradient in the context of ranking
and selection problems, which limits our discussion to offline problems with a small
number of alternatives to measure. However, as we develop the idea in later chapters, we show that this strategy can be applied to a wide range of online and offline
problems.

The idea works as follows. Assume that we have a finite number of discrete
alternatives with independent, normally distributed beliefs. After n measurements,
we let θ^n be our vector of means and β^n our vector of precisions (inverse variances).
We represent our state of knowledge as $S^n = (\theta^n, \beta^n)$. If we stop measuring now,
we would pick the best option, which we represent by

$$x^n = \max_{x \in \mathcal{X}} \theta_x^n.$$

Optimal Learning. By Warren B. Powell and Ilya O. Ryzhov

The value of being in state S^n is then given by

$$V^n(S^n) = \theta_{x^n}^n.$$

Now let $S^{n+1}(x)$ be the next state if we choose to measure $x^n = x$ right now, allowing us to observe W_x^{n+1}. This allows us to update θ_x^n and β_x^n, giving us an estimate θ_x^{n+1} for the mean and β_x^{n+1} for the precision (using equations (4.1) and (4.2)). Given that we choose to measure $x = x^n$, we transition to a new state of knowledge $S^{n+1}(x)$, and the value of being in this state is now given by

$$V^{n+1}(S^{n+1}(x)) = \max_{x' \in \mathcal{X}} \theta_{x'}^{n+1}.$$

At iteration n, θ_x^{n+1} is a random variable because we do not yet know what W^{n+1} is going to be. We would like to choose x at iteration n which maximizes the expected value of $V^{n+1}(S^{n+1}(x))$. We can think of this as choosing x^n to maximize the incremental value, given by

$$\nu_x^{KG,n} = \mathbb{E}[V^{n+1}(S^{n+1}(x)) - V^n(S^n)|S^n]. \tag{5.1}$$

The right–hand side of (5.1) can be viewed as the derivative (or gradient) of $V^n(S^n)$ with respect to the measurement x. Thus, we are choosing our measurement to maximize the gradient with respect to the knowledge gained from the measurement, hence the label "knowledge gradient." We write the knowledge gradient policy using

$$X^{KG,n} = \arg\max_{x \in \mathcal{X}} \nu_x^{KG,n}. \tag{5.2}$$

The knowledge gradient, $\nu^{KG,n}$, is the amount by which the solution improves if we choose to measure alternative x. This is illustrated in Figure 5.1, where the estimated mean of choice 4 is best, and we need to find the value from measuring choice 5. The estimated mean of choice 5 will move up or down according to a normal distribution (we assume with mean 0). The solid area under the curve that exceeds the estimate for choice 4 is the probability that measuring 5 will produce a value that is better than the current best, which means that V^{n+1} will increase. The knowledge gradient is the expected amount by which it will increase (we receive a value of 0 if it does not go up).

5.1   THE KNOWLEDGE GRADIENT FOR INDEPENDENT BELIEFS

For the case of independent normally distributed beliefs, the knowledge gradient is particularly easy to compute. When independence holds, we only change our beliefs about one alternative in every time period. Suppose that we are at time n, with estimates θ_x^n of the true values μ_x, after which we choose to measure a particular alternative x^n. Then, we will have $\theta_x^{n+1} = \theta_x^n$ for $x \neq x^n$. Furthermore, for $x = x^n$, while the exact value of θ_x^{n+1} is still unknown at time n, there is a very simple

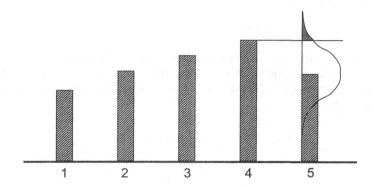

Figure 5.1 Illustration of the knowledge gradient if we were to measure choice 5.

expression for the *conditional* distribution of θ_x^{n+1} given the time-n beliefs. We are able to write

$$\theta_x^{n+1} \sim \mathcal{N}(\theta_x^n, \tilde{\sigma}_x^{2,n}), \tag{5.3}$$

where $\tilde{\sigma}_x^{2,n}$ is the same quantity that we saw in (2.9). The distribution in (5.3) is known as the *predictive distribution* of θ_x^{n+1}, because it represents our best prediction of the results of our next observation before the observation actually occurs.

Below we provide the calculations required to compute the knowledge gradient, and we follow this presentation with a discussion of some properties of this policy. The full derivation of the knowledge gradient policy is deferred to Section 5.9.1. For now, it is enough to keep in mind that (5.1) involves computing an expected value over the predictive distribution.

5.1.1 Computation

For the case of independent normally distributed beliefs, the knowledge gradient is particularly easy to compute. Recall that the precision is simply the inverse of the variance, which is given by $\sigma_x^{2,n}$. Making the transition from precisions to variances, let σ_W^2 be the variance of our measurement W. The updating formula for the variance of our belief $\sigma_x^{2,n}$, assuming we measure $x^n = x$, is given by

$$
\begin{aligned}
\sigma_x^{2,n} &= \left((\sigma_x^{2,n-1})^{-1} + (\sigma_W^2)^{-1}\right)^{-1} \\
&= \frac{(\sigma_x^{2,n-1})}{1 + \sigma_x^{2,n-1}/\sigma_W^2}.
\end{aligned} \tag{5.4}
$$

Now let $Var^n(\cdot)$ be the variance of a random variable given what we know about the first n measurements. For example, $Var^n \theta_x^n = 0$ since, given the first n measurements, θ_x^n is deterministic. Next we compute the change in the variance in our belief about

θ^{n+1} given θ^n, given by

$$\tilde{\sigma}_x^{2,n} \quad = \quad Var^n[\theta_x^{n+1} - \theta_x^n].$$

We need to remember that given what we know at iteration n, which means given θ^n, the only reason that θ^{n+1} is random (that is, with a variance that is not equal to zero) is because we have not yet observed W^{n+1}. As in Chapter 2, after some derivation, we can show that

$$\tilde{\sigma}_x^{2,n} \quad = \quad \sigma_x^{2,n} - \sigma_x^{2,n+1} \tag{5.5}$$

$$= \quad \frac{(\sigma_x^{2,n})}{1 + \sigma_W^2/\sigma_x^{2,n}} \tag{5.6}$$

$$= \quad (\beta_x^n)^{-1} - (\beta_x^n + \beta^W)^{-1}. \tag{5.7}$$

It is useful to compare the updating equation for the variance (5.4) with the change in the variance in (5.6). The formulas have a surprising symmetry to them. Equation (5.7) gives the expression in terms of the precisions.

We then compute ζ_x^n, which is given by

$$\zeta_x^n \quad = \quad - \left| \frac{\theta_x^n - \max_{x' \neq x} \theta_{x'}^n}{\tilde{\sigma}_x^n} \right|. \tag{5.8}$$

ζ_x^n is the *normalized influence* of decision x. It is the number of standard deviations from the current estimate of the value of decision x, given by θ_x^n, and the best alternative other than decision x. We always need to keep in mind that the value of information lies in its ability to change our decision. So, we are always comparing the value of a choice to the best of all the other alternatives. The quantity ζ_x^n captures the distance between a choice and the next best alternative, measured in units of standard deviations of the change resulting from a measurement.

We next compute

$$f(\zeta) = \zeta \Phi(\zeta) + \phi(\zeta), \tag{5.9}$$

where $\Phi(\zeta)$ and $\phi(\zeta)$ are, respectively, the cumulative standard normal distribution and the standard normal density. That is,

$$\phi(\zeta) = \frac{1}{\sqrt{2\pi}} e^{-\frac{\zeta^2}{2}},$$

and

$$\Phi(\zeta) = \int_{-\infty}^{\zeta} \phi(x)\, dx.$$

$\phi(\zeta)$ is, of course, quite easy to compute. $\Phi(\zeta)$ cannot be calculated analytically, but very accurate approximations are easily available. For example, MATLAB provides the function $\texttt{normcdf}(x, \mu, \sigma)$, while Excel provides $\texttt{NORMSDIST}(\zeta)$. Searching the Internet for "calculate cumulative normal distribution" will also turn up analytical approximations of the cumulative normal distribution.

Finally, the knowledge gradient is given by

$$\nu_x^{KG,n} = \tilde{\sigma}_x^n f(\zeta_x^n). \tag{5.10}$$

Table 5.1 illustrates the calculations for a problem with five choices. The priors θ^n are shown in the second column, followed by the prior precision. The precision of the measurement is $\beta^W = 1$.

Table 5.1 Calculations illustrating the knowledge gradient index

Choice	θ^n	β^n	β^{n+1}	$\tilde{\sigma}$	$\max_{x' \neq x} \theta_{x'}^n$	ζ	$f(\zeta)$	ν_x^{KG}
1	20.0	0.0625	1.0625	3.8806	28	-2.0616	0.0072	0.0279
2	22.0	0.1111	1.1111	2.8460	28	-2.1082	0.0063	0.0180
3	24.0	0.0400	1.0400	4.9029	28	-0.8158	0.1169	0.5731
4	26.0	0.1111	1.1111	2.8460	28	-0.7027	0.1422	0.4048
5	28.0	0.0625	1.0625	3.8806	26	-0.5154	0.1931	0.7493

Interestingly, the knowledge gradient formula in (5.10) is symmetric. This means that, if we are looking for the alternative with the lowest value (rather than the highest), we still have

$$\mathbb{E}\left[\min_{x'} \theta_{x'}^n - \min_{x'} \theta_{x'}^{n+1} \mid S^n, x^n = x\right] = \tilde{\sigma}_x^n f(\zeta_x^n),$$

with the only difference being that $\max_{x' \neq x} \theta_{x'}^n$ is replaced by $\min_{x' \neq x} \theta_{x'}^n$ in the definition of ζ_x^n. This symmetry is a consequence of our choice of a Gaussian learning model with Gaussian observations. Intuitively, the normal density is symmetric about its mean; and thus, whether we are looking for the largest or the smallest normal value, the computation consists of integrating over the tail probability of some normal distribution. This does not mean that alternative x has the exact same KG factor in both cases (we are changing the definition of ζ_x^n), but it does mean that the KG formula retains the same basic form and intuitive interpretation. Later on, we show that this does not hold when the learning model is not Gaussian.

5.1.2 Some Properties of the Knowledge Gradient

Recall that we provided a mathematical framework for an optimal learning policy in Section 4.2.2. It is important to keep in mind that the knowledge gradient policy is not optimal, in that it is not guaranteed to be the best possible policy for collecting information. But for ranking and selection problems, the knowledge gradient policy has some nice properties. These include

- Property 1: The knowledge gradient is always positive, $\nu_x^{KG,n} \geq 0$ for all x. Thus, if the knowledge gradient of an alternative is zero, that means we won't measure it.

- Property 2: The knowledge gradient policy is optimal (by construction) if we are going to make exactly one measurement.

- Property 3: If there are only two choices, the knowledge gradient policy is optimal for any measurement budget N.

- Property 4: If N is our measurement budget, the knowledge gradient policy is guaranteed to find the best alternative as N is allowed to be big enough. That is, if x^N is the solution we obtain after N measurements, and

$$x^* = \arg\max \mu_x$$

is the true best alternative, then $x^N \to x^*$ as $N \to \infty$. This property is known as asymptotic optimality.

- Property 5: There are many heuristic policies that are asymptotically optimal (for example, pure exploration, mixed exploration–exploitation, epsilon-greedy exploration and Boltzmann exploration). But none of these heuristic policies is myopically optimal. The knowledge gradient policy is the only pure policy (an alternative term would be to say it is the only stationary policy) that is both myopically and asymptotically optimal.

- Property 6: The knowledge gradient has no tunable algorithmic parameters. Heuristics such as the Boltzmann policy and interval estimation have tunable algorithmic parameters, such as ρ in (4.9) and z_α in (4.10). The knowledge gradient has no such parameters but, as with all Bayesian methods, does require a prior, which is sometimes used as a tunable parameter.

The knowledge gradient is not an optimal policy for collecting information, but these properties suggest that it is generally going to work well. But there are situations where it can work poorly, as we demonstrate in Section 5.2 below.

5.1.3 The Four Distributions of Learning

It is useful to identify four different distributions that arise in the learning process. We illustrate these in the context of learning problems where all random variables are represented by normal distributions.

1) The *prior distribution* $N(\theta_x^0, \sigma_x^{2,0})$, which gives our initial distribution of belief about the mean.

2) The *sampling distribution* of the random variable W, which is given by $N(\mu, \sigma_W^2)$, where μ is the true mean and σ_W^2 is the sample variance.

3) The *posterior distribution* (after n measurements), given by $N(\theta_x^n, \sigma_x^{2,n})$, which reflects the noise in the measurements W.

4) The conditional distribution of θ_x^{n+1} given our beliefs at time n. This is also known as the *predictive distribution*. If we have seen $W^1, ..., W^n$, then θ_x^{n+1} is random before we have observed W^{n+1}, which means that $\theta_x^{n+1} \sim \mathcal{N}(\theta_x^n, \tilde{\sigma}_x^{2,n})$.

An understanding of these distributions is useful because it helps to highlight the different types of uncertainties we are trying to resolve by collecting information.

5.2 THE VALUE OF INFORMATION AND THE S-CURVE EFFECT

The knowledge gradient computes the marginal value of information. What if we perform n_x observations of alternative x, rather than just a single measurement? In this section, we derive the value of n_x measurements to study the marginal value of information. Note that this can be viewed as finding the value of a single measurement with precision $n_x \beta^W$, so below we view this as a single, more accurate measurement.

As before, let θ_x^0 and β_x^0 be the mean and precision of our prior distribution of belief about μ_x. Now let θ_x^1 and β_x^1 be the updated mean and precision after measuring alternative x a total of n_x times in a row. As before, we let $\beta^W = 1/\sigma_W^2$ be the precision of a single measurement. This means that our updated precision after n_x observations of x is

$$\beta_x^1 = \beta_x^0 + n_x \beta^W.$$

In Section 2.2.1, we showed that

$$\tilde{\sigma}^{2,n} = Var^n[\theta^{n+1} - \theta^n],$$

where Var^n is the conditional variance given what we know after n iterations. We are interested in the total variance reduction over n measurements. We denote this by $\tilde{\sigma}^{2,0}$, and calculate

$$\tilde{\sigma}^{2,0}(n_x) = \sigma^{2,0} - \sigma^{2,1}$$
$$= (\beta^0)^{-1} - (\beta^0 + n_x \beta^W)^{-1}.$$

Recall that we have already seen this quantity in Section 3.2, where it had the same meaning.

We next take advantage of the same steps we used to create equation (2.13) and write

$$\theta_x^1 = \theta_x^0 + \tilde{\sigma}_x^0(n_x)Z,$$

where Z is a standard normal random variable and where $\tilde{\sigma}_x^0(n_x) = \sqrt{\tilde{\sigma}_x^{2,0}(n_x)}$ is the standard deviation of the conditional change in the variance of θ^1 given that we make n_x observations.

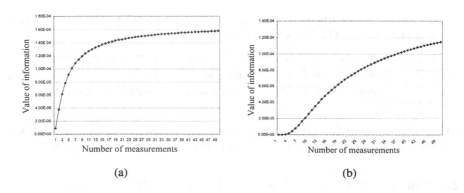

(a)   (b)

Figure 5.2 Value of making n measures. In (a), the value of information is concave, while in (b) the value of information follows an S-curve.

We are now ready to calculate the value of our n_x measurements. Assume we are measuring a single alternative x, so $n_x > 0$ and $n_{x'} = 0$ for $x' \neq x$. Then we can write

$$v_x(n_x) = \mathbb{E}\left[\max_{x'}(\theta_{x'}^0 + \tilde{\sigma}_{x'}^0(n_{x'})Z_{x'})\right] - \max_{x'}\theta_{x'}^0.$$

We can compute the value of n_x observations of alternative x using the knowledge gradient formula in equation (5.10),

$$v_x(n_x) = \tilde{\sigma}_x^0(n_x)f\left(\frac{\theta_x^0 - \max_{x' \neq x}\theta_{x'}^0}{\tilde{\sigma}_x^0(n_x)}\right),$$

where $f(\zeta)$ is given in equation (5.9).

Now we have what we need to study some properties of the value of information. Consider a problem where $\sigma_W = 1.0$, $\sigma^0 = 1.5$ and $\Delta = \mu_x - \max_{x' \neq x}\mu_{x'} = 5$. Figure 5.2(a) shows the value of n measurements as n ranges from 1 to 50. This plot shows that the value of information is concave, as we might expect. Each additional measurement brings value, but less than the previous measurement, a behavior that seems quite intuitive. Figure 5.2(b), however, gives the same plot for a problem where $\sigma_W = 2.5$. Note that when the measurement noise increases, the value of information forms an S-curve, with a very small value of information from the first few measurements, but then rising.

The S-curve behavior arises when a single measurement simply contains too little information. Imagine, for example, trying to find the best baseball hitter out of a group using the experience of a single at-bat. If you use the knowledge gradient to guide the choice of what to measure, it is possible that it would not evaluate an alternative simply because we are learning too little from a single observation. This is one setting where the knowledge gradient policy may not work well. In fact, it is fairly easy to create situations where the knowledge gradient produces numbers on the order of 10^{-10}.

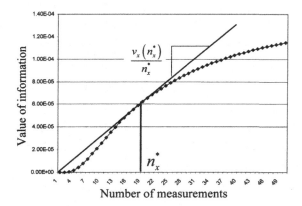

Number of measurements

Figure 5.3   The KG(*) policy, which maximizes the average value of a series of measurements of a single alternative.

A variant of the KG policy is called the KG(*) policy, which finds the number of measurements n_x^* which produces the highest *average* value of each observation, computed using

$$n_x^* = \arg \max_{n_x > 0} \frac{v_x(n_x)}{n_x}. \tag{5.11}$$

Figure 5.3 illustrates this policy. If the value of information is concave, then $n_x^* = 1$. If the function is nonconcave, then it is very easy to find the value of n_x^* that solves equation (5.11).

The KG(*) policy implicitly assumes that our measurement budget is large enough to sample alternative x roughly n_x^* times. There are many applications where this is just not going to be true. For example, a baseball coach needs about 100 at-bats to get a real sense of how well a player can hit. Coaches simply do not have the time to let every player have this many at-bats before deciding who should be on the team. Furthermore, it is not effective to simply spread a fixed measurement budget over a large number of alternatives, especially when the measurement noise is large.

Figure 5.4 illustrates the value of spreading a budget of 50 measurements uniformly over $N \leq 50$ alternatives. If the measurement noise λ is small (in the figure, this corresponds to $\lambda = 1/2$), we do the best if we can observe all 50 alternatives exactly once, after which we pick what appears to be the best choice. Since our evaluation is very accurate, we do the best when we have the most choices. As the measurement noise increases, we get more value if we focus our measurement budget over a subset of the alternatives. For example, if $\lambda = 16$, we do best if we arbitrarily choose eight alternatives out of our population of 50, and then focus on finding the best among these eight.

This behavior can help explain why some choices tend to be biased in favor of cosmetic differences. If you have to choose the best baseball player, there is a bias

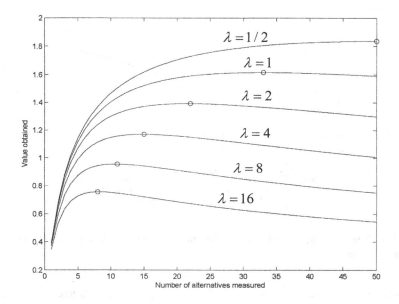

Figure 5.4   The value of spreading a budget of 50 measurements over $M \le 50$ alternatives (from Frazier & Powell 2010).

toward choosing tall, strong athletes, since these are attributes that are easy to identify and help to narrow the field. People are biased toward brands that are well known, and companies tend to work with established vendors. Simply landing near the top of an Internet search can also be a way of pruning a set of choices. However it is done, there are many examples where there is a need to use an initial filter to reduce the set of choices to one that can be evaluated using a fixed measurement budget.

5.3   KNOWLEDGE GRADIENT FOR CORRELATED BELIEFS

There are many problems where updating our belief about one alternative tells us something about other alternatives. Some examples include the following:

- We are trying to find the best set of features for a laptop. We try one laptop with 2G of RAM, a 2.4–GHz processor, a CD/DVD drive, and a 14–inch screen and weighing 5.2 pounds. We then offer a second laptop with 4G of RAM, but everything else the same. A third laptop has 4G of RAM, a 3.2–GHz processor, a CD/DVD drive, a solid–state internal disk drive, a 13–inch screen and weighs 3.7 pounds (with a much higher price). We start by selling the first laptop, and find that we are getting sales higher than expected. Given the similarities with the second laptop, it is reasonable to assume that the sales of this laptop will also be higher than expected.

- Now we are trying to find the best price for our laptop. We start with an initial guess of the sales volume we will for prices in $100 increments from 700 to 1200. We start at $1100, and sales are much lower than we expected. This would lower our beliefs about sales at $1000 and $1200.

- We are trying to find the best of several paths to use for routing a bus. Each time the bus finishes a trip, we record the start and end times, but not the times for individual components of the trip (the bus driver does not have the time for this). If the travel time for one path is higher than expected, this would increase our estimates for other paths that have shared links.

- We are estimating who has a disease. If a concentration of a disease is higher than expected in one part of the population, then we would expect that people who are nearby, or otherwise have a reason to interact, will also have a higher likelihood of having the disease.

Correlations are particularly important when the number of possible measurements is much larger than the measurement budget. The measurement might be continuous (choosing locations to evaluate the population for infections), or there may simply be a very large number of choices (such as websites relevant to a particular issue). The number of choices to measure may be far larger than our budget to measure them in a reliable way.

Perhaps one of the most important features of the knowledge gradient is its ability to handle correlated beliefs. We assume that we have a covariance matrix (or function) that tells us how measurements of x and x' are correlated. If x is a scalar, we might assume that the covariance of μ_x and $\mu_{x'}$ is given by

$$Cov(x, x') \propto e^{-\rho|x-x'|}.$$

Or, we just assume that there is a known covariance matrix Σ with element $\sigma_{xx'}$. For now, we continue to assume that the alternatives x are discrete, an assumption we relax later.

There is a way of updating our estimate of θ^n which gives us a more convenient analytical form than what is given in Section 2.2.3. To simplify the algebra a bit, we let $\lambda^W = \sigma_W^2 = 1/\beta^W$. As we did in Section 5.1, we need the *change* in the variance of our belief due to a measurement. Following the development in Chapter 2, let $\Sigma^{n+1}(x)$ be the updated covariance matrix given that we have chosen to measure alternative x, and let $\tilde{\Sigma}^n(x)$ be the change in the covariance matrix due to evaluating x, which is given by

$$
\begin{aligned}
\tilde{\Sigma}^n(x) &= \Sigma^n - \Sigma^{n+1} & (5.12)\\
&= \frac{\Sigma^n e_x (e_x)^T \Sigma^n}{\Sigma_{xx}^n + \lambda^W}, & (5.13)
\end{aligned}
$$

where e_x is a column vector of 0's with a 1 in the position corresponding to x. We note that (5.12) closely parallels (5.5) for the case of independent beliefs. Now define the column vector $\tilde{\sigma}^n(x)$, which gives the change in our belief about each alternative x' resulting from measuring alternative x. This is calculated using

$$\tilde{\sigma}^n(x) = \frac{\Sigma^n e_x}{\sqrt{\Sigma_{xx}^n + \lambda^W}}. \tag{5.14}$$

Also let $\tilde{\sigma}_i(\Sigma, x)$ be the component $(e_i)^T \tilde{\sigma}(x)$ of the vector $\tilde{\sigma}(x)$. Let $Var^n(\cdot)$ be the variance given what we know after n measurements. We note that if we measure alternative x^n, then

$$Var^n \left[W_{x^n}^{n+1} - \theta_{x^n}^n \right] = Var^n \left[\mu_{x^n} + \varepsilon^{n+1} \right]$$
$$= \Sigma_{x^n x^n}^n + \lambda^W. \tag{5.15}$$

It is useful to think about this for a moment. We index W^{n+1} by x^n to emphasize the fact that our observation is a scalar quantity. Since θ^n is deterministic given n (that is, given what we know about our first n measurements), $Var^n \left[W_{x^n}^{n+1} - \theta^n \right] = Var^n \left[W_{x^n}^{n+1} \right]$. Next recall that $W_{x^n}^{n+1} = \mu_{x^n} + \varepsilon^{n+1}$, where μ_{x^n} is random (because of our Bayesian prior) and ε^{n+1} is random (and independent of μ_{x^n}). So, we are doing a noisy measurement around an uncertain mean. Our estimate of the variance of μ_{x^n} after n measurements is $\Sigma_{x^n x^n}^n$. This reasoning gives us (5.15).

Next define the random variable

$$Z^{n+1} = (W_{x^n}^{n+1} - \theta_{x^n}^n) / \sqrt{Var^n[W_{x^n}^{n+1} - \theta_{x^n}^n]}.$$

We can now rewrite (2.22) as

$$\theta^{n+1} = \theta^n + \tilde{\sigma}(x^n) Z^{n+1}. \tag{5.16}$$

Equation (5.16) nicely brings out the definition of $\tilde{\sigma}(x^n)$ as the vector of variances of the future estimates θ_x^{n+1} given what we know at time n (which makes θ^n deterministic). Essentially, even though a single measurement may change our beliefs about every alternative, the "randomness" in this change comes from a scalar observation, so the conditional distribution is expressed in terms of the scalar Z^{n+1}. This is a useful way of representing θ^{n+1}, especially for problems with correlated beliefs, as we see next.

The knowledge gradient policy for correlated beliefs is computed using

$$X^{KG}(s) = \arg\max_x \mathbb{E}\left[\max_i \mu_i^{n+1} \mid S^n = s, x^n = x \right] \tag{5.17}$$
$$= \arg\max_x \mathbb{E}\left[\max_i \left(\theta_i^n + \tilde{\sigma}_i(x^n) Z^{n+1} \right) \mid S^n, x^n = x \right],$$

where Z is a one-dimensional standard normal random variable. The problem with this expression is that the expectation is hard to compute. We encountered the same expectation when measurements are independent, but in this case we just have to do an easy computation involving the normal distribution. When the measurements are correlated, the calculation becomes more difficult.

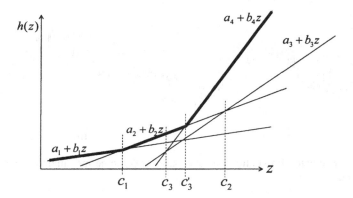

Figure 5.5 Regions of z over which different choices dominate. Choice 3 is always dominated.

There is a way to compute the expectation exactly. We start by defining

$$h(\theta^n, \tilde{\sigma}(x)) \;=\; \mathbb{E}\left[\max_i \left(\theta_i^n + \tilde{\sigma}_i(x^n)Z^{n+1}\right) \mid S^n, x^n = x\right]. \qquad (5.18)$$

Substituting (5.18) into (5.17) gives us

$$X^{KG}(s) = \arg\max_x h(\theta^n, \tilde{\sigma}(x)). \qquad (5.19)$$

Now let $h(a, b) = \mathbb{E}\max_i(a_i + b_i Z)$, where $a = \theta_i^n$, $b = \tilde{\sigma}_i(\Sigma^n, x^n)$ and Z is our standard normal deviate. Both a and b are M-dimensional vectors. We next sort the elements b_i so that $b_1 \leq b_2 \leq \ldots$ so that we get a sequence of lines with increasing slopes. If we plot the lines, we get a series of cuts similar to what is depicted in Figure 5.5. We see there are ranges for z over which the line for alternative 1 is higher than any other line; there is another range for z over which alternative 2 dominates. But there is no range over which alternative 3 dominates; in fact, alternative 3 is dominated by every other alternative for any value of z. What we need to do is to identify these dominated alternatives and drop them from further consideration.

To do this we start by finding the points where the lines intersect. If we consider the lines $a_i + b_i z$ and $a_{i+1} + b_{i+1} z$, we find they intersect at

$$z = c_i = \frac{a_i - a_{i+1}}{b_{i+1} - b_i}.$$

For the moment, we are going to assume that $b_{i+1} > b_i$ (that is, no ties). If $c_{i-1} < c_i < c_{i+1}$, then we can generally find a range for z over which a particular choice dominates, as depicted in Figure 5.5. We can identify dominated alternatives such as alternative 3 in the figure when $c_{i+1} < c_i$. When this happens, we simply drop it from the set. Thus, instead of using c_2, we would use c_3' to capture the intersection between the lines for choices 2 and 4.

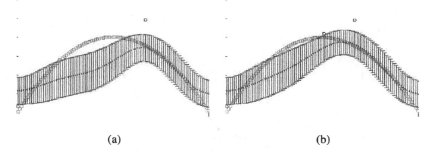

(a)   (b)

Figure 5.6   Estimate of a continuous function after (a) three measurements and (b) four measurements.

Once we have the sequence c_i in hand, we can compute (5.17) using

$$h(a,b) = \sum_{i=1}^{M}(b_{i+1} - b_i)f(-|c_i|),$$

where as before, $f(z) = z\Phi(z) + \phi(z)$. Of course, the summation has to be adjusted to skip any choices i that were found to be dominated.

Figure 5.6 illustrates the use of the correlated knowledge gradient algorithm when it is used to try to estimate a continuous function, starting with a constant confidence interval over the entire range. The logic begins by sampling the two endpoints. The third measurement is shown in Figure 5.6(a), which occurs roughly at the 2/3 point (even though our best estimate of the maximum is at the midpoint). Figure 5.6(b) shows the fourth measurement, which is to the left of the third measurement (which corresponds roughly to where the current estimate of the maximum lies). The measurements illustrate that we are not choosing points that correspond to the highest point on the curve, but instead we are choosing the points where we have the best chance of improving our estimate of the maximum of the function.

Figure 5.7 shows the log of the expected opportunity cost as a function of the number of measurements for three different situations. The curve labeled "correlated KG" represents the performance of the correlated knowledge gradient algorithm exactly as presented in this section. The curve labeled "hybrid KG" represents a situation where we have correlated beliefs, as in Section 2.2.3, but we do not consider the correlations when making decisions, and instead apply the independent KG formula from Section 5.1. Finally, "independent KG" indicates a situation where we do not account for correlations in any way, neither in our prior nor in our policy. Clearly, we do best when we incorporate correlations into our decision-making. Interestingly, we do the worst when our problem has correlated beliefs, but we do not consider them. It is not enough to use a more powerful learning model – we also need a good policy to match.

Handling correlations between measurements has tremendous practical value. When we assumed independent measurements, it was necessary to measure each

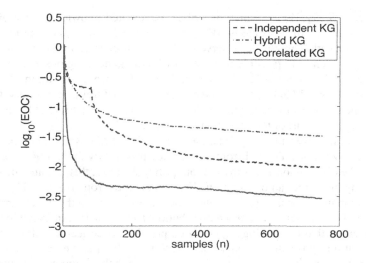

Figure 5.7   Expected opportunity cost as a function of the number of measurements, for each of three policies (from Frazier et al. (2008)).

option at least once. There are applications where the number of potential measurements is far greater than the number of measurements that we can actually make. If we have information about correlations, we can handle (in principle) a much larger number of measurements (even potentially infinite) by using correlations to fill in the gaps.

5.4 ANTICIPATORY VERSUS EXPERIENTIAL LEARNING

It is important to recognize learning in two different settings:

Anticipatory learning - This is the learning that we *anticipate* doing as part of our decision rule, when we are simulating decisions in the future.

Experiential learning - This is learning that we would do as we *experience* data from real observations from the field.

Anticipatory learning occurs within the decision function $X^\pi(S_t)$ when decisions require calculations that approximate in some way how new information may change our underlying belief. In our discussion of decision trees in Section 1.6, Section 1.6.1 presented a basic decision tree where outcomes did not change downstream probabilities. This is a case of a policy that does not use anticipatory learning. Sections 1.6.2 and 1.6.3 provided examples where observations of information were used to change downstream distributions within the decision tree used to make the decision. These sections provide examples of policies that use anticipatory learning.

Experiential learning is what happens after a decision has been made, and we then observe an actual outcome. Section 4.2.2 introduced the transition function

$S_{t+1} = S^M(S_t, x_t, W_{t+1})$ which describes the updated knowledge state given an observation W_{t+1}. In some cases, W_{t+1} comes from a real physical system where the new information might be observed sales, or an observed response to a drug treatment. It is entirely possible that the distribution generating observed values of W_{t+1} is quite different than a distribution $f_W(w)$ that we are assuming for W_{t+1} within our policy $X^\pi(S_t)$. If this is the case, then it is natural to use observations of W_{t+1} to update our estimate of the density $f_W(w)$. This is experiential learning, and it is assumed to be part of the transition function $S^M(\cdot)$.

There are, however, many applications where $S^M(\cdot)$ represents a computer simulation, and W_{t+1} is generated from an assumed probability distribution $f_W(w)$. If $f_W(w)$ is the distribution we use within our policy $X^\pi(S_t)$, then we are not actually learning anything from an observation W_{t+1} that comes from $f_W(w)$. This is quite common in many stochastic simulations, where the focus is on the uncertainty in W, and not on the uncertainty about the distribution $f_W(w)$ of W. An exception would be a simulation that is focusing explicitly on policies for collecting information. In such simulations, it is natural to make observations of W_{t+1} as we step forward using $S_{t+1} = S^M(S_t, x_t, W_{t+1})$ which are drawn from a truth that is unknown to the policy $X^\pi(S_t)$.

With these concepts in mind, we can describe three modeling strategies based on how new information is used.

1) No anticipatory or experiential learning. This is the default strategy in both deterministic and stochastic optimization, where we assume that observations of random variables are realizations from a known distribution and, as a result, would not change our belief about the distribution. This situation is analogous to the "independent KG" curve in Figure 5.7.

2) Experiential learning without anticipatory learning. Here we use actual observations of random variables to update our beliefs, but we ignore the fact that we are going to do this when we make a decision. Learning is thus reduced to a statistical problem, with no optimization component. This is exactly the "hybrid KG" situation in Figure 5.7.

3) Experiential learning with anticipatory learning. This is where we use learning within the policy, as in equation (5.19), and in the transition function, as in equations (2.22) and (2.23). Such a situation corresponds to the "correlated KG" curve in Figure 5.7.

We ignore the combination of using anticipatory learning without experiential learning, because it does not make sense to anticipate the effect of observations on beliefs, but then not use real observations to update beliefs.

Anticipatory learning spans any policy (optimal or heuristic) which uses some representation of the uncertainty in our belief about the value of each choice. We did this heuristically in Chapter 4 with policies such as mixed exploitation–exploration, epsilon–greedy exploration, and Boltzmann exploration. In this chapter, the knowledge gradient, expected improvement, and linear loss policies all represent methods which use some form of explicit learning within the measurement policy.

5.5 THE KNOWLEDGE GRADIENT FOR SOME NON-GAUSSIAN DISTRIBUTIONS

It is important to remember that, while the definition of the knowledge gradient given in (5.1) is general, and can be written down for almost any learning problem, the KG formula given in (5.10) is geared specifically to the ranking and selection problem with independent alternatives and a normal–normal learning model (normally distributed prior, normally distributed measurements). This is not particularly surprising, because the formula uses the functions ϕ and Φ that characterize a Gaussian distribution. We might expect other learning models to yield completely different formulas.

In this section, we give expressions for the marginal value of a single measurement for some of the non-Gaussian learning models presented in Chapter 2. They have certain notable differences from the normal–normal case, but all of them are based on the relationship between our beliefs about alternative x, and our beliefs about "the best alternative other than x." In all of these examples, we assume that the alternatives are independent, for the simple reason that there are no convenient Bayesian conjugate priors for any distribution other than normal.

5.5.1 The Gamma–Exponential Model

Suppose that we have M exponentially distributed random variables with means $\lambda_1, ..., \lambda_M$, and we wish to find the one with the largest rate. For instance, we might be looking at a number of servers, with the goal of finding the one with the highest (fastest) service rate. Alternately, we might have a number of products, each with an exponential daily demand. In this setting, we might want to discover which product has the lowest demand (corresponding to the largest exponential rate), so that we might take this product out of production. A final example is the problem of network routing, where we have a number of possible routes for sending packets, and the objective is to find the route with the lowest ping time. In all of these problems, the observations (service time, daily demand, network latency) are positive, which means that the normal–normal model is not a good fit.

Instead, we use the gamma–exponential model (which we first saw in Section 2.3.1). We start by assuming that the rate λ_x of each alternative is uncertain and follows a gamma distribution with parameters a_x^0 and b_x^0. When we choose to measure alternative x^n at time n, we make a random observation $W_{x^n}^{n+1} \sim \mathbb{E}\left(\lambda_{x^n}\right)$ and update our beliefs according to the equations

$$
a_x^{n+1} = \begin{cases} a_x^n + 1 & \text{if } x^n = x, \\ a_x^n & \text{otherwise,} \end{cases}
$$

$$
b_x^{n+1} = \begin{cases} b_x^n + W_x^{n+1} & \text{if } x^n = x, \\ b_x^n & \text{otherwise.} \end{cases}
$$

We still have an independent ranking and selection problem, as in (4.1) and (4.2), but the specific updating mechanism for the alternative we measure is taken from the gamma–exponential model in Chapter 2.

Recall that, given the beliefs a_x^n and b_x^n, our estimate of λ_x is a_x^n / b_x^n, the mean of the gamma distribution. The KG factor of alternative x at time n is now given by

$$
\nu_x^{KG,n} = \mathbb{E}\left[\max_{x'} \frac{a_{x'}^{n+1}}{b_{x'}^{n+1}} - \max_{x'} \frac{a_{x'}^n}{b_{x'}^n} \,\Big|\, S^n \right]. \tag{5.20}
$$

We omit the steps in the computation of this expectation; interested readers can follow the procedure in Section 5.9.1 and compute the formula for themselves. However, we point out one interesting detail. As in the normal–normal case, the computation of $\nu_x^{KG,n}$ requires us to find the conditional distribution, given S^n and $x^n = x$, of the parameter b_x^{n+1}. This is known as the *predictive distribution* of b_x^n. It is easy to see that

$$
\begin{aligned}
\mathbb{P}\big(b_x^{n+1} > y \mid S^n, x^n = x\big) &= \mathbb{P}\big(W^{n+1} > y - b_x^n \mid S^n, x^n = x\big) \\
&= \mathbb{E}\big[\mathbb{P}\big(W^{n+1} > y - b_x^n \mid \lambda_x\big) \mid S^n, x^n = x\big] \\
&= \mathbb{E}\big[e^{-\lambda_x(y - b_x^n)} \mid S^n, x^n = x\big] \\
&= \left(\frac{b_x^n}{y}\right)^{a_x^n}.
\end{aligned}
$$

We get from the second line to the third line by using the cumulative distribution (cdf) of the exponential distribution, since W_x^{n+1} is exponential given λ_x. The last line uses the moment-generating function of the gamma distribution. From this calculation, it follows that the density of b_x^{n+1} is given by

$$
f(y) = \frac{a_x^n \, (b_x^n)^{a_x^n}}{y^{a_x^n + 1}},
$$

which is precisely the Pareto density with parameters a_x^n and b_x^n.

Once we have this fact, computing (5.20) is a matter of taking an expectation of a certain function over the Pareto density. Define $C_x^n = \max_{x' \neq x} \frac{a_x^n}{b_x^n}$. As before, this quantity represents the "best of the rest" of our estimates of the rates. We then compute

$$
\tilde{\nu}_x^n = \frac{(b_x^n)^{a_x^n} (C_x^n)^{a_x^n + 1}}{(a_x^n + 1)^{a_x^n + 1}}, \tag{5.21}
$$

which is a sort of baseline knowledge gradient. However, depending on certain conditions, we may subtract an additional penalty from this quantity when we compute the final KG formula.

It can be shown that the KG formula is given by

$$
\nu_x^{KG,n} = \begin{cases}
\tilde{\nu}_x^n & \text{if } x = \arg\max_{x'} \frac{a_{x'}^n}{b_{x'}^n}, \\
\tilde{\nu}_x^n - \left(C_x^n - \frac{a_x^n}{b_x^n}\right) & \text{if } \frac{a_x^{n+1}}{b_x^n} > \max_{x'} \frac{a_{x'}^n}{b_{x'}^n}, \\
0 & \text{otherwise.}
\end{cases} \tag{5.22}
$$

Table 5.2 Calculations of the KG formula for the gamma–exponential model.

Choice	a^n	b^n	a^n/b^n	C^n	Too low?	$\tilde{\nu}_x^n$	Penalty	Final KG
1	1.0	7.2161	0.1386	0.3676	Yes			0
2	4.0	12.1753	0.3285	0.3676	No	0.0472	0.0390	0.0081
3	3.0	8.1802	0.3667	0.3676	No	0.0390	0.0008	0.0382
4	5.0	19.3574	0.2583	0.3676	Yes			0
5	2.0	5.4413	0.3676	0.3667	No	0.0540	0	0.0541

This particular formula is skewed toward exploitation. If $x \neq \arg\max_{x'} \frac{a_{x'}^n}{b_{x'}^n}$, we subtract an additional value $C_x^n - \frac{a_x^n}{b_x^n}$ from the KG factor of alternative x. However, if $x = \arg\max_{x'} \frac{a_{x'}^n}{b_{x'}^n}$, there is no additional penalty. Furthermore, if our estimate of λ_x is low enough that $\frac{a_x^n + 1}{b_x^n} \leq \max_{x'} \frac{a_{x'}^n}{b_{x'}^n}$, the KG factor is automatically zero. Since $\nu_x^{KG,n} \geq 0$ for all x, just like in the normal–normal case, this means that we won't even consider an alternative if our beliefs about it are too low. So, there is a more pronounced slant in favor of alternatives with high estimates than we saw in the normal–normal setting.

Table 5.2 illustrates this issue for a problem with five alternatives. Based on our beliefs, it seems that alternative 5 is the best. This alternative also has the highest KG factor. Although our estimate of λ_3 is very close to our estimate of λ_5 (they differ by less than 0.001), the KG factor for alternative 5 is nearly 1.5 times larger than the KG factor for alternative 3. Of the three remaining alternatives, only one is believed to be good enough to warrant a (very small) nonzero KG factor.

One should not take this numerical example too close to heart, however. If we keep the same numbers but let $a_3^n = 1$ and $b_3^n = 2.78$, then alternative 3 will have the largest KG factor, even though our estimate of it would actually be 0.3597, lower than in Table 5.2. Just as in the normal–normal problem, the KG method weighs our estimate of a value against the variance of our beliefs. All other things being equal, alternatives with lower values of a_x^n (that is, alternatives that we have measured fewer times) tend to have higher KG factors, so the KG method will occasionally be moved to explore an alternative that does not currently seem to be the best.

As a final detail in our discussion, let us consider the case where our goal is to find the alternative with the smallest rate, rather than the largest. So, instead of looking for the product with the lowest demand (highest rate) in order to take it out of production, we are now looking for the product with the highest demand (lowest rate) in order to determine the most promising line of research for new products. In this case, the KG factor of alternative x is defined to be

$$\nu_x^{KG,n} = \mathbb{E}\left[\min_{x'} \frac{a_{x'}^n}{b_{x'}^n} - \min_{x'} \frac{a_{x'}^{n+1}}{b_{x'}^{n+1}} \,\middle|\, S^n \right].$$

The predictive distribution of b_x^{n+1}, given S^n and $x^n = x$, is still Pareto with parameters a_x^n and b_x^n. As before, let $C_x^n = \min_{x' \neq x} \frac{a_{x'}^n}{b_{x'}^n}$, and let $\tilde{\nu}_x^n$ be as in (5.21). The KG formula resulting from taking the appropriate expectation over the Pareto density is

$$
\nu_x^{KG,n} = \begin{cases} \tilde{\nu}_x^n & \text{if } x \neq \arg\max_{x'} \frac{a_{x'}^n}{b_{x'}^n}, \\ \tilde{\nu}_x^n - \left(C_x^n - \frac{a_x^n}{b_x^n} \right) & \text{if } \frac{a_x^n + 1}{b_x^n} > C_x^n, \\ 0 & \text{otherwise.} \end{cases} \quad (5.23)
$$

This formula is the mirror image of (5.22). It has a similar appearance, but its effect is precisely the opposite: It rewards exploration. In this case, we never impose a penalty on any alternative *except* for $\arg\max_{x'} \frac{a_{x'}^n}{b_{x'}^n}$. In fact, if our beliefs about the best alternative are too good (i.e. our estimate of the corresponding λ_x is too small), the KG factor of this alternative is zero, and we do not measure it. Even if we believe that the best alternative is relatively close to the second best, we still penalize the one that we think is the best.

Recall that, in the normal–normal case, the KG formula was the same regardless if we were looking for the alternative with the largest or the smallest value. However, in the gamma–exponential model, there is a clear difference between minimizing and maximizing. The reason is because the gamma and exponential distributions are not symmetric. Our use of the gamma prior means that our estimate of λ_x could potentially take on any arbitrarily high value, but it can never go below zero. Thus, roughly speaking, our beliefs about λ_x are more likely to be low than to be high. To put it another way, the true value λ_x can always be higher than we think. Thus, if we are looking for the lowest value of λ_x, we need to push ourselves to explore more, so as not to get stuck on one alternative that seems to have a low rate. However, if we are looking for the highest value of λ_x, it is often enough to stick with an alternative that seems to have a high rate: If the rate is not really as high as we think, we will discover this quickly.

5.5.2 The Gamma–Poisson Model

Let us now consider a ranking and selection problem where our observations are discrete. For instance, we might consider the problem of finding the product with the highest average demand, assuming that the individual daily demands are integer-valued. Then, the daily demand for product x can be modeled as a Poisson random variable with rate λ_x. We assume that λ_x follows a gamma distribution with parameters a_x^0 and b_x^0. If we choose to measure the demand for product x^n after n days, our beliefs are updated according to the equations

$$
a_x^{n+1} = \begin{cases} a_x^n + N_x^{n+1} & \text{if } x^n = x, \\ a_x^n & \text{otherwise,} \end{cases}
$$

$$
b_x^{n+1} = \begin{cases} b_x^n + 1 & \text{if } x^n = x, \\ b_x^n & \text{otherwise,} \end{cases}
$$

where N_x^{n+1} is the number of units of product x ordered on the next day. We assume that $N_x^{n+1} \sim Poisson(\lambda_x)$. If we are looking for the largest rate, the definition of the KG factor is once again

$$\nu_x^{KG,n} = \mathbb{E}\left[\max_{x'} \frac{a_{x'}^{n+1}}{b_{x'}^{n+1}} - \max_{x'} \frac{a_{x'}^{n}}{b_{x'}^{n}} \,\middle|\, S^n\right].$$

The problem looks deceptively similar to the gamma–exponential problem. However, the first difference is that the predictive distribution of a_x^{n+1} is now discrete, since our observation is Poisson. In fact, it can be shown that

$$\mathbb{P}\big(a_x^{n+1} = a_x^n + k \mid S^n, x^n = x\big) = \frac{\Gamma\left(a_x^n + k\right)}{\Gamma\left(a_x^n\right)\Gamma\left(k+1\right)}\left(\frac{b_x^n}{b_x^n+1}\right)^{a_x^n}\left(\frac{1}{b_x^n+1}\right)^k$$

for $k = 0, 1, 2, \dots$. We can view this as a sort of generalization of the negative binomial distribution. In fact, if a_x^n is an integer, then

$$\frac{\Gamma\left(a_x^n + k\right)}{\Gamma\left(a_x^n\right)\Gamma\left(k+1\right)} = \binom{a_x^n + k - 1}{a_x^n - 1},$$

and the predictive distribution of a_x^{n+1} is the classic negative binomial distribution, with a_x^{n+1} representing the total number of Bernoulli trials that take place before a_x^n failures occur, with $\frac{1}{b_x^n+1}$ being the success probability.

There is no closed-form expression for the cdf of a negative binomial distribution; however, because the distribution is discrete, the cdf can always be evaluated exactly by computing and adding the appropriate terms of the probability mass function. Let $F_a(y) = \mathbb{P}(Y_a \le y)$, where Y_a has the negative binomial distribution for a failures, with success probability $\frac{1}{b_x^n+1}$. The basic KG quantity equals

$$\tilde{\nu}_x^n = C_x^n F_{a_x^n}\left(C_x^n\left(b_x^n+1\right)\right) - \frac{a_x^n}{b_x^n}F_{a_x^n+1}\left(C_x^n\left(b_x^n+1\right)+1\right),$$

and it can be shown that the KG factor is given by

$$\nu_x^{KG,n} = \begin{cases} \tilde{\nu}_x^n - \left(C_x^n - \frac{a_x^n}{b_x^n}\right) & \text{if } x \ne \arg\max_{x'} \frac{a_{x'}^n}{b_{x'}^n}, \\ \tilde{\nu}_x^n & \text{if } \frac{a_x^n}{b_x^n+1} \le C_x^n, \\ 0 & \text{otherwise.} \end{cases} \qquad (5.24)$$

Interestingly, if our estimate of λ_x is too high, we will not measure x at all, but if we measure an alternative that does not seem to have the highest rate, the KG factor has an extra penalty.

5.5.3 The Pareto-Uniform Model

Let us consider the problem of finding the product with the largest demand from a different angle. Suppose now that the demand for product x is uniformly distributed on the interval $[0, B_x]$, where B_x is unknown. We assume that B_x follows a Pareto distribution with parameters $\alpha^0 > 1$ and $b^0 > 0$. When we choose to perform a market study on product x at time n, we update our beliefs about x according to

$$b_x^{n+1} = \begin{cases} \max\left(b_x^n, W_x^{n+1}\right) & \text{if } x^n = x, \\ b_x^n & \text{otherwise,} \end{cases}$$

$$\alpha_x^{n+1} = \begin{cases} \alpha_x^n + 1 & \text{if } x^n = x, \\ \alpha_x^n & \text{otherwise.} \end{cases}$$

The random variable W_x^{n+1} is our observation of the demand for product x and is uniform on $[0, B_x]$.

The goal is to find the product with the largest possible demand $\max_x B_x$. Recall that $\mathbb{E}[B_x \mid S^n] = \frac{\alpha_x^n b_x^n}{\alpha_x^n - 1}$. Then, the knowledge gradient is defined to be

$$\nu_x^{KG,n} = \mathbb{E}\left[\max_{x'} \frac{\alpha_{x'}^{n+1} b_{x'}^{n+1}}{\alpha_{x'}^{n+1} - 1} - \max_{x'} \frac{\alpha_{x'}^n b_{x'}^n}{\alpha_{x'}^n - 1} \,\middle|\, S^n\right].$$

The procedure for computing this expectation is the same as in the other models that we have considered. One thing that makes it a bit messier than usual is that the predictive distribution of b_x^{n+1}, given S^n and $x^n = x$, is neither discrete nor continuous, but a mixture of both. This is because if we measure x, then b_x^{n+1} is the maximum of a constant and our observation, which has the effect of "folding" part of the continuous density of the observation. It can be shown that

$$\mathbb{P}\left(b_x^{n+1} = b_x^n \mid S^n, x^n = x\right) = \frac{\alpha_x^n}{\alpha_x^n + 1}.$$

For $y > b_x^n$, however, the predictive distribution has a scaled Pareto density

$$f\left(y \mid S^n, x^n = x\right) = \frac{1}{\alpha_x^n + 1} \frac{\alpha_x^n \left(b_x^n\right)^{\alpha_x^n}}{y^{\alpha_x^n + 1}}.$$

This mixed distribution complicates the computation slightly, but the principle is the same. As usual, let

$$C_x^n = \max_{x' \neq x} \frac{\alpha_{x'}^n b_{x'}^n}{\alpha_{x'}^n - 1}$$

denote the "best of the rest." In this case, the basic KG quantity is

$$\tilde{\nu}_x^n = \frac{1}{\alpha_x^n \left(\alpha_x^n - 1\right)} \frac{\left(\alpha_x^n + 1\right)^{\alpha_x^n - 1} \left(b_x^n\right)^{\alpha_x^n}}{\left(\alpha_x^n\right)^{\alpha_x^n - 1} \left(C_x^n\right)^{\alpha_x^n - 1}},$$

and the KG formula is

$$\nu_x^{KG,n} = \begin{cases} \tilde{\nu}_x^n & \text{if } x \neq \arg\max_{x'} \frac{\alpha_{x'}^n b_{x'}^n}{\alpha_{x'}^n - 1}, \\ \tilde{\nu}_x^n - \left(\frac{\alpha_x^n b_x^n}{\alpha_x^n - 1} - C_x^n\right) & \text{if } \frac{(\alpha_x^n + 1) b_x^n}{\alpha_x^n} \leq C_x^n, \\ 0 & \text{otherwise.} \end{cases} \qquad (5.25)$$

Thus, if we think that x has the largest possible demand and our estimate of B_x is too much larger than our other estimates, we do not measure x. If x seems to have the largest demand but the actual estimate is relatively close to our other estimates, the KG factor of x is nonzero but has an extra penalty term. Otherwise, the KG factor is equal to $\tilde{\nu}_x^n$, with no additional penalty. This particular model promotes exploration.

5.5.4 The Beta–Bernoulli Model

The last model that we consider in this section is the beta–Bernoulli model. In this model, we assume that our observations can only take on the values 0 and 1. The unknown parameter of each alternative is the corresponding success probability.

In fact, it is this setting, rather than the normal–normal setting, that motivated much of the pioneering work on optimal learning. A popular application arose in the setting of clinical trials. Our alternatives may correspond to different experimental medical treatments. A treatment can result in either success or failure, and we are interested in finding the treatment with the highest success probability. The challenge comes when we have a relatively small number of clinical trials that we can perform. We must decide which treatments to test in order to find the highest success probability most efficiently.

As in Chapter 2, we assume that the success probability ρ_x of treatment x follows a beta distribution with parameters α_x^0 and β_x^0. If we decide to test treatment x^n at time n, the result $W_{x^n}^{n+1}$ of the test is 1 with probability ρ_{x^n} and 0 with probability $1 - \rho_{x^n}$. Our beliefs about the treatment are updated using the equations

$$
\alpha_x^{n+1} = \begin{cases} \alpha_x^n + W_x^{n+1} & \text{if } x^n = x, \\ \alpha_x^n & \text{otherwise,} \end{cases}
$$

$$
\beta_x^{n+1} = \begin{cases} \beta_x^n + \left(1 - W_x^{n+1}\right) & \text{if } x^n = x, \\ \beta_x^n & \text{otherwise.} \end{cases}
$$

Our estimate of ρ_x given S^n is $\mathbb{E}\left(\rho_x \mid S^n\right) = \frac{\alpha_x^n}{\alpha_x^n + \beta_x^n}$. Recall that α_x^n roughly corresponds to the number of successes that we have observed in n trials, whereas β_x^n represents the number of failures. As usual, the KG factor is defined as

$$
\nu_x^{KG,n} = \mathbb{E}\left[\max_{x'} \frac{\alpha_{x'}^{n+1}}{\alpha_{x'}^{n+1} + \beta_{x'}^{n+1}} - \max_{x'} \frac{\alpha_{x'}^n}{\alpha_{x'}^n + \beta_{x'}^n} \,\middle|\, S^n \right].
$$

Unlike some of the other models discussed in this section, the predictive distribution in the beta–Bernoulli model is very simple. Given S^n and $x^n = x$, the conditional distribution of α_x^{n+1} is essentially Bernoulli. There are only two possible outcomes, whose probabilities are given by

$$
\mathbb{P}\left(\alpha_x^{n+1} = \alpha_x^n + 1\right) = \frac{\alpha_x^n}{\alpha_x^n + \beta_x^n}, \quad \mathbb{P}\left(\alpha_x^{n+1} = \alpha_x^n\right) = \frac{\beta_x^n}{\alpha_x^n + \beta_x^n}.
$$

The predictive distribution of β_x^{n+1} is also Bernoulli, but the probabilities are reversed. That is, $\mathbb{P}\left(\beta_x^{n+1} = \beta_x^n + 1\right)$ is now $\frac{\beta_x^n}{\alpha_x^n + \beta_x^n}$.

Table 5.3 Calculations of the KG formula for the beta–Bernoulli model.

Choice	α^n	β^n	$\frac{\alpha^n}{\alpha^n+\beta^n+1}$	$\frac{\alpha^n}{\alpha^n+\beta^n}$	$\frac{\alpha^n+1}{\alpha^n+\beta^n+1}$	C^n	KG
1	1	13	0.0667	0.0714	0.1333	0.8874	0
2	2	11	0.1429	0.1538	0.2143	0.8874	0
3	1	20	0.0455	0.0476	0.0909	0.8874	0
4	67	333	0.1671	0.1675	0.1696	0.8874	0
5	268	34	0.8845	0.8874	0.8878	0.1675	0

Letting $C_x^n = \max_{x' \neq x} \frac{\alpha_{x'}^n}{\alpha_{x'}^n+\beta_{x'}^n}$ as usual, we can derive the KG formula

$$
\nu_x^{KG,n} = \begin{cases} \frac{\alpha_x^n}{\alpha_x^n+\beta_x^n}\left(\frac{\alpha_x^n+1}{\alpha_x^n+\beta_x^n+1} - C_x^n\right) & \text{if } \frac{\alpha_x^n}{\alpha_x^n+\beta_x^n} \leq C_x^n < \frac{\alpha_x^n+1}{\alpha_x^n+\beta_x^n+1}, \\ \frac{\beta_x^n}{\alpha_x^n+\beta_x^n}\left(C_x^n - \frac{\alpha_x^n}{\alpha_x^n+\beta_x^n+1}\right) & \text{if } \frac{\alpha_x^n}{\alpha_x^n+\beta_x^n+1} \leq C_x^n < \frac{\alpha_x^n}{\alpha_x^n+\beta_x^n}, \\ 0 & \text{otherwise.} \end{cases} \tag{5.26}
$$

Observe that

$$
\frac{\alpha_x^n}{\alpha_x^n + \beta_x^n + 1} \leq \frac{\alpha_x^n}{\alpha_x^n + \beta_x^n} \leq \frac{\alpha_x^n + 1}{\alpha_x^n + \beta_x^n + 1},
$$

and the KG factor depends on where C_x^n falls in relation to these quantities. In this case, we do not measure x if we believe ρ_x to be too low or too high. We will only benefit from measuring x if our beliefs about ρ_x are reasonably close to our beliefs about the other success probabilities. There is a certain symmetry to this formula (both low and high estimates result in knowledge gradients of zero); and in fact, it has the same symmetric quality of the KG formula for the normal–normal model. If our objective is to find the smallest success probability rather than the largest, (5.26) remains the same; the only change is that we replace the maximum in the definition of C_x^n by a minimum.

An unexpected consequence of this structure is that it is entirely possible for all the KG factors to be zero in the beta–Bernoulli problem. Table 5.3 illustrates one possible instance where this might happen. Intuitively, it can occur when we are already quite certain about the solution to the problem. In the problem shown in the table, it seems clear that alternative 5 is the best, with $268 + 34 = 302$ trials yielding a very high success probability. Among the other alternatives, there is some competition between 2 and 4, but neither is close to alternative 5. As a result, our beliefs about 2 and 4 are too low for one measurement to change their standing with respect to 5. Similarly, our beliefs about alternative 5 are too high for one measurement to change its standing with respect to any of the others. Thus, all the KG factors are zero, which essentially means that the KG method does not really care which alternative to measure (we can choose any one at random).

Of course, it is conceivable (though it may seem unlikely) that alternative 5 is not really the best. For instance, if we could measure alternative 2 several hundred

Table 5.4 Table showing the distinctions between sampling, prior, and predictive distributions for different learning models.

Sampling Distribution	Prior Distribution	Predictive Distribution
Normal	Normal	Normal
Exponential	Gamma	Pareto
Poisson	Gamma	Negative binomial
Uniform	Pareto	Mixed discrete/Pareto
Bernoulli	Beta	Bernoulli
Multinomial	Dirichlet	Multinomial
Normal	Normal–gamma	Student's t-distribution

times, we might find that it actually has a higher success probability, and we were simply unlucky enough to observe 11 failures in the beginning. Unfortunately, the KG method only looks ahead one time step into the future and thus is unable to consider this possibility. This hints at possible limitations of the knowledge gradient approach when our observations are discrete, similar to the S-curve effect of Section 5.2.

5.5.5 Discussion

Our examination of ranking and selection with non-Gaussian learning models underscores several interesting issues that were not as clear in the basic normal–normal problem. For one thing, there is now a real distinction between the prior, sampling and predictive distributions. In the normal–normal model, all three of these distributions were normal; but in other problems, all three can come from different families.

So far, we have managed to derive knowledge gradient formulas in all of these cases. It is clear, however, that a knowledge gradient formula depends not only on the particular type of learning model that we are using, but also on the objective function. In the gamma–exponential model, the KG algorithm uses two different formulas depending on whether we are looking for the largest exponential parameter or the smallest. Even in the basic normal–normal model, we can change the KG formula completely by changing the objective function.

The power and appeal of the KG method come from our ability to write the definition of the knowledge gradient in (5.1) in terms of some arbitrary objective function. Every new objective function requires us to recompute the KG formula, but as long as we are able to do this, we can create algorithms for problems where the objective function is very complicated. This allows us to go beyond the simple framework of ranking and selection, where the goal is always to pick the alternative with the highest value. Later on in this book, we will create knowledge gradient methods for problems where the alternatives are viewed as components that make up a large system, and the objective is a complicated function of our beliefs about the alternatives that somehow expresses the value of the entire system.

5.6 RELATIVES OF THE KNOWLEDGE GRADIENT

The fundamental idea underlying KG is that a single piece of information contributes economic value by improving our ability to optimize (e.g. allowing us to choose a better alternative). This basic concept is quite intuitive, and it is no surprise that it has attracted a lot of attention in the scientific literature. However, there are different ways to plausibly define the value of information as a mathematical quantity, leading to decision-making policies that are closely related to KG, but not exactly identical. This section examines two well-known variations on the theme of value of information, known respectively as "expected improvement" and "linear loss."

5.6.1 Expected Improvement

The expected improvement (EI) policy grew out of a class of learning problems where, instead of dealing with a finite set of alternatives, we have a continuous spectrum. For example, a semiconductor manufacturer uses liquid argon to provide an environment for sputtering, or depositing thin layers of metal, on semiconductor wafers. Argon is expensive and has to be purchased from a chemical manufacturing company, so the exact amount x that is needed may require some fine-tuning. We study this problem class in much more detail in Chapter 16. For now, we can examine EI in the context of the basic ranking and selection problem with M alternatives.

EI defines improvement in the following way. At time n, our current estimate of the largest value is given by $\max_{x'} \theta_{x'}^n$. We would like to find alternatives x whose true value μ_x is greater than this estimated quantity. To put it another way, we prefer x such that μ_x is more likely to exceed (or improve upon) our current estimate. We then define the EI factor of x as

$$\nu_x^{EI,n} = \mathbb{E}\left[\max\left\{ \mu_x - \max_{x'} \theta_{x'}^n, 0 \right\} \,\middle|\, S^n, x^n = x \right]. \tag{5.27}$$

This expression is exactly identical to the one we derived in Section 3.2. We can interpret the EI factor as the value of collecting a single observation about x versus doing nothing. According to EI, collecting information about x is only valuable if $\mu_x > \max_{x'} \theta_{x'}^n$, that is, if μ_x brings about an improvement. Otherwise, the value is zero. Since μ_x is unknown at time n (but $\max_{x'} \theta_{x'}^n$ is known), we take an expected value of the improvement over the distribution of μ_x. At time n, this is $\mathcal{N}\left(\theta_x^n, (\sigma_x^n)^2 \right)$. Like KG, the EI expectation in (5.27) leads to an explicit formula

$$\nu_x^{EI,n} = \sigma_x^n f\left(\frac{\theta_x^n - \max_{x'} \theta_{x'}^n}{\sigma_x^n} \right), \tag{5.28}$$

which closely resembles (5.10), with some differences that we highlight below. The policy then makes the decision using

$$X^{EI,n} = \arg\max_x \nu_x^{EI,n}.$$

There are two main differences between EI and KG. First, $\nu_x^{KG,n}$ is calculated based on how likely θ_x^{n+1} is to exceed $\max_{x' \neq x} \theta_{x'}^n$; that is, x is compared to the best of the other alternatives. On the other hand, EI simply uses the current best estimate $\max_{x'} \theta_{x'}^n$, the maximum over all alternatives, as the reference point. Second, EI uses the prior variance σ_x^n instead of the one-period variance reduction $\tilde{\sigma}_x^n$. One way to interpret this is that EI essentially assumes $\sigma_W^2 = 0$. That is, EI considers what would happen if we could learn μ_x exactly in a single measurement.

We can still use EI in problems with noisy measurements, but the noise is not explicitly considered by the EI calculation. The literature on EI sometimes goes so far as to assume that $\sigma_W^2 = 0$ in the underlying learning problem. This is not very interesting in ranking and selection, because we could find the exact optimal solution just by measuring every alternative once. However, in problems with continuous decisions, finding the best x is still challenging even when observations are exact. We return to EI in Chapter 16, where it is known as efficient global optimization.

5.6.2 Linear Loss*

Two policies have been proposed in the literature under the names $LL(1)$ and LL_1, which might leave the impression that they are the same policy. Despite having very similar names, $LL(1)$ and LL_1 are actually two different policies. Both are examples of "linear-loss" or LL methods, close analogs to KG that were developed for ranking and selection with unknown means and variances. We discussed learning in this setting in Section 2.3.5. Supposing that alternative x has value μ_x, and our observations of μ_x also have an unknown precision β_x^W, we can represent our beliefs about the mean and precision using a normal–gamma prior. Recall that, when we say that (μ_x, β_x^W) follows a normal–gamma distribution with parameters $\theta_x^0, \tau_x^0, a_x^0, b_x^0$, we mean that β_x^W is gamma with parameters a_x^0 and b_x^0, and the conditional distribution of μ_x given that $\beta^W = r$ is normal with mean θ_x^0 and precision $\tau_x^0 r$. For each observation of μ_x that we collect, we use (2.37)-(2.40) to update our beliefs.

Both $LL(1)$ and LL_1 originate from a technique called $LL(N)$. This procedure was designed for a version of ranking and selection where, instead of collecting observations one at a time, we collect N of them at once, in a batch. Instead of making sequential decisions, where we first choose an alternative, then collect an observation, update our beliefs, and make a new decision based on the new information, we make only one decision. Before collecting the batch of information, we choose how many observations (out of N total) should be collected for each alternative. For example, if we have three alternatives and $N = 10$, we might make a decision to collect $k_1 = 3$ observations of μ_1, $k_2 = 5$ observations of μ_2, and $k_3 = 2$ observations of μ_3. We make this decision based on the beliefs we have prior to collecting the batch. For example, if we have a lot of uncertainty about μ_2 (corresponding to a low value of τ_2^0), we might want to assign more observations to alternative 2.

Our goal is to choose an allocation vector $k = (k_1, ..., k_M)$ with $\sum_{i=1}^{M} k_i = N$ and $k_i \geq 0$ to maximize the usual objective function (4.7). In this setting, we can write the objective as

$$\max_{k} \mathbb{E} F^k(\mu, \beta^W, W) \tag{5.29}$$

with $F^k(\mu, \beta^W, W) = \max_x \theta_x^N$. Notice that β^W is now included as part of the "truth," since it is also a vector of unknown parameters and affects the observations we get. The name "linear loss" is due to the fact that maximizing $\mathbb{E} F^k$ is the same as minimizing an equivalent expression $\mathbb{E}\left[\mathbb{E}^N\left(\max_x \mu_x - \max_x \theta_x^N\right)\right]$, the expected loss or difference between the true best value and the estimated best value.

We cannot compute (5.29) directly, but we can approximate it. Let $x^* = \arg\max_x \theta_x^0$ be the alternative that seems to be the best initially, before any information is collected. Because x^* does not depend on the choice of allocation k, we can rewrite (5.29) as

$$\max_{k} \mathbb{E}\left(\max_x \theta_x^N - \theta_{x^*}^N\right). \tag{5.30}$$

Although (5.29) and (5.30) have different optimal values, they have the same optimal solution (that is, the same k maximizes both functions). Define an indicator function

$$I_x^N = \begin{cases} 1 & \text{if } x = \arg\max_{x'} \theta_{x'}^N, \\ 0 & \text{otherwise,} \end{cases}$$

that equals 1 if and only if x is believed to be the best alternative after N measurements have been made. Also define a second indicator

$$\tilde{I}_x^N = \begin{cases} 1 & \text{if } \theta_x^N \geq \theta_{x^*}^N, \\ 0 & \text{otherwise,} \end{cases}$$

to show whether x is believed to be better than x^* (but not necessarily better than all other alternatives) at time N. Then,

$$
\begin{aligned}
\mathbb{E}\left(\max_x \theta_x^N - \theta_{x^*}^N\right) &= \mathbb{E} \sum_{x=1}^{M} I_x^N \cdot \left(\theta_x^N - \theta_{x^*}^N\right) \\
&\leq \mathbb{E} \sum_{x=1}^{M} \tilde{I}_x^N \cdot \left(\theta_x^N - \theta_{x^*}^N\right) \\
&= \sum_{x=1}^{M} P\left(\theta_x^N \geq \theta_{x^*}^N\right) \mathbb{E}\left(\theta_x^N - \theta_{x^*}^N \mid \theta_x^N \geq \theta_{x^*}^N\right). \tag{5.31}
\end{aligned}
$$

If we replace our objective function by the upper bound in (5.31), we have replaced the objective function $\max_x \theta_x^N$ with a sum of pairwise comparisons between individual θ_x^N and a single value $\theta_{x^*}^N$. This is more tractable because we can express the predictive distribution of θ_x^N at time 0, given that we will allocate k_x observations to alternative x, in terms of Student's t-distribution. Specifically, $\sqrt{\frac{\tau_x^0 a_x^0 (\tau_x^0 + k_x)}{b_x^0 k_x}} \left(\theta_x^N - \theta_x^0\right)$ follows a t-distribution with $2a_x^0$ degrees of freedom. We can then compute the expectation in (5.31) for a particular choice of k, and maximize k by setting the derivative with respect to k_x equal to zero. After a few more approximations, the LL method arrives at the allocation

$$k_x = \frac{N + \sum_{x'=1}^{M} \tau_{x'}^0}{\sum_{x'=1}^{M} \sqrt{\frac{b_{x'}^0 \eta_{x'}^0}{b_x^0 \eta_x^0}}} - \tau_x^0, \tag{5.32}$$

where

$$\eta_x^0 = \begin{cases} \sqrt{\lambda_{x,x^*}^0} \frac{2a_x^0 + \lambda_{x,x^*}^0 \left(\theta_{x^*}^0 - \theta_x^0\right)^2}{2a_x^0 - 1} \phi_{2a_x^0}\left(\sqrt{\lambda_{x,x^*}^0}\left(\theta_{x^*}^0 - \theta_x^0\right)\right), & x \neq x^*, \\ \sum_{x' \neq x^*} \eta_{x'}^0, & x = x^*, \end{cases}$$

the function $\phi_{2\alpha}$ is the density of the t-distribution with 2α degrees of freedom, and

$$\lambda_{x,x^*}^0 = \left(\frac{b_x^0}{\tau_x^0 a_x^0} + \frac{b_{x^*}^0}{\tau_{x^*}^0 a_{x^*}^0}\right)^{-1}. \tag{5.33}$$

The procedure uses several approximations, such as allowing k_x to be continuous when computing the optimal allocation, so it is possible for (5.32) to produce negative numbers. In this case, we need to adjust the allocation manually by rounding k_x up or down while ensuring that the allocations still add up to N.

What happens when we leave the batch setting and return to our usual world of sequential measurements? One easy solution is to run the $LL(N)$ method for $N = 1$ at every time step $n = 0, 1, 2, ..., N - 1$. This is precisely what is meant by the $LL(1)$ technique. As a result, we will obtain a vector k^n of the form e_{x^n}, a vector of zeroes with a single 1 corresponding to the alternative that we need to measure. We then measure the alternative, update our beliefs, and repeat the same procedure assuming a batch of size 1.

The similarly named LL_1 procedure works along the same lines. The main difference is that, in the batch setting, LL_1 assumes that all N samples will be allocated to a single alternative, and then it finds the most suitable alternative under these conditions. This significantly simplifies computation, because the batch is guaranteed to change only a single set of beliefs. Setting $N = 1$ inside the policy then produces a straightforward rule for making decisions at time n,

$$X^{LL_1,n} = \arg\max_x \nu_x^{LL_1,n},$$

where

$$\nu_x^{LL_1,n} = \sqrt{\lambda_{x,\tilde{x}}^n} \Psi_{2a_x^n}\left(\sqrt{\lambda_{x,\tilde{x}}^n} |\theta_x^n - \theta_{\tilde{x}}^n|\right), \tag{5.34}$$

for $\tilde{x} = \arg\max_{x' \neq x} \theta_x^n$, and $\Psi_d(z) = \frac{d+z^2}{d-1}\phi_d(z) - z\Phi_d(-z)$, with ϕ_d, Φ_d being the pdf and cdf of the standard t-distribution with d degrees of freedom. The quantity $\lambda_{x,\tilde{x}}^n$ is computed exactly as in (5.33), but using the time-n beliefs. Notice that this procedure compares our beliefs about x to our beliefs about $\tilde{x}$, the best alternative other than x. This recalls the behavior of the KG policy; and indeed, LL_1 is the exact analog of KG for ranking and selection with unknown measurement precision.

The computation required for LL_1 is simpler and requires fewer approximations than $LL(1)$. It is less suitable for the batch setting, where we would most likely choose to divide our observations among many alternatives, but better suited to the sequential setting considered in this chapter. Additionally, some experiments in the simulation literature suggest that we can learn more effectively and discover better alternatives by using LL_1 to make decisions sequentially or with very small batches, rather than by running $LL(N)$ to allocate our entire learning budget at once.

5.7 THE PROBLEM OF PRIORS

Perhaps one of the most important yet difficult challenges in learning is constructing a reasonable prior. Sometimes a domain expert may have a reasonable understanding about the shape of a function. The domain expert may refer to a person or to information derived from the Internet or prior studies. In the worst case, we may have to resort to doing an initial sample using randomly chosen observations. If we insist on using a Bayesian model, this last strategy is referred to as *empirical Bayes*.

The biggest problem arises when we have a prior, but it is not a very good one. Furthermore, we may be willing to assume that the precision of our prior belief is higher than it really is. Figure 5.8 presents a thought experiment with several possible scenarios. In Figure 5.8(a), we have an unknown function, a constant prior over the interval, and a confidence interval that reasonably captures the spread of the function. This is a fairly accurate statement that we have no idea where the optimum is, but we do have an idea of the range of the function. Figure 5.8(b) is our estimate of the function after a series of measurements using the knowledge gradient, which shows that we do a very good job of finding the true optimum, with a very high level of confidence (note that we do not produce a precise estimate of the value of the function at the optimum).

Figure 5.8(c) illustrates a prior that is too low, and where we do not accurately represent the precision of our belief. Figure 5.8(d) then shows that we start by sampling the function at a random point, and not surprisingly the observed value is much higher than our prior. As a result, we tend to focus our search near this point, since our prior suggests that there is no value in sampling points far from the initial sample. If we were to repeat the exercise with a different random starting point, we would have focused our entire search close to that point. The result is an unreliable estimate of the optimum.

Figure 5.8(e) then depicts a prior that is too high. When this is the case, Figure 5.8(f) shows that we end up sampling the entire function, since every observation produces an estimate that is very low relative to our prior estimate for the rest of the function. We then proceed to sample any point that has not been sampled before (and which is not close to a previously sampled point). This works if our budget is large enough to sample the entire function, but it would work poorly if this were not the case.

Perhaps the most visible lesson from this illustration is that it is important to be honest about the uncertainty in the prior. Figures 5.8(c) and 5.8(e) both display

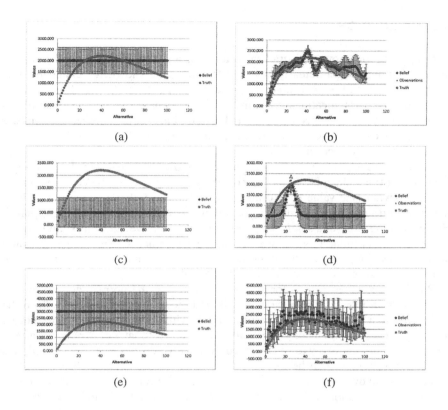

Figure 5.8 The effect of the prior on the search process: (a) Unbiased prior. (b) Resulting measurements. (c) Prior that is biased low. (d) A low prior produces measurements focused around whatever point is chosen first. (e) Prior that is biased high. (f) A high prior produces a search that evaluates the entire function.

confidence intervals that do not cover the function. A narrow confidence bound is particularly problematic if the prior is biased low, because we tend to focus all of our energy around any point that we sample.

Of course, from the Bayesian point of view, there is no one fixed function. Rather, our prior encodes our beliefs about a wide range of possible true functions. The message of the preceding example can be taken to mean that we should make sure that this range is in some way "wide enough."

Care should be used to ensure that the prior does not exclude portions of the function that may be quite good. For this reason, it is better to be biased high and to use care not to overstate the precision of your prior (that is, where your confidence interval is too narrow). However, using a prior that is too high, and/or a confidence interval that is too large, may simply result in a lot of unnecessary exploration. In other words, the better your prior is, the better your search will be. This may be a way of saying that there is no free lunch in information collection.

5.8 DISCUSSION

The appeal of the knowledge gradient is that it is a simple idea that can be applied to many settings. A particularly powerful feature of the knowledge gradient is that it can capture the important dimension of correlated beliefs. In fact, it is useful to review the list of applications given in Section 1.2 where you will see that almost all of these are characterized by correlated beliefs. Later in the volume, we consider more complex sets of alternatives such as finding the best subset, or the best value of a continuous, multidimensional parameter vector.

The knowledge gradient (as with any policy based on the expected value of a single measurement) is vulnerable to the nonconcavity of information. Indeed, if you have a problem where the value of information is nonconcave, then you have to address the issues discussed in Section 5.2, regardless of your choice of learning policy. However, if the value of a single measurement is nonconcave, then this simply means that you have to think about taking repeated measurements. This behavior would almost always be present if the information W^n is binomial, which means that we should be thinking about the value of multiple trials.

We have presented results for situations other than the normal–normal model, but we suspect that most applications will lend themselves reasonably well to the normal–normal model, for two reasons. First, while the initial prior may not be normal, the central limit theorem generally means that estimates of parameters after a few observations are likely to be described by a normal distribution. Second, while a single observation W^n may be non–normal, if we have to use repeated observations (in a single trial) to overcome the nonconcavity of information, then the combined effect of multiple observations is likely to be accurately described by a normal distribution.

5.9 WHY DOES IT WORK?*

5.9.1 Derivation of the Knowledge Gradient Formula

It is not necessary to know how to derive the knowledge gradient formula in order to be able to use it effectively, but sometimes it is nice to go past the "trust me" formulas and see the actual derivation. The presentation here is more advanced (hence the * in the section title), but it is intended to be tutorial in nature, with additional steps that would normally be excluded from a traditional journal article.

The knowledge gradient method is characterized by simplicity and ease of use. In every time step, we can compute $\nu_x^{KG,n}$ for each alternative x by plugging the current values of θ_x^n and $\bar{\sigma}_x^{2,n}$ into the formulas (5.7) and (5.8), and then applying (5.10). After that, our measurement decision is given by $X^{KG,n} = \arg\max_x \nu_x^{KG,n}$, the alternative with the largest knowledge gradient value.

However, it is worthwhile to go through the derivation of the knowledge gradient formula at least once. Not only does this make the KG formulas look less unwieldy, by showing how they originate from the definition of the knowledge gradient, it also gives a sense of how we might go about creating a knowledge gradient method in other optimal learning problems. Later on in this chapter, we derive KG formulas for ranking and selection problems that use some of the non-Gaussian learning models from Chapter 2. This can be done using the same approach as for the independent normal case, but the resulting KG formulas will be very different from the formulas for the normal–normal model.

The expression in (5.2) gives a generic definition of a KG policy, in terms of the expected improvement made by making a measurement. We can write down this expectation in many different settings, but the way we compute it (if, indeed, we can compute it at all) will vary from problem to problem. Often, deriving a computable form for the KG policy poses a computational challenge in research. Thus, the KG approach is a very general idea, but the algorithmic realization of that idea is heavily problem-specific.

We now show how $\nu_x^{KG,n}$ is derived for a particular choice of alternative x at time n. At time n, the estimates θ_x^n and $\bar{\sigma}_x^{2,n}$ are known to us for all x. However, the future estimates θ_x^{n+1} are still random, because we have not yet decided on the $(n+1)$st measurement. It is important to remember that, because the problem is sequential, each new time step changes what is random and what is known. For example, the nth estimate θ_x^n is a random variable from the point of view of any time $n' < n$, but it is a constant from the point of view of time n, when the first n observations have been irrevocably made.

Our goal is to compute (5.1), which requires us to find

$$\mathbb{E}\big[V^{n+1}\left(S^n\right)\big|\ S^n, x^n = x\big] = \mathbb{E}\Big[\max_{x'} \theta_{x'}^{n+1} \,|\, S^n, x^n = x\Big].$$

We assume that we measure x at time n, and we examine how this measurement will affect our beliefs about the best alternative. Fortunately, we can simplify our expression for $\max_{x'} \theta_{x'}^{n+1}$. Recall from the updating equations (4.1) and (4.2) that $\theta_{x'}^{n+1} = \theta_{x'}^n$ for any $x' \neq x^n$. Thus, we can rewrite

$$\mathbb{E}\Big[\max_{x'} \theta_{x'}^{n+1}\,\Big|\ S^n, x^n = x\Big] = \mathbb{E}\Big[\max\Big(\max_{x' \neq x}\theta_{x'}^n, \theta_x^{n+1}\Big)\,\Big|\ S^n, x^n = x\Big]. \quad (5.35)$$

From the point of view of time n, the quantity $\max_{x'} \theta_{x'}^{n+1}$ is merely a maximum of a single random variable and a constant. It is typically much easier to compute the expected value of such a quantity than the maximum of multiple random variables.

Next, we consider the conditional distribution of θ_x^{n+1} given S^n and $x^n = x$. From the updating equations, we know that

$$\theta_x^{n+1} = \frac{\beta_x^n}{\beta_x^n + \beta_x^W}\theta_x^n + \frac{\beta_x^W}{\beta_x^n + \beta_x^W}W_x^{n+1}, \quad (5.36)$$

a weighted average of a constant θ_x^n and a random variable W_x^{n+1}. Given our beliefs at time n, the conditional distribution of the true value μ_x of x is normal with mean θ_x^n and variance $\bar{\sigma}_x^{2,n}$. Then, given μ_x, the observation W_x^{n+1} is itself conditionally normal with mean μ_x and variance σ_x^2. What we need, however, is the conditional distribution of W_x^{n+1} given our beliefs at time n, but not given the true value μ_x, and we can find this distribution by computing the moment-generating function

$$
\begin{aligned}
\mathbb{E}\left(e^{-rW_x^{n+1}}\right) &= \mathbb{E}\left(\mathbb{E}\left(e^{-rW_x^{n+1}} \mid \mu_x\right)\right) \\
&= \mathbb{E}\left(e^{-r\mu_x}e^{\frac{1}{2}\sigma_x^2 r^2}\right) \\
&= e^{\frac{1}{2}\sigma_x^2 r^2}\mathbb{E}\left(e^{-r\mu_x}\right) \\
&= e^{\frac{1}{2}\sigma_x^2 r^2}e^{-r\theta_x^n}e^{\frac{1}{2}\bar{\sigma}_x^{2,n}r^2} \\
&= e^{-r\theta_x^n}e^{\frac{1}{2}(\sigma_x^2+\bar{\sigma}_x^{2,n})r^2}.
\end{aligned}
$$

This is clearly the moment-generating function of the normal distribution with mean θ_x^n and variance $\sigma_x^2 + \bar{\sigma}_x^{2,n}$. The variance can also be written in precision notation as $1/\beta_x^W + 1/\beta_x^n$.

It follows that the conditional distribution of θ_x^{n+1} given S^n and $x^n = x$ is also normal, since θ_x^{n+1} is a linear function of W_x^{n+1} by (5.36). We can find the mean and variance of this distribution by computing

$$
\begin{aligned}
\mathbb{E}\left[\theta_x^{n+1} \mid S^n, x^n = x\right] &= \frac{\beta_x^n}{\beta_x^n + \beta_x^W}\theta_x^n + \frac{\beta_x^W}{\beta_x^n + \beta_x^W}\mathbb{E}\left[W_x^{n+1} \mid S^n, x^n = x\right] \\
&= \frac{\beta_x^n}{\beta_x^n + \beta_x^W}\theta_x^n + \frac{\beta_x^W}{\beta_x^n + \beta_x^W}\theta_x^n \\
&= \theta_x^n,
\end{aligned}
$$

and

$$
\begin{aligned}
Var\left[\theta_x^{n+1} \mid S^n, x^n = x\right] &= \left(\frac{\beta_x^W}{\beta_x^n + \beta_x^W}\right)^2 Var\left[W_x^{n+1} \mid S^n, x^n = x\right] \\
&= \left(\frac{\beta_x^W}{\beta_x^n + \beta_x^W}\right)^2 \left(\frac{1}{\beta_x^W} + \frac{1}{\beta_x^n}\right) \\
&= \frac{\beta_x^W}{\beta_x^n\left(\beta_x^W + \beta_x^n\right)}.
\end{aligned}
$$

Using the definition of the precision again, we can write

$$
\begin{aligned}
\frac{\beta_x^W}{\beta_x^n\left(\beta_x^W + \beta_x^n\right)} &= \frac{\bar{\sigma}_x^{2,n}}{\sigma_x^2\left(\frac{1}{\sigma_x^2} + \frac{1}{\bar{\sigma}_x^{2,n}}\right)} \\
&= \frac{\bar{\sigma}_x^{2,n}}{1 + \sigma_x^2/\bar{\sigma}_x^{2,n}},
\end{aligned}
$$

which is precisely $\tilde{\sigma}_x^{2,n}$ by (5.6).

We have found that the conditional distribution of θ_x^{n+1}, given S^n and $x^n = x$, is normal with mean θ_x^n and variance $\tilde{\sigma}_x^{2,n}$. In words, when we measure x at time n, we expect that the next observation will be equal to θ_x^n, that is, our beliefs θ_x^n are accurate on average. As a consequence of this measurement, the variance of our beliefs about μ_x will decrease by an amount equal to the variance of the next observation.

Now, we can return to (5.35) and rewrite the right-hand side as

$$\mathbb{E}\left[\max\left(\max_{x'\neq x}\theta^n_{x'},\theta^{n+1}_x\right)\,\bigg|\,S^n,x^n=x\right]=\mathbb{E}\left[\max\left(\max_{x'\neq x}\theta^n_{x'},\theta^n_x+\tilde{\sigma}^n_x Z\right)\right],$$

where Z is a standard normal (mean 0, variance 1) random variable. Our goal is now to compute an expectation of a function of Z, which looks far more tractable than the expression we started out with.

Let $C^n_x=\max_{x'\neq x}\theta^n_{x'}$, and observe that

$$\max\left(C^n_x,\theta^n_x+\tilde{\sigma}^n_x Z\right)=\begin{cases}C^n_x & \text{if }Z\leq\frac{C-\theta^n_x}{\tilde{\sigma}^n_x},\\ \theta^n_x+\tilde{\sigma}^n_x & \text{otherwise.}\end{cases}$$

Thus,

$$\mathbb{E}\left[\max\left(C^n_x,\theta^n_x+\tilde{\sigma}^n_x Z\right)\right]=\int_{-\infty}^{\frac{C^n_x-\theta^n_x}{\tilde{\sigma}^n_x}}C^n_x\phi(z)dz$$
$$+\int_{\frac{C^n_x-\theta^n_x}{\tilde{\sigma}^n_x}}^{\infty}\left(\theta^n_x+\tilde{\sigma}^n_x z\right)\phi(z)dz,$$

where

$$\int_{-\infty}^{\frac{C^n_x-\theta^n_x}{\tilde{\sigma}^n_x}}C^n_x\phi\left(z\right)dz=C^n_x\Phi\left(\frac{C^n_x-\theta^n_x}{\tilde{\sigma}^n_x}\right)$$

and

$$\int_{\frac{C^n_x-\theta^n_x}{\tilde{\sigma}^n_x}}^{\infty}\left(\theta^n_x+\tilde{\sigma}^n_x z\right)\phi\left(z\right)dz=\theta^n_x\Phi\left(-\frac{C^n_x-\theta^n_x}{\tilde{\sigma}^n_x}\right)+\tilde{\sigma}^n_x\phi\left(\frac{C^n_x-\theta^n_x}{\tilde{\sigma}^n_x}\right).$$

The first term in the second equation uses the symmetry of the normal distribution, and the second term is due to the fact that

$$\int_y^\infty z\phi(z)\,dz=\phi(y),$$

which can be verified by a back-of-the-envelope calculation.

Observe now that, due to the symmetry of the normal density,

$$\phi\left(\frac{C^n_x-\theta^n_x}{\tilde{\sigma}^n_x}\right)=\phi\left(-\frac{C^n_x-\theta^n_x}{\tilde{\sigma}^n_x}\right)=\phi(\zeta^n_x),$$

which gives us one of the terms that make up the KG formula. To obtain the other term, we consider two cases. If $C^n_x\leq\theta^n_x$, then $\theta^n_x=\max_{x'}\theta^n_{x'}$ and

$$C^n_x\Phi\left(\frac{C^n_x-\theta^n_x}{\tilde{\sigma}^n_x}\right)+\theta^n_x\Phi\left(-\frac{C^n_x-\theta^n_x}{\tilde{\sigma}^n_x}\right)=C^n_x\Phi(\zeta^n_x)+\theta^n_x-\theta^n_x\Phi(\zeta^n_x)$$
$$=\theta^n_x+\tilde{\sigma}^n_x\left(\frac{C^n_x-\theta^n_x}{\tilde{\sigma}^n_x}\right)\Phi\left(\zeta^n_x\right)$$
$$=\theta^n_x+\tilde{\sigma}^n_x\zeta^n_x\Phi\left(\zeta^n_x\right).$$

However, if $\theta_x^n < C_x^n$, then $C_x^n = \max_{x'} \theta_{x'}^n$ and

$$C_x^n \Phi\left(\frac{C_x^n - \theta_x^n}{\tilde{\sigma}_x^n}\right) + \theta_x^n \Phi\left(-\frac{C_x^n - \theta_x^n}{\tilde{\sigma}_x^n}\right) = C_x^n - C_x^n \Phi(\zeta_x^n) + \theta_x^n \Phi(\zeta_x^n)x)$$

$$= C_x^n + \tilde{\sigma}_x^n \zeta_x^n \Phi(\zeta_x^n).$$

Either way,

$$C_x^n \Phi\left(\frac{C_x^n - \theta_x^n}{\tilde{\sigma}_x^n}\right) + \theta_x^n \Phi\left(-\frac{C_x^n - \theta_x^n}{\tilde{\sigma}_x^n}\right) = \max_{x'} \theta_{x'}^n + \tilde{\sigma}_x^n \zeta_x^n \Phi(\zeta_x^n).$$

Putting all the terms together,

$$\mathbb{E}[\max(C_x^n, \theta_x^n + \tilde{\sigma}_x^n Z)] = \max_{x'} \theta_{x'}^n + \tilde{\sigma}_x^n (\zeta_x^n \Phi(\zeta_x^n) + \phi(\zeta_x^n))$$

$$= \max_{x'} \theta_{x'}^n + \tilde{\sigma}_x^n f(\zeta_x^n),$$

whence

$$\nu_x^{KG,n} = \mathbb{E}\left[\max_{x'} \theta_{x'}^{n+1} - \max_{x'} \theta_{x'}^n \;\middle|\; S^n, x^n = x\right]$$

$$= \mathbb{E}\left[\max\left(\max_{x' \neq x} \theta_{x'}^n, \theta_x^{n+1}\right) \;\middle|\; S^n, x^n = x\right] - \max_{x'} \theta_{x'}^n$$

$$= \tilde{\sigma}_x^n f(\zeta_x^n),$$

which is precisely the KG formula that we introduced earlier.

The above derivation relies on our assumption of independent alternatives, as well as our use of a normal–normal learning model. However, in the process, we followed a set of steps that can be used to derive knowledge gradient formulas for other learning problems, as well. In particular, we use the exact same approach to derive KG formulas for ranking and selection problems with non-Gaussian learning models later in this chapter. The steps are:

1) We calculate the conditional distribution, given S^n and $x^n = x$, of the value function $V^{n+1}(S^{n+1}(x))$. In ranking and selection with independent alternatives, this reduces to the problem of finding the conditional distribution of θ_x^{n+1} (in the above derivation, this was normal). Recall that this is called the predictive distribution of θ_x^{n+1}, because it is what we predict about time $n+1$ at time n.

2) We calculate the conditional expectation of $V^{n+1}(S^{n+1}(x))$ over the predictive distribution. This is especially simple in ranking and selection with independent alternatives, because it is simply an expected value of a function of a one-dimensional random variable (in the above derivation, this was the standard normal random variable Z).

3) We subtract the quantity $V^n(S^n)$, which is viewed as deterministic at time n.

In most of the problems discussed in this book, the real challenge lies in the second step. For example, if we introduce correlations into the problem, the predictive distribution of the vector θ^{n+1} is still fairly simple, but it is a bit more difficult to calculate the expectation of the value function $V^{n+1}\left(S^{n+1}(x)\right)$ over this distribution. However, there are many learning problems that seem complicated, but that actually yield simple closed-form expressions for the knowledge gradient. Later on, we encounter many more varied examples of such problems.

5.10 BIBLIOGRAPHIC NOTES

Section 5.1 - The idea of collecting information based on the expected value of a single measurement was first introduced by Gupta & Miescke (1994) and Gupta & Miescke (1996). The concept of the knowledge gradient was developed in greater depth in Frazier et al. (2008). We connect the KG concept to the idea of information economics from Chapter 3; in fact, Chick & Gans (2009) describes a KG-like approach as the "economic approach to simulation selection."

Section 5.2 - The material on the nonconcavity of information is based on Frazier & Powell (2010). Additional discussion on the nonconcavity of information can be found in Weibull et al. (2007), which also presents conditions where it is not always optimal to choose the alternative that appears to be best.

Section 5.3 - The knowledge gradient for correlated beliefs was first introduced in Frazier et al. (2009).

Section 5.5 - The development of the knowledge gradient for non-Gaussian distributions in this section is mostly new. A version of KG for the gamma–exponential model was presented in Ryzhov & Powell (2011c).

Section 5.6 - The expected improvement algorithm was proposed by Jones et al. (1998) for problems with continuous decisions. This algorithm is also sometimes known as efficient global optimization or EGO in this setting. Some more recent work on EI can be found in Gramacy & Lee (2011). The $LL(N)$ methodology was originally put forth by Chick & Inoue (2001), with extensive empirical validation undertaken by Inoue et al. (1999) and Branke et al. (2005). A general overview of the LL approach is available in Chick (2006). The LL_1 procedure is a more recent development laid out in Chick et al. (2010).

PROBLEMS

5.1 Your estimate of the long-run performance of a mutual fund was that it returns 8 percent, but your distribution of belief around this number is normally distributed with a standard deviation of 3 percent. You update this each year (assume that successive years are independent), and from history you estimate that the standard deviation of the return in a particular year is 6 percent. At the end of the most recent year, the mutual fund returned -2 percent.

a) Use Bayesian updating to update the mean and standard deviation.

b) What do we mean when we say that the "normal distribution is conjugate"?

c) What will be the precision of my estimate of the long-run performance in four years (after four measurements, starting with the current state of knowledge)?

5.2 You have three places that serve takeout food around your area and you want to maximize the quality of the total food you intake over time. The three restaurants are:

1. Vine Garden (VG)

2. Wise Sushi (WS)

3. Food Village (FV)

You assume a normal prior with (μ_x, β_x) on the quality of the food in these places and the measurements are normally distributed with precision $\beta^W = 1$.

1. Define the expected opportunity cost (EOC) for this setting (assume that the discount factor, $\gamma = 1$).

2. For a single ω, Table 5.5 below contains your prior, the truth, and the outcomes of your observations until the third time step $n = 3$. Write down the empirical opportunity cost after the third observation (note that you will need to update your priors).

Table 5.5 Priors and Observations

Iteration	VG	WS	FV
Prior (μ_x, β_x)	$(6, 2)$	$(7, 1)$	$(8, 1)$
Truth for μ_x	5	8	7
1	4		
2		9	
3			8.5

3. If you are valuing your current utility from food higher than future time periods ($\gamma < 1$), how would you expect the behavior of the optimal policy to change as opposed to having $\gamma = 1$?

5.3 Table 5.8 shows the priors θ^n and the standard deviations σ^n for five alternatives.

a) Compute the knowledge gradient for each alternative in a spreadsheet. Create a plot with the mean, the standard deviation, and the knowledge gradient for each alternative.

b) Three of the alternatives have the same standard deviation, but with increasing priors. Three have the same prior, but with increasing standard deviations. From these two (overlapping) sets of alternatives, describe how the knowledge gradient changes as we vary priors and the standard deviation of our belief.

Table 5.6   Calculations Illustrating the Knowledge Gradient Index

Choice	θ^n	σ^n
1	3.0	8.0
2	4.0	8.0
3	5.0	8.0
4	5.0	9.0
5	5.0	10.0

5.4   Assume that we have a standard normal prior about a true parameter μ which we assume is normally distributed with mean μ^0 and variance $(\sigma^0)^2$.

a) Given the observations $W^1, \ldots, W^n$, is θ^n deterministic or random?

b) Given the observations $W^1, \ldots, W^n$, what is $\mathbb{E}(\mu|W^1, \ldots, W^n)$ (where μ is our truth)? Why is μ random given the first n measurements?

c) Given the observations $W^1, \ldots, W^n$, what is the mean and variance of θ^{n+1}? Why is θ^{n+1} random?

5.5   As a venture capitalist specializing in energy technologies, you have to decide to invest in one of three strategies for converting solar power to electricity. A major concern is the efficiency of a solar panel, which tends to run around 11 to 12 percent. You are at the point where you are running field experiments, but each field experiment produces an estimate of the efficiency which has a standard deviation of 4.0. The first technology appears to have an efficiency of 11.5 percent, but the standard deviation in your distribution of belief around this number is 2.0. The second technology has an estimated efficiency of 11.0 with a standard deviation of 3.5, while the third has an estimated efficiency of 12.0, with a standard deviation of 1.5. You want to choose the technology with the highest efficiency.

a) Use the knowledge gradient to tell you which technology you should experiment with next.

b) If you are only going to do one last experiment, is this the optimal choice? Explain.

c) If you did not make this last investment, you would choose technology 3 with an efficiency of 12.0. What is the expected efficiency of the technology that would be chosen as best after the last investment?

5.6   Consider the problem of finding the best person to serve as the lead-off hitter on a baseball team. The lead-off hitter is evaluated primarily for his ability to get on base. If x is the hitter, his outcome would be recorded as a Bernoulli random variable W_x^n, where $W_x^n = 1$ if he gets on base, and $W_x^n = 0$ otherwise. We are going to conduct these experiments during spring training, where we are primarily focused on finding the best lead-off hitter (we do not really care about his performance while we are collecting the information). Learning is accomplished using the beta–Bernoulli model. Each alternative x has an unknown success probability ρ_x, and our goal is to find the alternative with the highest success probability. We begin with a prior belief $\rho_x \sim Beta\left(\alpha_x^0, \beta_x^0\right)$. Supposing that we measure alternative x^n at time n, our beliefs are updated according to the equations

$$\alpha_x^{n+1} = \begin{cases} \alpha_x^n + W_x^{n+1} & \text{if } x^n = x, \\ \alpha_x^n & \text{otherwise,} \end{cases}$$

$$\beta_x^{n+1} = \begin{cases} \beta_x^n + \left(1 - W_x^{n+1}\right) & \text{if } x^n = x, \\ \beta_x^n & \text{otherwise,} \end{cases}$$

where the observation W^{n+1} is equal to 1 with probability ρ_x and 0 with probability $1 - \rho_x$. Recall that, under this model, our estimate of the uncertain truth ρ_x at time n is $\mathbb{E}\left(\rho_x \mid S^n\right) = \frac{\alpha_x^n}{\alpha_x^n + \beta_x^n}$.

a) Suppose that we measure alternative x at time n. Show (by conditioning on the truth) that

$$\mathbb{P}\left(W_x^{n+1} = 1 \mid S^n\right) = \frac{\alpha_x^n}{\alpha_x^n + \beta_x^n}, \qquad \mathbb{P}\left(W_x^{n+1} = 0 \mid S^n\right) = \frac{\beta_x^n}{\alpha_x^n + \beta_x^n}.$$

b) Use the definition of the knowledge gradient to write out an expression for the knowledge gradient for this problem. You do not have to reduce the expression in any way (for example, you will have an expectation, but you do not have to reduce it to a convenient expression).

5.7   Garrett Jones was a minor leaguer in baseball trying to break into the major leagues. He was called up to play in a few major league games, where he made one hit in eight at-bats. After this weak performance, he was sent back to the minor leagues. The major league club that was evaluating him is looking for someone who can hit at a certain level against an existing major league hitter. Think of this as choosing between an uncertain minor leaguer, and a more certain major leaguer (so this is a case with two alternatives).

a) It is reasonable to assume that no one would ever make a decision based on a single at-bat. Assume that our minor leaguer will be given at least 10 at-bats and that we will now assume that our prior belief about his batting average is normally distributed with mean 0.250 and standard deviation 0.20. Further assume that our belief about the major leaguer is also normally distributed with mean 0.267 and standard deviation of 0.10. Finally assume that we are going to approximate the observed batting average from at least 10 at-bats as normally distributed with mean:

$$W_{minor} = \frac{H}{m},$$

where H is the number of hits and m is the number of at-bats. The variance of W_{minor} is given by

$$\sigma_W^2 = \rho_{minor}(1 - \rho_{minor})/m,$$

where $\rho_{minor} = .235$ is the expected batting average of the minor leaguer (the true batting average is a random variable). Give the expression for the knowledge gradient resulting from m at-bats and compute the knowledge gradient.

b) Assume that the knowledge gradient for a single at-bat is very small. Without actually computing the knowledge gradient, plot what is likely the general shape of the value of observing m at-bats as a function of m. If you were going to use the KG(*) policy, what would you do? What are the implications of this shape in terms of how a coach should evaluate different minor leaguers?

5.8 Consider a ranking and selection problem with exponential observations and gamma priors. That is, if we choose to measure alternative x at time n, we observe $W_x^{n+1} \sim \exp(\lambda_x)$. The rate λ_x is unknown, but we start with the assumption that $\lambda_x \sim Gamma(\alpha_x^0, \beta_x^0)$ and update these beliefs as we make measurements. Our beliefs about the alternatives are independent. Thus, if we measure alternative x at time n, we update $\alpha_x^{n+1} = \alpha_x^n + 1$ and $\beta_x^{n+1} = \beta_x^n + W_x^{n+1}$ while keeping $\alpha_y^{n+1} = \alpha_y^n$ and $\beta_y^{n+1} = \beta_y^n$ for all $y \neq x$.

a) Suppose that our objective is to find the largest rate λ_x. Define the knowledge gradient of alternative x at time n as

$$\nu_x^{KG,n} = \mathbb{E}\left[\max_y \frac{\alpha_y^{n+1}}{\beta_y^{n+1}} - \max_y \frac{\alpha_y^n}{\beta_y^n} \mid S^n, x^n = x\right].$$

Argue that

$$\mathbb{E}\left[\max_y \frac{\alpha_y^{n+1}}{\beta_y^{n+1}} \mid S^n, x^n = x\right] = \mathbb{E}\left[\max\left(C_x^n, \frac{\alpha_x^n + 1}{Y}\right)\right],$$

where $C_x^n = \max_{y \neq x} \frac{\alpha_y^n}{\beta_y^n}$ and $Y \sim Pareto(\alpha_x^n, \beta_x^n)$.

(Remember that the $Pareto(a, b)$ density is $g(t) = \frac{ab^a}{t^{a+1}}$ for $t > b$ and zero elsewhere.)

b) Suppose that $\frac{\alpha_x^n + 1}{\beta_x^n} \leq C_x^n$. Show that

$$\mathbb{E}\left[\max\left(C_x^n, \frac{\alpha_x^n + 1}{Y}\right)\right] = C_x^n.$$

c) Suppose that $\frac{\alpha_x^n + 1}{\beta_x^n} > C_x^n$. Show that

$$\mathbb{E}\left[\max\left(C_x^n, \frac{\alpha_x^n + 1}{Y}\right)\right] = \frac{\alpha_x^n}{\beta_x^n} + \frac{(\beta_x^n)^{\alpha_x^n}(C_x^n)^{\alpha_x^n + 1}}{(\alpha_x^n + 1)^{\alpha_x^n + 1}}.$$

d) Based on parts b) and c), show that

$$\nu_x^{KG,n} = \begin{cases} \frac{(\beta_x^n)^{\alpha_x^n}(C_x^n)^{\alpha_x^n + 1}}{(\alpha_x^n + 1)^{\alpha_x^n + 1}} & \text{if } x = \arg\max_y \frac{\alpha_y^n}{\beta_y^n}, \\ \frac{(\beta_x^n)^{\alpha_x^n}(C_x^n)^{\alpha_x^n + 1}}{(\alpha_x^n + 1)^{\alpha_x^n + 1}} - \left(\max_y \frac{\alpha_y^n}{\beta_y^n} - \frac{\alpha_x^n}{\beta_x^n}\right) & \text{if } x \neq \arg\max_y \frac{\alpha_y^n}{\beta_y^n}, \\ & \frac{\alpha_x^n + 1}{\beta_x^n} > \max_y \frac{\alpha_y^n}{\beta_y^n}, \\ 0 & \text{otherwise.} \end{cases}$$

5.9   We again consider a ranking and selection problem with exponential observations and gamma priors. There are five alternatives, and the belief state for a certain time step n is given in Table 5.7. The objective is to find the largest rate.

Table 5.7   Priors for Exercise 5.9

x	α_x^n	β_x^n
1	2	18
2	3	17
3	1	7
4	2	15
5	3	14

5.10   Consider a ranking and selection problem with independent alternatives, exponential observations and gamma priors.

a) Suppose that we want to find the alternative with the highest rate. Derive the KG formula given in (5.22). (Hint: Consider the three cases given in the formula separately before trying to take integrals.)

b) Repeat your analysis for the case where our goal is to find the lowest rate. Show that the KG formula is given by (5.23).

5.11   Table 5.8 shows the priors θ^n and the standard deviations σ^n for five alternatives.

a) Compute the knowledge gradient for each alternative in a spreadsheet. Create a plot with the mean, standard deviation and the knowledge gradient for each alternative.

b) Three of the alternatives have the same standard deviation, but with increasing priors. Three have the same prior, but with increasing standard deviations.

From these two (overlapping) sets of alternatives, describe how the knowledge gradient changes as we vary priors and the standard deviation of our belief.

Table 5.8 Calculations Illustrating the Knowledge Gradient Index

Choice	θ^n	σ^n
1	3.0	8.0
2	4.0	8.0
3	5.0	8.0
4	5.0	9.0
5	5.0	10.0

a) Compute $\nu_x^{KG,n}$ for all x. Which alternative will the KG policy measure at time n?

b) Suppose now that $\alpha_2^n = 1$ and $\beta_2^n = 6$, while our beliefs about the other alternatives remain unchanged. How does this change your answer to part a)? Why did KG change its decision, even though the estimate $\frac{\alpha_2^n}{\beta_2^n}$ is actually smaller now than what it was in part a)?

5.12 Table 5.9 shows the priors θ^n and the standard deviations σ^n for five alternatives.

a) Compute the knowledge gradient for each in a spreadsheet.

b) You should observe that the knowledge gradients are fairly small. Provide a plain English explanation for why this would be the case.

Table 5.9 Priors for Exercise 5.12

Choice	θ^n	σ^n
1	3.0	4.0
2	4.0	6.0
3	20.0	3.0
4	5.0	5.0
5	6.0	7.0

5.13 In Section 5.9.1, we showed that $\mathbb{E}\left[\theta_x^{n+1} \mid S^n, x^n = x\right] = \theta_x^n$ in the normal–normal model. Verify that this also holds for our estimates of the unknown parameters in other learning models:

a) Show that $\mathbb{E}\left[\frac{\alpha_x^{n+1}}{\beta_x^{n+1}} \mid S^n, x^n = x\right] = \frac{\alpha_x^n}{\beta_x^n}$ in the gamma–exponential model.

b) Repeat part a) for the gamma–Poisson model.

c) Show that $\mathbb{E}\left[\frac{\alpha_x^{n+1}}{\alpha_x^{n+1}-1} b_x^{n+1} \mid S^n, x^n = x\right] = \frac{\alpha_x^n}{\alpha_x^n - 1} b_x^n$ in the Pareto-uniform model.

d) Show that $\mathbb{E}\left[\frac{\alpha_x^{n+1}}{\alpha_x^{n+1}+\beta_x^{n+1}} \mid S^n, x^n = x\right] = \frac{\alpha_x^n}{\alpha_x^n + \beta_x^n}$ in the beta–Bernoulli model.

In each of these cases, our estimates of the unknown parameters (the rate λ_x, the upper endpoint B_x, and the success probability ρ_x) are expected to stay the same on average. That is, given that we are at time n, we expect that the estimate will not change on average between time n and time $n + 1$. This is called the *martingale property*.

5.14 Consider a ranking and selection problem with independent alternatives, Poisson observations and gamma priors. Suppose that the objective is to find the alternative with the highest Poisson rate. Show that the KG formula is given by (5.24).

5.15 Consider a ranking and selection problem with independent alternatives, uniform observations and Pareto priors. Suppose that the objective is to find the largest upper endpoint among the uniform distributions.

a) Show that the predictive distribution of b_x^{n+1} given S^n and $x^n = x$ is a mixed discrete/continuous distribution given by

$$\mathbb{P}\left(b_x^{n+1} = b_x^n \mid S^n, x^n = x\right) = \frac{\alpha_x^n}{\alpha_x^n + 1}$$

and

$$f\left(y \mid S^n, x^n = x\right) = \frac{1}{\alpha_x^n + 1} \frac{\alpha_x^n (b_x^n)^{\alpha_x^n}}{y^{\alpha_x^n+1}}, \quad y > b_x^n.$$

b) Show that the KG formula for alternative x in this problem is given by (5.25).

5.16 Consider a ranking and selection problem with independent alternatives, Bernoulli observations and beta priors. Suppose that the objective is to find the alternative with the largest success probability. Show that the KG formula is given by (5.26), and verify that this formula remains the same (aside from changing the maximum in the definition of C_x^n to a minimum) if we change the objective to finding the lowest success probability instead of the highest.

5.17 Consider a ranking and selection problem with independent alternatives, normal observations and normal priors. However, instead of the usual objective function $F^\pi = \max_x \theta_x^N$, we use

$$F^\pi = \sum_x \mathbb{E}\left[(\mu_x - \theta_x^N)^2 \mid S^N\right].$$

That is to say, instead of trying to find the alternative with the highest value, our goal is now to make measurements in such a way as to reduce (on average) the sum of squared errors of our final estimates at time N of all the values. Derive the KG formula for this problem.

5.18 Suppose that we have four different products. The profit margin of product x, $x = 1, 2, 3, 4$ is represented by μ_x. We can choose to perform a market study on product x to get an observation W_x of its profit margin. The observation is normally distributed with mean μ_x and variance $(\sigma_x^W)^2$. Table 5.10 below gives the true values of μ_x and $(\sigma_x^W)^2$:

Table 5.10   Priors for Exercise 5.18.

x	μ_x	$(\sigma_x^W)^2$
1	15	4
2	10	3
3	12	5
4	11	2

However, we do not know the true values of μ_x. We describe our beliefs about them using a multivariate Gaussian prior with the following mean vector and covariance matrix:

$$\theta^0 = \begin{bmatrix} 12 \\ 14 \\ 13 \\ 10 \end{bmatrix}, \quad \Sigma^0 = \begin{bmatrix} 12 & 0 & 6 & 3 \\ 0 & 7 & 4 & 2 \\ 6 & 4 & 9 & 0 \\ 3 & 2 & 0 & 8 \end{bmatrix}.$$

Assume that we choose to observe product 3 and we observe $W_3^1 = 15$. Show how our beliefs would change using the updating equations

$$\theta^1 = \theta^0 + \frac{W_x^1 - \theta_x^0}{(\sigma_x^W)^2 + \Sigma_{xx}^0} \Sigma^0 e_x, \quad \Sigma^1 = \Sigma^0 - \frac{\Sigma^0 e_x e_x^T \Sigma^0}{(\sigma_x^W)^2 + \Sigma_{xx}^n}, \tag{5.37}$$

where $x = 3$ is the particular product that you are considering, and e_x is a column vector of zeroes with a single 1 in the xth coordinate. Report the resulting values of θ^1 and Σ^1. (Equation (5.37) gives the "convenient" version of the updating equations, where you don't have to compute an inverse. You do not have to derive these equations.)

5.19 Consider a ranking and selection problem with independent alternatives, normal observations and normal priors. However, instead of the usual objective function $F^\pi = \max_x \theta_x^N$, we use

$$F^\pi = \max_x \left| \theta_x^N \right|$$

That is, we want to find the alternative with the largest absolute value. Show that the KG factor of alternative x is given by

$$\nu_x^{KG,n} = \tilde{\sigma}_x^n \left(f\left(\zeta_x^n \right) + f\left(\delta_x^n \right) \right),$$

where

$$\zeta_x^n = -\left|\frac{\theta_x^n - \max_{x' \neq x}|\theta_{x'}^n|}{\bar{\sigma}_x^n}\right|, \qquad \delta_x^n = -\frac{\theta_x^n + \max_{x' \neq x}|\theta_{x'}^n|}{\bar{\sigma}_x^n}.$$

5.20   The revenue generated by an online advertisement has an exponential distribution with parameter λ (thus the mean revenue is $\frac{1}{\lambda}$). We do not know λ, so we use a gamma prior and assume $\lambda \sim Gamma(a, b)$ for $a > 1$. Recall that the density of the gamma distribution is given by

$$f(x) = \frac{b(bx)^{a-1}e^{-bx}}{\Gamma(a)},$$

where $\Gamma(a) = (a - 1)!$ if a is integer. The mean of $f(x)$ is a/b and the variance is a/b^2.

a) What is the current belief about the value of λ? That is, if you had to guess the value of λ, what would you say, and why? If you assume that λ is exactly equal to this belief, what is the mean revenue generated by the advertisement?

b) Now take an expectation of the mean revenue over the entire distribution of belief. That is, compute $\mathbb{E}(\frac{1}{\lambda})$ for $\lambda \sim Gamma(a, b)$.

c) Why are your answers to (a) and (b) different? Which one should you actually use as your estimate of the mean reward, and why?

5.21   Consider a ranking and selection problem with normal observations and normal priors (and independent beliefs).

a) Create a MATLAB file called kg.m which implements the KG policy. As a template, you can use the code that was first introduced in exercise 4.2 which can be downloaded from

   http://optimallearning.princeton.edu/exercises/exploration.m
   http://optimallearning.princeton.edu/exercises/explorationRun.m

b) Set $N = 5000$, $M = 50$ and report the confidence interval.

c) Set $N = 1$, $M = 1$ and run the policy 100 times. How often does KG find the best alternative?

5.22   You would like to find the price of a product that maximizes revenue. Unknown to you, the demand for the product is given by

$$D(p) = 100e^{-.02p}.$$

Total revenue is given by $R(p) = pD(p)$. Assume prices are integers between 1 and 100.

You set a price and watch it for a week. Assume that the observed revenue R^n in week n is given by

$$R^n = R(p) + \epsilon^n,$$

where $\epsilon^n \sim \mathcal{N}(0, 400^2)$. However, since you believe that the function is continuous, you realize that your beliefs are correlated. Assume that your belief about $R(p)$ and $R(p')$ is correlated with covariance function

$$Cov^0(R(p), R(p')) = 400^2 e^{-0.03|p-p'|}.$$

So, $Cov^0(R(p), R(p)) = Var^0(R(p)) = 400^2$, and $Cov^0(R(20), R(30)) = 400^2 \times 0.7408$. Use this to create your prior covariance matrix Σ^0. Assume that your initial estimate of $R(p)$ is $\theta_p^0 = 2000$ for each p (this is known as a "uniform prior," and represents a situation where you have no idea which price is the best.) Note that we are using online learning for this exercise.

a) Write out the updating formulas for updating your estimate θ_p^n giving the estimated revenue when you charge price p, and the updating formula for the covariance matrix Σ^n.

b) Implement the algorithm for computing the knowledge gradient in the presence of correlated beliefs (call this algorithm KGCB), using as a starting point

http://optimallearning.princeton.edu/exercises/KGCorrBeliefs.m

An example illustration of the KGCB algorithm is given in

http://optimallearning.princeton.edu/exercises/KGCorrBeliefsEx.m

Verify your algorithm first by running it with a diagonal covariance matrix and showing that the independent and correlated KG algorithms give you the same numbers. Note that you may find that for some prices, the knowledge gradient is too small to compute (for example, you get a very negative exponent).

c) Next use the initial covariance matrix described above. Plot the log of the knowledge gradient for prices between 1 and 100 (again, be careful with large negative exponents), and compare your results to the log of the knowledge gradient assuming independent beliefs. How do they compare?

d) Now we are going to compare policies. Please do the following:

   i) Run your KGCB algorithm for 10 measurements, and plot after each measurement the opportunity cost, which means you take what you think is the best price based on your current set of estimates and compare it to the revenue you would get if you knew the best price (hint: it is $50). Repeat this exercise 20 times, and report the average opportunity cost, averaged over the 20 iterations. The goal here is to get a sense of the variability of the performance of a learning policy.

ii) We want to compare KGCB against pure exploration, pure exploitation, and interval estimation using $z_\alpha = 1.96$ (a standard default). For each policy, perform 20 sample paths and plot the average opportunity cost over all 20 sample paths. Compare all three policies. [Please make sure that you are resetting the covariance matrix to its initial structure after you are done with a sample path (once you have gone over a single truth). If you skip this part, the KG algorithm will assume it has very precise priors for the second run, which of course is not true.]

e) The knowledge gradient policy can be quite sensitive to the prior. Instead of an initial prior of 2000, now assume that we start with a uniform prior of 500 (same standard deviation). If you edit the prior (column B), the package regenerates the truth. You are going to have to re-enter the formula for the truth after changing the prior. After doing this, perform 3 repetitions of KG, pure exploration and pure exploitation, and contrast their performance.

5.23 This exercise uses the optimal learning library. Download the zip file from

http://optimallearning.princeton.edu/exercises/OptimalLearningCalculator.zip

Download and extract all the files, which will consist of an Excel spreadsheet KGCalculator.xls and a folder called lib which needs to be in the same location as the spreadsheet.

You would like to find the price of a product that maximizes revenue. Unknown to you, the demand for the product is given by

$$D(p) = 100e^{-.02p}.$$

Total revenue is given by $R(p) = pD(p)$. Assume prices are integers between 1 and 100.

You set a price and watch it for a week. Assume that the observed revenue R^n in week n is given by

$$R^n = R(p) + \epsilon^n$$

where $\epsilon^n \sim \mathcal{N}(0, 400^2)$. However, since you believe that the function is continuous, you realize that your beliefs are correlated. Assume that your belief about $R(p)$ and $R(p')$ is correlated with covariance function

$$Cov^0(R(p), R(p')) = 400^2 e^{-0.03|p-p'|}.$$

So, $Cov^0(R(p), R(p)) = Var^0(R(p)) = 400^2$, and $Cov^0(R(20), R(30)) = 400^2 \times 0.7408$. Use this to create your prior covariance matrix Σ^0. Assume that your initial estimate of $R(p)$ is $\theta_p^0 = 1400$ for each p (this is known as a "uniform prior" and represents a situation where you have no idea which price is the best).

a) Write out the updating formulas for updating your estimate θ_p^n giving the estimated revenue when you charge price p, and also write out the updating formula for the covariance matrix Σ^n.

b) Using the Optimal Learning calculator, set the package for the problem class "Correlated" on the home screen (do not use "Independent" or "On a line"). Enter 100 under M (cell A3). Then hit the "edit covariance" button, and use the formula above to enter the covariance matrix. Returning to the home screen, hit "Edit settings" and set "measurement error" to 400 (this is the standard deviation of the measurement noise). Finally, change the "range end" from 5 to 100 (lower left–hand corner of the edit settings screen). Hit the "exit and save changes" button to exit. Enter the prior of 2000 in column B. Using the formula for the true demand given above, compute the true revenue for each price and enter this in column D. Verify that the highest revenue occurs at $50. Now we are ready to try learning this true function.

c) Before making any measurements, hit the "graphs" button and make a copy of the "current beliefs" graph and the log of the knowledge gradient, and include this in your writeup. Is the true belief within the confidence interval of your prior?

d) Now we are going to compare policies. Please do the following:

  i) Enter the number 50 next to the field "number of iterations" in column H. Run your KGCB algorithm for 50 measurements (by hitting the "implement policy" button). Make a copy of the graphs showing the current belief about the function, number of measurements by alternative, and the opportunity cost. Repeat this three times (so your report should have three sets of three graphs). Discuss how well your KG policy performed. Contrast the ability of KG to identify a near optimal price to its ability to estimate the true function.

  ii) We want to compare KGCB against pure exploitation, pure exploration, and interval estimation using the best value of z_α that you obtained in problem set 3. For each policy, perform 10 measurements, and we want to perform 20 iterations for each policy. To avoid doing this manually, go to "edit settings" and enter 20 in the field next to "budget" (this is your number of measurements - it should read "Infinity" when you first see this field). When you hit save, notice that the field "number of iterations" now reads "number of samples." Now enter 20 in this box (so we repeat each simulation 20 times). Now you are set up to do competitions between KG and other policies. First hit the "other" button (column I), click on the pull down menu, and choose pure exploitation. Save and exit, and then hit "implement policy." When the comparison is done, you will see a summary. At the top it shows the opportunity cost (with confidence intervals) for both KG and the competing policy. Report these numbers. Repeat this five times, for each of the three competing policies. Be sure to hit the "clear history" button before each run. You want to be sure that cell B3 is zero before you hit the implement policy button. Discuss the performance of each policy, covering not only how well one policy seems to outperform the others, but also the variability of this performance.

e) The knowledge gradient policy can be quite sensitive to the prior. Instead of an initial prior of 2000, now assume that we start with a uniform prior of 500 (same standard deviation). If you edit the prior (column B), the package regenerates the truth. You are going to have to re-enter the formula for the truth after changing the prior. After doing this, perform three repetitions of KG, pure exploration, and pure exploitation, and contrast their performance.

CHAPTER 6

BANDIT PROBLEMS

The *multi-armed bandit problem* is a venerable topic in optimal learning and has inspired some of the pioneering work in the field. The story that was originally used to motivate the problem (and gave the problem its name) is not really an important application, but is useful for understanding the basic idea behind the problem. The term "one-armed bandit" refers to a slot machine operated by pulling a lever (or "arm") on its side. A multi-armed bandit, then, is a collection of slot machines, each with a different winning probability (or different average winnings).

Suppose that we have M slot machines, and we have enough money to play N times. If we play slot machine x at time n, we receive random winnings W_x^{n+1}. Suppose that the expected value of these winnings, given by μ_x, is unknown. We would like to estimate μ_x, but the only way to do this is by putting money into the machine and collecting a random observation. For this reason, we must balance our desire to find the slot machine with the highest value with our desire to achieve good results on every play.

It is easy to see the similarity between this problem and the ranking and selection problem from Chapter 4. In the multi-armed bandit problem, we have a clear set of alternatives, namely the different arms. Every arm has an unknown value, namely the average winnings. We can create and update estimates of these values using the exact same model laid out in Section 4.1. At each time step, we choose an arm x, observe the winnings W_x^{n+1}, and update our beliefs.

The difference lies in the objective function. In ranking and selection, we choose a measurement policy to optimize the offline objective function

$$\max_{\pi \in \Pi} \mathbb{E}^{\pi} \mu_{x^N},$$

where $x^N = \arg\max_x \theta_x^N$. This objective is designed for problems in which our measurements are only important insofar as they give us a good final estimate. For example, we could be running tests on various experimental drug compounds in a laboratory setting. We do not incur a penalty from a poor outcome of a single experiment. In fact, such an outcome could actually be desirable, because it may give us important information about the problem.

By contrast, the multi-armed bandit problem is online. The idea is that we are playing the slot machines in real time, and our wealth depends directly on the outcome of each individual play. The correct objective function for this problem is

$$\max_{\pi \in \Pi} \mathbb{E}^{\pi} \sum_{n=0}^{N} \gamma^n \mu_{x^n}, \qquad (6.1)$$

where x^n is the decision we make at time n, and $\gamma \in (0, 1]$ is a discount factor. This objective function causes us to play more conservatively than the offline objective. It is possible that a poor outcome of a single play may give us valuable information, but it would also penalize our objective value, and we might not necessarily be able to use the information we gained to make up for our immediate losses later on.

The rest of the model is the same as before; the objective function is the only real difference between ranking and selection and multi-armed bandits. However, this difference can be very important in practical applications. Although the traditional slot machine example is not in itself a practical application, it is possible to think of many applications where the online objective is a crucial feature of the problem. For example, suppose that we are performing clinical trials on a set of experimental medical treatments. The treatments have already passed the laboratory trials, and we have moved to testing them on human patients. In this case, it is much more important to be mindful of the outcome of each individual trial and to try to ensure the best possible outcome for every patient while still learning something about which treatment is the most effective.

For another example, suppose that we are running an online advertising system. Every day, we can choose one company's advertisement to display on our website. We would like to experiment with different companies' advertisements to figure out which is the most interesting and profitable. However, we must also keep an eye on the revenue we collect from the advertisements that we display each day. Here, too, we have to try to achieve a good outcome (high profitability) for every advertisement that we display while also learning in the process.

In the multi-armed bandit problem, the difficulty of balancing between exploration (pulling an arm that may turn out to have high winnings) and exploitation (pulling an arm that already seems to have the highest winnings) is highlighted especially starkly. Perhaps for this reason, there is a deep literature on bandit problems. The crowning result of this literature is the creation of a policy that is optimal as long as $N \to \infty$; that is, we have infinitely many chances to measure. This policy is based on a clever shortcut credited to J.C. Gittins. Gittins found that instead of solving the dynamic program from Section 4.2.2 with the multidimensional state variable, it was possible to characterize the optimal solution using something known as an "index policy." This works by computing an index $I_x^n (\theta_x^n, \beta_x^n)$ for each option x at each iteration n. The index I_x^n is computed by solving M single-dimensional problems. The index policy chooses to measure $x^n = \arg\max_x I_x^n$, which is to say that we play the machine x^n which corresponds to the largest index. Thus, instead of solving a single M-dimensional problem, we have to solve M one-dimensional problems.

6.1 THE THEORY AND PRACTICE OF GITTINS INDICES

The idea behind Gittins indices works as follows. Assume that we are playing a single slot machine, and that we have the choice of continuing to play the slot machine or stopping and switching to a process that pays a reward r. If we choose not to play, we receive r, and then find ourselves in the same state (since we did not collect any new information). If we choose to play, we earn a random amount W, plus we earn $\mathbb{E}\{V(S^{n+1}, r)|S^n\}$, where S^{n+1} represents our new state of knowledge resulting from our observed winnings. For reasons that will become clear shortly, we write the value function as a function of the state S^{n+1} and the stopping reward r.

The value of being in state S^n, then, can be written as

$$V(S^n, r) \quad = \quad \max\left[r + \gamma V(S^n, r), \mathbb{E}\left\{W^{n+1} + \gamma\, V(S^{n+1}, r)\big|\, S^n\right\}\right].$$

The first choice represents the decision to receive the fixed reward r, while in the second choice we get to observe W^{n+1} (which is random when we make the decision). When we have to choose x^n, we will use the expected value of our return if we continue playing, which is computed using our current state of knowledge. For example, in the Bayesian normal–normal model, $\mathbb{E}\{W^{n+1}|S^n\} = \theta^n$, which is our estimate of the mean of W given what we know after the first n measurements.

If we choose to stop playing at iteration n, then S^n does not change, which means we earn r and face the identical problem again for our next play. In this case, once we decide to stop playing, we will never play again, and we will continue to receive r (discounted) from now on. For this reason, r is called the retirement reward. The infinite horizon, discounted value of retirement is $r/(1 - \gamma)$. This means that we can rewrite our optimality recursion as

$$V(S^n, r) = \max\left[\frac{r}{1 - \gamma}, \mathbb{E}\{W^{n+1} + \gamma\, V(S^{n+1}, r)\big|\, S^n\}\right], \qquad (6.2)$$

Here is where we encounter the magic of Gittins indices. We compute the value of r that makes us indifferent between stopping and accepting the reward r (forever), versus continuing to play the slot machine. That is, we wish to solve the equation

$$\frac{r}{1-\gamma} = \mathbb{E}\left\{W^{n+1} + \gamma \, V(S^{n+1}, r) \big| \, S^n\right\} \tag{6.3}$$

for r. The Gittins index $I^{Gitt,n}$ is the particular value of r that solves (6.3). This index depends on the state S^n. If we use a Bayesian perspective and assume normally distributed rewards, we would use $S^n = (\theta^n, \beta^n)$ to capture our distribution of belief about the true mean μ. If we use a frequentist perspective, our state variable would consist of our estimate $\bar{\theta}^n$ of the mean, our estimate $\hat{\sigma}^{2,n}$ of the variance, and the number N^n of observations (this is equal to n if we only have one slot machine).

If we have multiple slot machines, we consider every machine separately, as if it were the only machine in the problem. We would find the Gittins index $I_x^{Gitt,n}$ for every machine x. Gittins showed that, if $N \to \infty$, meaning that we are allowed to make infinitely many measurements, it is optimal to play the slot machine with the highest value of $I_x^{Gitt,n}$ at every time n. Notice that we have not talked about how exactly (6.3) can be solved. In fact, this is a major issue, but for now, assume that we have some way of computing $I_x^{Gitt,n}$.

Recall that, in ranking and selection, it is possible to come up with trivial policies that are asymptotically optimal as the number of measurements goes to infinity. For example, the policy that measures every alternative in a round-robin fashion is optimal for ranking and selection: If we have infinitely many chances to measure, this policy will measure every alternative infinitely often, thus discovering the true best alternative in the limit. However, in the multi-armed bandit setting, this simple policy is likely to work extremely badly. It may discover the true best alternative in the limit, but it will do poorly in the early iterations. If $\gamma < 1$, the early iterations are more important than the later ones, because they contribute more to our objective value. Thus, in the online problem, it can be more important to pick good alternatives in the early iterations than to find the true best alternative. The Gittins policy is the only policy with the ability to do this optimally.

6.1.1 Gittins Indices in the Beta–Bernoulli Model

The Gittins recursion in (6.2) cannot be solved using conventional dynamic programming techniques. Even in the beta–Bernoulli model, one of the simplest learning models we have considered, the number of possible states S^n is uncountably infinite. In other models like the normal–normal model, S^n is also continuous. However, in some models, the expectation in the right-hand side of (6.2) is fairly straightforward, allowing us to get a better handle on the problem conceptually.

Let us consider the beta–Bernoulli model for a single slot machine. Each play has a simple 0/1 outcome (win or lose), and the probability of winning is ρ. We do know this probability exactly, so we assume that ρ follows a beta distribution with parameters α^0 and β^0. Recall that the beta–Bernoulli model is conjugate, and the updating equations are given by

$$\begin{aligned}
\alpha^{n+1} &= \alpha^n + W^{n+1}, \\
\beta^{n+1} &= \beta^n + \left(1 - W^{n+1}\right),
\end{aligned}$$

where the distribution of W^{n+1} is Bernoulli with success probability ρ. After n plays, the distribution of ρ is beta with parameters α^n and β^n. The knowledge state for a single slot machine is simply $S^n = (\alpha^n, \beta^n)$. Consequently,

$$
\begin{aligned}
\mathbb{E}\left(W^{n+1} \mid S^n\right) &= \mathbb{E}\left[\mathbb{E}\left(W^{n+1} \mid S^n, \rho\right) \mid S^n\right] \\
&= \mathbb{E}\left(\rho \mid S^n\right) \\
&= \frac{\alpha^n}{\alpha^n + \beta^n}.
\end{aligned}
$$

Then, writing $V(S^n, r)$ as $V(\alpha^n, \beta^n, r)$, we obtain

$$
\begin{aligned}
\mathbb{E}\left\{W^{n+1} + \gamma\, V(S^{n+1}, r) \mid S^n\right\} &= \frac{\alpha^n}{\alpha^n + \beta^n} + \gamma \frac{\alpha^n}{\alpha^n + \beta^n} V\left(\alpha^n + 1, \beta^n, r\right) \\
&\quad + \gamma \frac{\beta^n}{\alpha^n + \beta^n} V\left(\alpha^n, \beta^n + 1, r\right). \qquad (6.4)
\end{aligned}
$$

For fixed α and β, the quantity $V(\alpha, \beta, r)$ is a constant. However, if the observation W^{n+1} is a success, we will transition to the knowledge state $(\alpha^n + 1, \beta^n)$; and if it is a failure, the next knowledge will be $(\alpha^n, \beta^n + 1)$. Given S^n, the conditional probability of success is $\frac{\alpha^n}{\alpha^n + \beta^n}$.

From (6.4), it becomes clear why Gittins indices are difficult to compute exactly. For any value of r and any α, β, we need to know $V(\alpha + 1, \beta, r)$ as well as $V(\alpha, \beta + 1, r)$ before we can compute $V(\alpha, \beta, r)$. But there is no limit on how high α and β are allowed to go. These parameters represent roughly the tallies of successes and failures that we have observed, and these numbers can take on any integer value if we assume an infinite horizon.

However, it is possible to compute $V(\alpha, \beta, r)$ approximately. For all α and β such that $\alpha + \beta$ is "large enough," we could assume that $V(\alpha, \beta, r)$ is equal to some value, perhaps zero. Then, a backwards recursion using these terminal values would give us approximations of $V(\alpha, \beta, r)$ for small α and β.

The quality of such an approximation would depend on how many steps we would be willing to perform in the backwards recursion. In other words, the larger the value of $\alpha + \beta$ for which we cut off the recursion and set a terminal value, the better. Furthermore, the approximation would be improved if these terminal values were themselves as close to the actual value functions as possible.

One way of choosing terminal values is the following. First, fix a value of r. If $\alpha + \beta$ is very large, it is reasonable to suppose that

$$
V(\alpha, \beta, r) \approx V(\alpha + 1, \beta, r) \approx V(\alpha, \beta + 1, r).
$$

Then, we can combine (6.2) with (6.4) to approximate the Gittins recursion as

$$
V(\alpha, \beta, r) = \max\left[\frac{r}{1 - \gamma}, \frac{\alpha}{\alpha + \beta} + \gamma V(\alpha, \beta, r)\right]. \qquad (6.5)
$$

In this case, it can be shown that (6.5) has the solution

$$
V(\alpha, \beta, r) = \frac{1}{1 - \gamma} \max\left(r, \frac{\alpha}{\alpha + \beta}\right),
$$

Table 6.1 Gittins indices for the beta–Bernoulli model with $\alpha, \beta = 1, ..., 7$ for $\gamma = 0.9$.

β \ α	1	2	3	4	5	6	7
1	0.7029	0.8001	0.8452	0.8723	0.8905	0.9039	0.9141
2	0.5001	0.6346	0.7072	0.7539	0.7869	0.8115	0.8307
3	0.3796	0.5163	0.6010	0.6579	0.6996	0.7318	0.7573
4	0.3021	0.4342	0.5184	0.5809	0.6276	0.6642	0.6940
5	0.2488	0.3720	0.4561	0.5179	0.5676	0.6071	0.6395
6	0.2103	0.3245	0.4058	0.4677	0.5168	0.5581	0.5923
7	0.1815	0.2871	0.3647	0.4257	0.4748	0.5156	0.5510

and the solution to (6.3) in this case is simply

$$r^* = \frac{\alpha}{\alpha + \beta}.$$

We can use this result to approximate Gittins indices for a desired α^n and β^n. First, we choose some large number N. If $\alpha + \beta \geq N$, we assume that $V(\alpha, \beta, r) = \frac{1}{1-\gamma}\frac{\alpha}{\alpha+\beta}$ for all r. Then, we can use (6.2) and (6.4) to work backwards and compute $V(\alpha^n, \beta^n, r)$ for a particular value of r. Finally, we can use a search algorithm to find the particular value r^* that makes the two components of the maximum in the expression for $V(\alpha^n, \beta^n, r)$ equal.

The computational cost of this method is high. If N is large, the backwards recursion becomes more expensive for each value of r, and we have to repeat it many times to find the value r^*. However, the recursion (for fixed r) is simple enough to be coded in a spreadsheet, and r can then be varied through trial and error (see Exercise 6.3). Such an exercise allows one to get a sense of the complexity of the problem.

When all else fails, the monograph by Gittins (1989) provides tables of Gittins indices for several values of γ and $\alpha, \beta = 1, 2, ..., 40$. A few of these values are given in Tables 6.1 and 6.2 for $\gamma = 0.9, 0.95$. The tables allow us to make several interesting observations. First, the Gittins indices are all numbers in the interval $[0, 1]$. In fact, this is always true in the beta–Bernoulli model (but not for other models, as we shall see in the next section). Second, the indices are increasing in the number of successes and decreasing in the number of failures. This is logical; if the number of successes is low, and the number of trials is high, we can be fairly sure that the success probability of the slot machine is low, and therefore the fixed-reward process should give us smaller rewards.

Table 6.2 Gittins indices for the beta–Bernoulli model with $\alpha, \beta = 1, ..., 7$ for $\gamma = 0.95$.

β \\ α	1	2	3	4	5	6	7
1	0.7614	0.8381	0.8736	0.8948	0.9092	0.9197	0.9278
2	0.5601	0.6810	0.7443	0.7845	0.8128	0.8340	0.8595
3	0.4334	0.5621	0.6392	0.6903	0.7281	0.7568	0.7797
4	0.3477	0.4753	0.5556	0.6133	0.6563	0.6899	0.7174
5	0.2877	0.4094	0.4898	0.5493	0.5957	0.6326	0.6628
6	0.2439	0.3576	0.4372	0.4964	0.5440	0.5830	0.6152
7	0.2106	0.3172	0.3937	0.4528	0.4999	0.5397	0.5733

6.1.2   Gittins Indices in the Normal–Normal Model

Let us now switch to the normal–normal model. Instead of winning probabilities, we deal with average winnings, and we assume that the winnings in each play follow a normal distribution. The quantity μ_x represents the unknown average winnings of slot machine x. Every observation W_x^n is normal with mean μ_x and known precision β^W, and every unknown mean μ_x is normally distributed with prior mean θ_x^0 and prior precision β_x^0. In every time step, we select an alternative x, observe a random reward W_x^{n+1}, and apply (4.1) and (4.2) to obtain a new set of beliefs $\left(\theta^{n+1}, \beta^{n+1}\right)$.

The Gittins index of slot machine x at time n can be written as the function $I_x^{Gitt,n}\left(\theta_x^n, \sigma_x^n, \sigma_W, \gamma\right)$. Observe that this quantity only depends on our beliefs about slot machine x, and not on our beliefs about any other slot machines $y \neq x$. This is the key feature of any index policy. However, the index does depend on the problem parameters σ_W and γ. We find it convenient to write the index in terms of the variance rather than the precision, for reasons that will become clear below.

Gittins showed that $I_x^{Gitt,n}$ can be simplified using

$$I_x^{Gitt,n}\left(\theta_x^n, \sigma_x^n, \sigma_W, \gamma\right) = \theta_x^n + \sigma_W \cdot I_x^{Gitt,n}\left(0, \frac{\sigma_x^n}{\sigma_W}, 1, \gamma\right). \qquad (6.6)$$

This equation is reminiscent of the well-known property of normal random variables. Just as any random variable can be written as a function of a standard normal random variable, so a Gittins index can be written in terms of a "standard normal" Gittins index, as long as we are using a normal–normal learning model.

For notational convenience, we can write

$$I_x^{Gitt,n}\left(0, \frac{\sigma_x^n}{\sigma_W}, 1, \gamma\right) = G\left(\frac{\sigma_x^n}{\sigma_W}, \gamma\right).$$

Table 6.3 Gittins indices $G(s, \gamma)$ for the case where $s = 1/\sqrt{k}$. Source: Gittins (1989).

	Discount factor					
k	0.5	0.7	0.9	0.95	0.99	0.995
1	0.2057	0.3691	0.7466	0.9956	1.5758	1.8175
2	0.1217	0.2224	0.4662	0.6343	1.0415	1.2157
3	0.0873	0.1614	0.3465	0.4781	0.8061	0.9493
4	0.0683	0.1272	0.2781	0.3878	0.6677	0.7919
5	0.0562	0.1052	0.2332	0.3281	0.5747	0.6857
6	0.0477	0.0898	0.2013	0.2852	0.5072	0.6082
7	0.0415	0.0784	0.1774	0.2528	0.4554	0.5487
8	0.0367	0.0696	0.1587	0.2274	0.4144	0.5013
9	0.0329	0.0626	0.1437	0.2069	0.3608	0.4624
10	0.0299	0.0569	0.1313	0.1899	0.3529	0.4299
20	0.0155	0.0298	0.0712	0.1058	0.2094	0.2615
30	0.0104	0.0202	0.0491	0.0739	0.1520	0.1927
40	0.0079	0.0153	0.0375	0.0570	0.1202	0.1542
50	0.0063	0.0123	0.0304	0.0464	0.0998	0.1292
60	0.0053	0.0103	0.0255	0.0392	0.0855	0.1115
70	0.0045	0.0089	0.0220	0.0339	0.0749	0.0983
80	0.0040	0.0078	0.0193	0.0299	0.0667	0.0881
90	0.0035	0.0069	0.0173	0.0267	0.0602	0.0798
100	0.0032	0.0062	0.0156	0.0242	0.0549	0.0730

Thus, we only have to compute Gittins indices for fixed values of the prior mean and measurement noise, and then use (6.6) to translate it to our current beliefs. Table 6.3 gives the values of $G(s, \gamma)$ for $\gamma = 0.95, 0.99$ and $s = 1/\sqrt{k}$ for $k = 1, 2, \ldots$. This corresponds to a case where the measurement noise is equal to 1, and our beliefs about alternative x have the variance $1/k$ if we make k measurements of x.

The table reveals several interesting facts about Gittins indices. First, $G\left(1/\sqrt{k}, \gamma\right)$ is increasing in γ. If γ is larger, this means that we have a larger effective time horizon. Essentially, a larger portion of our time horizon "matters"; more and more of the rewards we collect remain large enough after discounting to have a notable effect on our objective value. This means that we can afford to do more exploration, because one low reward early on will not harm our objective value as much. Hence, the Gittins indices are higher, encouraging us to explore more.

Second, $G\left(1/\sqrt{k},\gamma\right)$ is decreasing in k. This is fairly intuitive. The standard Gittins index $G(s,\gamma)$ represents the uncertainty bonus and does not depend on our estimate of the value of an alternative. As the variance of our beliefs goes down, the uncertainty bonus should go down as well, and we should only continue to measure the alternative if it provides a high reward.

Unfortunately, even the seminal work by Gittins does not give the values of $G(s,\gamma)$ for all possible s and γ. In general, computing these values is a difficult problem in and of itself. For this reason, a minor literature on Gittins approximation has arisen in the past ten years. To obtain a practical algorithm, it is necessary to examine this literature in more depth.

6.1.3 Approximating Gittins Indices

Finding Gittins indices is somewhat like finding the cdf of the standard normal distribution. It cannot be done analytically, and requires instead a fairly tedious numerical calculation. We take for granted the existence of nice functions built into most programming languages for computing the cumulative standard normal distribution, for which extremely accurate polynomial approximations are available. In Excel, this is available using the function NORMINV.

As of this writing, such functions do not exist for Gittins indices. However, in the case of the normal–normal model, there is a reasonably good approximation that results in an easily computable algorithm. First, it can be shown that

$$G(s,\gamma) = \sqrt{-\log\gamma} \cdot b\left(-\frac{s^2}{\log\gamma}\right),$$

where the function b must be approximated. The best available approximation of Gittins indices is given by

$$\tilde{b}(s) = \begin{cases} \frac{s}{\sqrt{2}}, & s \le \frac{1}{7}, \\ e^{-0.02645(\log s)^2 + 0.89106 \log s - 0.4873}, & \frac{1}{7} < s \le 100, \\ \sqrt{s}\,(2\log s - \log\log s - \log 16\pi)^{\frac{1}{2}}, & s > 100. \end{cases}$$

Thus, the approximate version of (6.6) is

$$I_x^{Gitt,n} \approx \theta_x^n + \sigma_W\sqrt{-\log\gamma} \cdot \tilde{b}\left(-\frac{\sigma_x^{2,n}}{\sigma_W^2 \log\gamma}\right). \tag{6.7}$$

Figure 6.1.3 gives us an idea of the quality of this approximation. In a few select cases where the exact Gittins indices are known (see Table 6.3), we can evaluate the approximation against the exact values. We see that the approximation gives the most

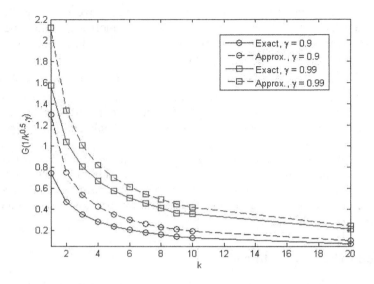

Figure 6.1   Comparison of approximate and exact Gittins indices for $\gamma = 0.9, 0.99$.

error for small values of k, but steadily improves as k increases. The approximation for $\gamma = 0.9$ tends to be slightly more accurate than the one for $\gamma = 0.99$. In general, the approximation is more accurate for lower values of γ.

6.2 VARIATIONS OF BANDIT PROBLEMS

The research into online learning problems has grown to consider a wide range of different variations. A sampling of this literature, with associated references, is summarized below.

Restless bandits - The standard bandit model assumes that the truth μ_x for arm x remains the same over time. Restless bandits describe a model where the means are allowed to vary over time (see Whittle 1988, Weber & Weiss 1990). Bertsimas & Nino-Mora (2000) provides a mathematically rigorous analysis of this problem class.

Continuous-armed bandits - The first author to have considered the online optimization of continuous functions (at least within the "bandit" literature) is Mandelbaum (1987). Bubeck et al. (2011) analyzes different policies for problems where the arm x may be a continuous, multidimensional variable using hierarchical discretization of the measurement space. Kleinberg (2004) provides tight minimax bounds on the regret for problems with continuous arms, but these are exponential in the number of dimensions.

Response-surface bandits - There are many applications where our beliefs about the value of different arms are correlated. Rusmevichientong & Tsitsiklis (2010) considers a problem where the value of arm x is given by a linear function $f(x) = a + bx$, so that as we learn about different arms, we learn the parameters a and b. This type of model was originally studied by Ginebra & Clayton (1995) under the name of "response-surface bandits."

Finite horizon bandits - Nino-Mora (2010) derives an index policy for finite horizon problems with discrete states, which occurs, for example, when we use the beta–Bernoulli model.

Intermittent bandits - Dayanik et al. (2008) derives an index policy for the problem where a bandit may not always be available. This is also known as "sleeping bandits" (Kleinberg et al. 2010).

A thorough review of bandit problems can be found in Bubeck (2010). In all these variations, the term "bandit" refers to a measurement x, which may be discrete or continuous. In this community, the use of the term bandit has the effect of putting a stochastic optimization problem in a learning setting, where we generally are looking for a policy where we would make a measurement in part for the value of information.

6.3 UPPER CONFIDENCE BOUNDING

A class of policies that has received considerable interest is known as *upper confidence bounding* or UCB policies. These policies are quite simple to implement, and different variants have been developed using this approach for many types of reward distributions. For example, imagine we have a problem where all the rewards are in the interval $[0, 1]$ (e.g. if we are using a beta–Bernoulli model). In this setting, one possible UCB policy defines the index of alternative x to be

$$I_x^{UCB1,n} = \theta_x^n + \sqrt{\frac{2 \log n}{N_x^n}}, \qquad (6.8)$$

where N_x^n is the number of times we have played arm x up to and including time n. The policy is somewhat analogous to interval estimation: We take our current estimate of μ_x and add an uncertainty bonus. Just as in interval estimation, this particular uncertainty bonus represents the half-width of a confidence interval. In a sense, (6.8) represents a probabilistic guess of the largest possible value that μ_x could realistically take on. We choose to measure, not the alternative with the highest estimated value, but rather the alternative that *could* potentially have the largest value.

The policy in (6.8) is geared toward the specific case where the rewards are in $[0, 1]$. A UCB policy designed for the normal–normal model defines the index of x

as

$$I_x^{UCB1-Normal,n} = \theta_x^n + 4\sigma_W \sqrt{\frac{\log n}{N_x^n}}. \tag{6.9}$$

There are also many other versions in the literature for both beta–Bernoulli and normal–normal problems, but (6.8) and (6.9) are two of the most prominent. Problems with bounded rewards turn out to be particularly attractive for the UCB approach because, in these settings, a particular proof technique can be applied to create UCB policies with provable bounds on the regret. There is also a UCB policy for the gamma–exponential model, where W_x follows an exponential distribution with parameter λ_x, and our beliefs about λ_x are represented by a gamma distribution with parameters a_x^n and b_x^n. In this case, we compute

$$I_x^{UCB-Exp,n} = \frac{b_x^n}{a_x^n - 1} + z \min\left(\sqrt{2 \frac{\log n + 2 \log \log n}{N_x^n}}, 1 \right), \tag{6.10}$$

with z being a tunable parameter.

The main reason why UCB policies have attracted attention is because they have an optimality property of sorts. If we are able to make N measurements, and we make them by following a UCB-type policy, then the average number of times we will play a suboptimal machine (a machine with $\mu_x < \max_{x'} \mu_{x'}$) can be bounded above by $C \log N$, where C is some constant. Thus, the number of times that we choose any particular suboptimal machine is on the order of $\log N$, known as a regret bound. It has been proven that this is the best possible bound (up to the choice of C) on the number of times a suboptimal machine is played. Both of the UCB policies given above have this property. In fact, it can even be shown that the epsilon-greedy policy from Section 17.2 has the UCB optimality property, although this randomized method is not what we typically think of as an index policy.

As is often the case, bounds can be loose, and the bound on the expected number of times that we may visit an incorrect arm given above can share this quality in some cases. For example, suppose that $C = 8$ for a particular UCB-type policy, and $N = 20$. Then, $C \log N = 11.8170$. This means that we can play any suboptimal machine up to eleven times on average. If we have many machines, this does not tell us much. Furthermore, the bound grows to infinity as $N \to \infty$, which once again indicates that an optimal policy can converge to a suboptimal arm (this is generally true of any discounted, online learning policy).

Still, UCB policies are a noteworthy alternative to Gittins approximations (based on a fundamentally different style of thinking), especially for some finite-horizon problems where Gittins indices are no longer optimal. One particularly attractive feature is their ease of computation. UCB policies tend to explore more than is really necessary. This can be effective for problems with small action spaces, and where we have relatively little information about the performance of each arm. UCB policies can be improved for problems where we feel we have a good prior that describes the population of potential truths. When this is the case, we can introduce a scaling factor for the second term in (6.8) or (6.9) which can be tuned using the population of truths. However, this would put a UCB policy into the same class as heuristics such as interval estimation.

In the next section, we present an adaptation of the knowledge gradient for bandit problems, and present some comparisons between the knowledge gradient and upper confidence bounding.

6.4 THE KNOWLEDGE GRADIENT FOR BANDIT PROBLEMS

The knowledge gradient approach that we introduced in Chapter 5 is a particularly simple and elegant strategy for collecting information. In the ranking and selection setting, it produces an easily computable algorithm. What is more, it can be adapted to handle correlated measurements, as well as non-Gaussian learning models. A natural question, then, is whether it can be adapted for the multi-armed bandit problem, which is the online version of ranking and selection.

In this section, we develop a simple relationship between the knowledge gradient for offline and online settings, which also allows us to consider problems with correlated beliefs. We present a few experimental comparisons that seem to suggest that this works quite well for online problems. We then discuss applications to problems with non–normal belief models.

6.4.1 The Basic Idea

Once again, consider the normal–normal Bayesian learning model. Suppose that we can make N measurements and that $\gamma = 1$. Furthermore, suppose that we have already made n measurements and have constructed estimates θ_x^n and β_x^n for each alternative x. Now, as a thought experiment, let us imagine that we will suddenly cease learning, starting at time n. We will still continue to collect rewards, but we will no longer be able to use the updating equations (4.1) and (4.2) to change our beliefs. We are stuck with our time-n beliefs until the end of the time horizon.

If this were to occur, the best course of action would be to choose $x^{n'} = \arg\max_x \theta_x^n$ for all times $n \le n' \le N$. Since we cannot change our beliefs anymore, all we can really do is choose the alternative that seems to be the best, based on the information that we managed to collect up to this point. The expected total reward that we will collect by doing this, from time n to time N, is given by

$$V^{Stop,n}\left(S^n\right) = (N - n + 1)\max_x \theta_x^n, \tag{6.11}$$

simply because there are $N - n + 1$ rewards left to collect. Because $\gamma = 1$, each reward is weighted equally. For instance, in the example given in Table 6.4, this quantity is $V^{Stop,n}\left(S^n\right) = 6 \cdot 5.5 = 33$.

Consider a different thought experiment. We are still at time n, but now our next decision will change our beliefs as usual. However, starting at time $n + 1$, we will cease to learn, and from there on we will be in the situation described above. This means that, starting at time $n + 1$, we will always measure the alternative given by

$\arg\max_x \theta_x^{n+1}$. The problem thus reduces to choosing one single decision x^n to maximize the expected total reward we collect, starting at time n.

This idea is essentially the knowledge gradient concept from a slightly different point of view. In ranking and selection, we chose each decision to maximize the incremental improvement (obtained from a single measurement) in our estimate of the best value. Essentially, we treated each decision as if it were the last time we were allowed to learn. We made each decision in such a way as to get the most benefit out of that single measurement. In the online setting, we do the same thing, only "benefit" is now expressed in terms of the total reward that we can collect from time n to the end of the time horizon.

The KG decision for the bandit problem is given by

$$X^{KG,n} = \arg\max_x \mathbb{E}\left[\mu_x + V^{Stop,n+1}\left(S^{n+1}\right) \mid S^n, x^n = x\right] \tag{6.12}$$

$$= \arg\max_x \theta_x^n + (N-n)\,\mathbb{E}\left[\max_{x'} \theta_{x'}^{n+1} \mid S^n, x^n = x\right] \tag{6.13}$$

$$= \arg\max_x \theta_x^n + (N-n)\,\mathbb{E}\left[\max_{x'} \theta_{x'}^{n+1} - \max_{x'} \theta_{x'}^n \mid S^n, x^n = x\right] \tag{6.14}$$

$$= \arg\max_x \theta_x^n + (N-n)\,\nu_x^{KG,n}, \tag{6.15}$$

where $\nu_x^{KG,n}$ is simply the knowledge gradient for ranking and selection, given by (5.10). We start with the basic Bellman equation in (6.12). The downstream value is given by $V^{Stop,n+1}$ because we assume that we will cease to learn starting at time $n+1$. Next, we use the fact that $\mathbb{E}\left(\mu_x \mid S^n\right) = \theta_x^n$, together with the definition of $V^{Stop,n+1}$ from (6.11) to obtain (6.13). Because the quantity $\max_{x'} \theta_{x'}^n$ is constant given S^n, and does not depend on x, we can put it into the expression without changing the arg max, thus arriving at (6.14). Finally, we apply the definition of the knowledge gradient from (5.1) to obtain (6.15).

This line of reasoning has given us a simple and easily computable algorithm for the multi-armed bandit problem. At first, the expression $\theta_x^n + (N-n)\,\nu_x^{KG,n}$ that we compute for alternative x may look very similar to the index policies we discussed earlier. Like interval estimation, Gittins indices, and other methods, KG takes θ_x^n and adds an uncertainty bonus $(N-n)\,\nu_x^{KG,n}$. Just as in the other index policies, the uncertainty bonus gets smaller as σ_x^n gets smaller: thus, if the level of uncertainty is zero, the uncertainty bonus is zero as well. Furthermore, all other things being equal, the uncertainty bonus is larger if n is smaller, reflecting the fact that it is more important to learn in the early stages of the problem, while we have more remaining time steps in which we can potentially use the information we collect.

However, the KG policy is not an index policy. Crucially, the knowledge gradient $\nu_x^{KG,n}$ depends not only on θ_x^n, but also on $\max_{x' \neq x} \theta_{x'}^n$. This cannot happen in an index policy, where the index of x is only allowed to depend on our beliefs about x. The knowledge gradient policy does not decompose the multi-armed bandit problem into many one-armed bandit problems. It considers each alternative relative to the others.

Table 6.4 shows the computations performed by the online KG policy for a particular problem with five alternatives and $N-n = 5$ measurements remaining in the

Table 6.4   Calculations for the online KG policy in a bandit problem with $M = 5$, $N - n = 5$, and $\beta_x^W = 1$ for all x.

Choice	θ^n	β^n	$\tilde{\sigma}^n$	ζ^n	$\nu^{KG,n}$	$\theta^n + 5 \cdot \nu^{KG,n}$
1	2	1/2	1.1547	-3.0311	0.0004	2.0020
2	4	1/2	1.1547	-1.2990	0.0527	4.2634
3	3	2/3	0.9487	-2.6352	0.0012	3.0062
4	5.5	1/2	1.1547	-0.8660	0.1234	6.1168
5	4.5	1/3	1.5	-0.6667	0.2267	5.6333

time horizon. This example illustrates the distinction between the online KG policy and the offline KG policy from Chapter 5. If this were a ranking and selection problem, we would measure the alternative with the highest KG factor, namely alternative 5. However, even though alternative 4 has a smaller KG factor, our estimate θ_4^n is sufficiently larger than θ_5^n to make the online KG policy choose alternative 4. Thus, the online KG policy favors exploitation more than the offline KG policy.

At the same time, if $N - n = 50$ in the same example, then the online KG policy would prefer alternative 5 to alternative 4, thus agreeing with the offline KG policy. Unlike the offline KG policy, online KG is what is known as a *nonstationary policy*. This means that the decision made by online KG depends on n as well as on S^n. The exact same belief state can lead online KG to measure different alternatives depending on the current time.

The formulation of KG for bandit problems is quite versatile. Suppose that we have a discount factor $\gamma < 1$. It is a simple matter to repeat the above reasoning and arrive at the decision rule

$$X^{KG,n} = \arg\max_x \theta_x^n + \gamma \frac{1 - \gamma^{N-n}}{1 - \gamma} \nu_x^{KG,n}.$$

Taking $N \to \infty$, we obtain the knowledge gradient rule for infinite-horizon bandit problems,

$$X^{KG,n} = \arg\max_x \theta_x^n + \frac{\gamma}{1 - \gamma} \nu_x^{KG,n}.$$

This is substantially easier to compute than Gittins indices. Keep in mind also that Gittins indices are only designed for infinite-horizon problems. If N is finite, the Gittins policy becomes a heuristic with γ serving as a tunable parameter. On the other hand, the KG policy can be defined for both finite- and infinite-horizon problems and requires no tuning in either case. Of course, it is not an optimal policy, but it is able to respond to these different environments without the need for any tunable parameters.

6.4.2 Some Experimental Comparisons

It is useful to see how the knowledge gradient adapted for online problems compares against some of the popular policies that have been proposed for this problem class.

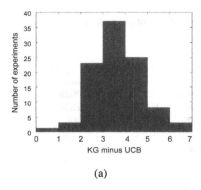
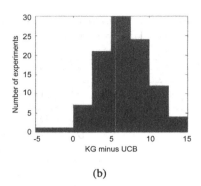

(a)   (b)

Figure 6.2 Histogram showing the average difference in the total discounted rewards collected by KG and approximate Gittins indices across 100 bandit problems with $M = 100$ and $\gamma = 0.9$. (a) Histogram using truth from the prior, and (b) histogram using truth from an unbiased uniform distribution (from Ryzhov et al. 2011).

Figure 6.2 shows a histogram of the performance of infinite-horizon KG minus the performance of a policy based on the Gittins approximation from Section 6.1.3 across 100 bandit problems with 100 alternatives. The numbers represent differences in the total discounted rewards (with $\gamma = 0.9$) collected by the two policies. We can see that all the numbers are positive, meaning that KG outperformed the approximate policy in every problem. To be sure, one can create problems where this is not the case, but it does indicate that KG can be competitive with the best existing approximation of the optimal policy.

The main advantage of the KG method, however, is that its nature as a nonindex policy makes it well suited to the case where our beliefs about the alternatives are correlated. Suppose that we begin with a multivariate normal prior $\mu \sim N\left(\theta^0, \Sigma^0\right)$, and use (2.22) and (2.23) to update our beliefs, but we keep the bandit objective function from (6.1). In this setting, index policies are inherently unsuitable: An index policy depends on our ability to decompose the problem and consider every alternative as if it were the only alternative in the problem, but the whole point of correlations is that the alternatives are inextricably related. Thus, while we can still use index policies such as Gittins and UCB as heuristics, they automatically lose their nice optimality properties in the correlated setting.

However, we can still define a knowledge gradient method in the correlated case. In fact, the KG decision rule is still given by (6.15), with the only change that we replace $\nu_x^{KG,n}$ by $h(\theta^n, \tilde{\sigma}(\Sigma^n, x))$ from (5.18). This quantity is then computed exactly as in Chapter 5. As of this writing, KG is the first algorithm that is able to consider bandit problems with multivariate normal priors.

Table 6.5 summarizes the results of a series of experiments on online problems with 100 alternatives with correlated beliefs. The knowledge gradient outperformed approximate Gittins, UCB, UCB1 and pure exploration for all 100 sample realizations. Only interval estimation proved to be competitive, and actually outperformed the tb

Table 6.5 Comparison between knowledge gradient and competing policies for 100 truth-from-prior experiments (source: Ryzhov et al. (2011)).

Comparison	Average Difference	Standard Error
KG minus approximate Gittins	0.7076	0.0997
KG minus interval estimation	-0.0912	0.0857
KG minus UCB	44.4305	0.6324
KG minus UCB1	1.2091	0.1020
KG minus pure exploitation	5.5413	0.1511

knowledge gradient policy 77 percent of the time (although the difference in the average performance was not statistically significant). Recall that interval estimation uses the policy of maximizing the index

$$I^{IE,n} = \theta_x^n + z_\alpha \sigma_x^n,$$

where θ_x^n is our current estimate of the value of alternative x and σ_x^n is the standard deviation of our estimate θ_x^n.

The performance of interval estimation seems surprising, given that the knowledge gradient policy is taking advantage of a covariance structure (which we assume is known in advance), while IE has no such ability. However, IE has a tunable parameter z_α, and it is important that this parameter be tuned carefully. Figure 6.3 shows the behavior of interval estimation as a function of z_α, along with the knowledge gradient

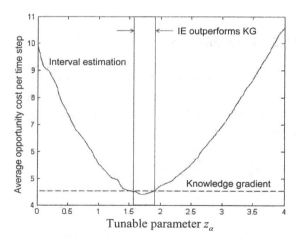

Figure 6.3 Expected opportunity cost for interval estimation as a function of z_α along with the knowledge gradient (from Ryzhov et al. 2011).

(which of course is a constant with respect to z_α). Note that IE outperforms KG only over a narrow range. Even more important, notice the severe degradation of IE as z_α moves away from its best setting.

This experiment demonstrates that there is a tremendous amount of information in a tunable parameter. Statements have been made in the literature that z_α can be safely chosen to be around 2 or 3, but problems have been found where the optimal value of z_α ranges anywhere from 0.5 to 5.0. Furthermore, the solution can be quite sensitive to the choice of z_α, suggesting that tuning has to be performed with care.

6.4.3 Non–Normal Models

Just as in ranking and selection, KG can also be extended to non–normal learning models, although we should take care to define the knowledge gradient in accordance with our reward structure. Suppose that we are working with the gamma–exponential model, where W_x follows an exponential distribution with parameter λ_x, and each λ_x has a gamma prior distribution with parameters a_x^0, b_x^0. Thus, our prior estimate of every reward is

$$
\begin{aligned}
\mathbb{E}\left(W_x\right) &= \mathbb{E}\left[\mathbb{E}\left(W_x \mid \lambda_x\right)\right] \\
&= \mathbb{E}\left[\frac{1}{\lambda_x}\right] \\
&= \frac{b_x^0}{a_x^0 - 1},
\end{aligned}
$$

where $b_x^0 > 0$ and $a_x^0 > 1$. Suppose that our objective function is to maximize the sum of the expected rewards,

$$
\max_{\pi \in \Pi} \sum_{n=0}^{N} \gamma^n \frac{1}{\lambda_{x^n}}.
$$

Then, the online KG decision rule for $\gamma = 1$ is given by

$$
X^{KG,n} = \arg\max_x \frac{b_x^n}{a_x^n - 1} + (N - n)\nu_x^{KG,n}.
$$

To write the knowledge gradient $\nu_x^{KG,n}$, let $C_x^n = \max_{x' \neq x} \frac{b_{x'}^n}{a_{x'}^n - 1}$ and define the baseline KG quantity

$$
\tilde{\nu}_x^n = \frac{(b_x^n)^{a_x^n}}{(a_x^n C_x^n)^{a_x^n - 1}} \left(\frac{1}{a_x^n - 1} - \frac{1}{a_x^n}\right).
$$

Then,

$$
\nu_x^{KG,n} = \begin{cases} \tilde{\nu}_x^n & \text{if } x \neq \arg\max_{x'} \frac{b_{x'}^n}{a_{x'}^n - 1}, \\ \tilde{\nu}_x^n - \left(\frac{b_x^n}{a_x^n - 1} - C_x^n\right) & \text{if } C_x^n \geq \frac{b_x^n}{a_x^n}, \\ 0 & \text{otherwise.} \end{cases} \tag{6.16}
$$

Once again, we have an easily computable KG algorithm, for a problem where Gittins approximations are not easily available. Note the presence of a penalty term in (6.16), much like in the expression for gamma–exponential KG in ranking and selection that was presented in Section 5.5.

6.5 BIBLIOGRAPHIC NOTES

Section 6.1 - Gittins index theory is due to Gittins & Jones (1974), Gittins (1979) and Gittins (1989). The term "retirement" was used by Whittle (1980) to explain the Gittins index. Berry & Fristedt (1985) also provides a rigorous analysis of bandit problems. This research has launched an entire field of research into the search for index policies for variations on the basic bandit problems. Glazebrook (1982) analyzes policies for variations of the basic bandit model. Glazebrook & Minty (2009) presents a generalized index for bandit problems with general constraints on information collection resources. Bertsimas & Nino-Mora (2000) show how an index policy can be computed using linear programming for a certain class of bandit problems. See the updated version of Gittins' 1989 book, Gittins et al. (2011), for a modern treatment of bandit problems and a much more thorough treatment of this extensive literature. The approximation of Gittins indices is due to Chick & Gans (2009), building on the diffusion approximation of Brezzi & Lai (2002).

Section 6.3 - Lai & Robbins (1985) and Lai (1987) provide the seminal research that shows that the number of times an upper confidence bound policy chooses a particular suboptimal machine is on the order of $\log N$, and that this is the best possible bound. Auer et al. (2002) derives finite-time regret bounds on the UCB1 and UCB1-normal policies and reports on comparisons against variations of UCB policies and epsilon-greedy on some small problems (up to 10 arms). The UCB policy for exponential rewards comes from Agrawal (1995).

Section 6.4 - The online adaptation of the knowledge gradient is due to Ryzhov et al. (2011). Some additional experimental comparisons can be found in Ryzhov & Powell (2009a). Ryzhov & Powell (2011c) presents the KG policy for the gamma–exponential model. Rates of convergence for KG-type policies are still an open question, but Bull (2011) is an interesting first step in this direction.

PROBLEMS

6.1 You have three materials, A, B, and C that you want to test for their ability to convert solar energy to electricity, and you wish to find which one produces the highest efficiency. Table 6.6 shows your initial beliefs (which we assume are independent) summarized as the mean and precision. Your prior belief is normal, and testing alternative x produces a measurement W_x which is normally distributed with precision $\beta^W = 1$.

Table 6.6   Three observations, for three alternatives, given a normally distributed belief, and assuming normally distributed observations.

Iteration	A	B	C
Prior (μ_x, β_x)	(5,.05)	(3,.02)	(4,.01)
1	3	-	-
2	-	2	-
3	-	-	6

a) You follow some measurement policy that has you first evaluating A, then B and finally C, obtaining the observations W_x^n shown in Table 6.6 (for example, $W_A^1 = 3$). Give the updated belief (mean and precision) for μ_A^1 given the observation $W_A^1 = 3$. Also compute the updated means only (not the precisions) for μ_B^2 and μ_C^3.

b) Give the objective function (algebraically) to find the best policy after N measurements if this is an offline learning problem. Compute a sample realization of the objective function for this example.

c) Give the objective function (algebraically) to find the best policy if this is an online learning problem. Compute a sample realization of the objective function for this example.

6.2   Consider a classic multi-armed bandit problem with normally distributed rewards.

a) Is the multi-armed bandit problem an example of an online or offline learning problem?

b) Let $R^n(x^n)$ be the random variable giving the reward from measuring bandit $x^n \in (1, 2, \ldots, M)$ in iteration n. Give the objective function we are trying to maximize (define any other parameters you may need).

c) Let $\Gamma(n)$ be the Gittins index when rewards are normally distributed with mean 0 and variance 1, and let $\nu_x(\mu_x, \sigma_x^2)$ be the Gittins index for a bandit where the mean reward is μ with variance σ^2. Write $\nu_x(\mu_x, \sigma_x^2)$ as a function of $\Gamma(n)$.

6.3   Consider a bandit problem with $\gamma < 1$ where we use the beta–Bernoulli learning model.

a) Suppose that, for a particular choice of α, β and r, we have $V(\alpha, \beta, r) \approx V(\alpha + 1, \beta, r) \approx V(\alpha, \beta + 1, r)$. Show that the Gittins recursion is solved by

$$V(\alpha, \beta, r) = \frac{1}{1 - \gamma} \max\left(r, \frac{\alpha}{\alpha + \beta}\right).$$

b) In a spreadsheet, choose values for r and γ (these should be stored in two cells of the spreadsheet, so that we can vary them), and create a table that compute the values of $V(\alpha, \beta, r)$ for all $\alpha, \beta = 1, 2, \ldots$ with $\alpha + \beta < 200$. When $\alpha + \beta = 200$, use $V(\alpha, \beta, r) = \frac{1}{1-\gamma}\frac{\alpha}{\alpha+\beta}$ as a terminal condition for the recursion.

c) The spreadsheet from part b) can now be used to compute Gittins indices. The Gittins index r^* for a particular α and β with $\alpha+\beta < 200$ is the smallest value of r for which $\frac{r}{1-\gamma}$ is equal to the entry for $V(\alpha, \beta, r)$ in the table. Use trial and error to find r^* for $\alpha, \beta = 1, \ldots, 5$ with $\gamma = 0.9$. Report the values you find, and compare them to the exact values of the Gittins indices in Table 6.1.

6.4 Consider a bandit problem with $\gamma < 1$. Repeat the derivation from Section 6.4 to show that the KG decision rule is given by

$$X^{KG,n} = \arg\max_x \theta_x^n + \gamma \frac{1-\gamma^{N-n}}{1-\gamma}\nu_x^{KG,n}$$

for finite N.

6.5 Consider a finite-horizon bandit problem with $\gamma = 1$ and a gamma–exponential learning model. Show that $\nu_x^{KG,n}$ is given by (6.16).

6.6 This exercise needs the spreadsheet:

http://optimallearning.princeton.edu/exercises/FiveAlternative.xls

available on the optimal learning website. You are going to have to construct a measurement policy to choose the best of five options, using the problems that are described in the attached spreadsheet. You are welcome to solve the problem directly in the accompanying spreadsheet. But this is an exercise that will be easier for some of you to solve using a programming environment such as MATLAB, Java, or perhaps Visual Basic in Excel.

The spreadsheet illustrates the calculations. Each time you choose a path, the spreadsheet will show you the time for the path. It is up to you to update your estimate of the average travel time and the variance of the estimate. Use a Bayesian framework for this exercise. You can repeat the exercise different times on the same set of random realizations. If you wish to use a fresh set of random numbers, hit the "Refresh button." You can see the data on the "Data" tab. The data tab uses the data in columns A–F, which will not change until you hit the refresh button. The data in columns H–L use the Rand() function, and will change each time there is a recompute (which can be annoying). If you click on the cells in columns H–L, you will see how the random numbers are being generated. The true means are in row 2, and the numbers in row 3 control the spread. You should use a prior estimate of the standard deviation equal to 10 for all your analyses.

The problem set requires testing a number of exploration policies. For each policy, compute two objective functions (averaged over 100 random number seeds):

1) The online objective function, which is the discounted sum of your measurement (for the chosen option) over all 100 measurements, with a discount factor of $\gamma = 0.80$. If $C^n(\omega)$ is the observed value of the measured option for the nth measurement for random number seed ω, your objective function would be

$$F^\pi = \frac{1}{100} \sum_{\omega=1}^{100} \sum_{n=0}^{100} \gamma^n C^n(\omega).$$

2) The final measurement is given by

$$G^\pi = \frac{1}{100} \sum_{\omega=1}^{100} C^{100}(\omega).$$

Here, F^π is our online objective function, while G^π is our offline objective function. Also let

$$F = \sum_{n=0}^{100} \gamma^n \mu^*,$$
$$G = \mu^*,$$

where μ^* is the true mean for the best choice, if we knew the true means. Where necessary, you may assume that the standard deviation of a measurement is 10. You have to test the following policies:

1) Pure exploitation.

2) Boltzmann exploration, using scaling factor $\rho = 1$.

3) Epsilon-greedy exploration, where the exploration probability is given by $1/n$.

4) Interval estimation. Test the performance of $z_\alpha = 1.0, 2.0, 3.0$ and 4.0 and select the one that performs the best.

5) Gittins indices (use the numerical approximation of Gittins indices given in Section 6.1.3.

Do the following

a) In a graph, report F^π and G^π for each of the policies below. Also show F and G to provide a measure of how well we are doing.

b) Discuss your results. Compare the performance of each policy in terms of the two different objective functions.

6.7   Consider a multi-armed bandit problem with independent normal rewards. In this exercise, you will implement a few online policies.

a) How should the opportunity cost be expressed in the online problem?

b) In exercise 4.4 you implemented the interval estimation policy in an offline setting. Now take your code from before, and adapt it to online problems by changing the objective function appropriately. The parameters of the problem should be the same as before (use the same priors and assume $\gamma = 1$), but now you need to compute opportunity cost differently. How does the best value of the tunable parameter z_α change when the problem becomes online? After you tune z_α, report the confidence interval for the opportunity cost using 200 simulations.

c) Now implement the Gittins index approximation for the independent normal–normal model. You cannot solve the Gittins recursion – just use the approximation function $\tilde{b}$. Our problem has a finite horizon, so you can treat the parameter γ in the Gittins index calculation as another tunable parameter. What value of γ gives you the best results?

d) Now implement the online KG policy. Compare the confidence intervals for KG, Gittins, and interval estimation.

CHAPTER 7

ELEMENTS OF A LEARNING PROBLEM

By now we have covered some elementary classes of learning problems. Fortunately, these describe a very broad class of applications. But there are some important problem classes that we have not yet discussed, and which we cannot solve using the tools presented so far. However, before we launch into what can seem to be a list of scattered applications, it is useful to lay out a more comprehensive modeling framework that at least hints at how we might deal with the major classes of unsolved problems. This presentation will motivate the problems we consider in the remainder of the volume, but will also serve potentially as areas for further research.

The dimensions of any stochastic, dynamic problem can be organized along five core dimensions:

1) States

2) Actions (or decisions or controls)

3) Exogenous information

4) The transition function

5) The objective function

In this chapter, we are not going to design learning policies. Instead, the goal is to describe the different dimensions of learning problems to provide a sense of the richness of this general problem area. We are not going to design policies for all of the problems that this chapter will identify, but offer this description as a potential source of research opportunities.

7.1 THE STATES OF OUR SYSTEM

The state of a system can be described as consisting of all the information needed to make a decision, compute the objective (contributions and rewards), and compute the transition to the next state. Elements of a state variable can include the following types of information:

- The physical state - These are variables that describe the locations, quantities and status of people, equipment, and goods.

- The information state - This is information separate from the physical state needed to make a decision, compute the objective and calculate the transition function. In most applications, this information arrives exogenously to the system, such as weather, prices, customer demands or changes in technology. Note that the physical state is a form of information, but it is useful to separate the physical state from exogenous information used to make decisions.

- The knowledge (or belief) state - This captures our distribution of belief about quantities that we do not know perfectly. For example, if we are using an independent normal–normal model, the knowledge state consists of a vector of means and a vector of variances (or precisions).

Up to now, our problems (ranking and selection and the bandit problem) have had only a belief state without a physical state. There is, of course, a vast range of problems in stochastic optimization where we are managing physical resources where we have to describe the state of our physical system.

The information state is best illustrated by example. One illustration arises when routing a driver through a network. As a driver arrives to node i, he is able to see the actual travel times on each link out of node i, but cannot see the actual travel time on any other link of the network. His physical state would be node i, while the elements of the information state would be the times on links out of node i. The knowledge state (or the belief state) is always a probability distribution describing parameters or quantities about which we are uncertain, but where we feel that we can describe these quantities using probability distributions. In our network example, we could describe the uncertain times on links that we cannot observe using probability distributions, where we may be willing to change our estimates of these distributions as we observe actual times.

We let R^n be the physical (or "resource") state at time n; I^n is the information state; and finally K^n is our state of knowledge (we could also let B^n be the "belief state"). Given the range and complexity of systems involving physical resources, we are not going to attempt to characterize the physical state other than as a generic variable R^n. Most of our attention focuses on problems with a pure knowledge state. But there is a growing literature that addresses the problem of optimal learning in the presence of a physical state, often under the broad umbrella of partially observable Markov decision processes (POMDPs). An example of an information problem with a physical state arises in sensor management, where observing information about, say, the disease in a population requires moving a sensor (e.g., a doctor) to an area to collect information. Once we do this, it changes our ability to make other observations moving forward.

There are generally two interpretations of K^n. The first is the Bayesian view, where K^n captures the probability distribution representing our belief about the parameters. This view requires that we start with some sort of initial distribution (the "prior") before we have collected any data. The second perspective is the frequentist view, which captures our state of knowledge through a set of statistics computed from observations. We have already discussed Bayesian and frequentist perspectives in Chapter 2. These approaches often learn in very similar ways (aside from the Bayesian ability to capture prior information), but the Bayesian view imposes a probability distribution on the possible values of the unknown parameters in the problem.

We have already seen problems with correlations in the belief model. This is where $Cov(\mu_x, \mu_{x'}) \neq 0$. We have only begun to explore the depth and richness of covariance structures for correlated beliefs. Correlated beliefs allow us to address problems with large (or infinite) numbers of potential alternatives with relatively small measurement budgets. Some examples of correlated beliefs include:

- Measurements are made of a continuous (and possibly scalar) function. We might be sampling disease within a population, the response due to a particular drug dosage, or the demand response to the price of a product. Measurements are correlated inversely proportional to the distance between two measurements.

- Measurements of multiattribute entities. We might be predicting the importance of a document (based on the attributes of the document) or the likelihood that someone will default on a loan (as a function of an individual's financial characteristics).

- Estimating the time on a path in a network. The observed travel time over one path will be correlated with other paths that share common links.

- Effectiveness of a drug compound. Different drug compounds will share common atomic subsequences which interact and determine the effectiveness of a drug. Drug compounds sharing common atoms (typically in specific locations) may exhibit correlations in their effectiveness.

The distinction between a physical (or resource) state and a belief (or knowledge) state has caused confusion in the dynamic programming community since the 1950s. It is common for people to equate "state" and "physical state," which is problematic when we only have a belief state. When it became clear that our belief about parameters could also evolve, Bellman & Kalaba (1959) used the term "hyperstate" to refer to the combination of physical state (an imperfect term) and state of knowledge. In Chapter 17 we address the problem of learning in the presence of a physical state. Rather than use terms like hyperstate, we prefer to use terms like physical (or resource) state, information state, and belief/knowledge state.

7.2 TYPES OF DECISIONS

The complexity of a problem depends in large part on the nature of the decision we have to make in order to make a measurement. Major problem classes include:

- Binary decisions - We can continue to collect information, or stop; we can decide to show a document to an expert, or not.

- Discrete choice - Here, we have a set of discrete choices (not too large - dozens, hundreds, perhaps thousands), where at each point in time we have to make a decision to collect information about one of the choices. A discrete choice could be a person to do a job, a technology, a drug compound, or a path through a network. So far, we have considered only discrete choice problems under the heading of ranking and selection or bandit problems.

- A scalar, continuous variable - We have to choose a quantity, price, location of a facility, or concentration of a chemical that we need to optimize.

- A vector, where the elements can come in several flavors:

  a) Binary - 0s and 1s, which we would use to choose subsets or portfolios of discrete items.

  b) Reals - We might have to set a multidimensional set of continuous parameters, such as testing a series of settings on a physical device such as an engine, or parameters governing the performance of a simulator.

  c) Integers - Similarly, we may have a vector of discrete variables (but other than 0 or 1), such as testing an allocation of ambulances or locomotives, or a mix of different types of aircraft.

  d) Multiattribute - We might choose a complex item (a document, a molecule, a person) characterized by a vector of categorical attributes.

We let x represent a generic "decision." We might have $x \in (0,1)$, or $x = (1, 2, \ldots, M)$, or $x = (x_d)_{d \in \mathcal{D}}$ where $d \in \mathcal{D}$ is a type of decision ("fly to Chicago," "try a particular drug compound") where x_d can be binary, discrete, or continuous. We let $\mathcal{X}$ be the set of feasible decisions. It is very important from a computational perspective to understand the nature of x, since there are problems where we assume that we can easily enumerate the elements of $\mathcal{X}$. In the list above, we open the door to problem classes where the size of $\mathcal{X}$ is infinite (if x is continuous) or much too large to enumerate.

A different dimension arises when we separate measurement decisions from implementation decisions. Some examples include:

- We may fund research (measurement) to build a device (the implementation).

- We may observe the performance of links in a network (measurement) to choose a path (implementation).

- We may wish to visit a company to talk to its management team (measurement) before deciding to invest in the company (implementation).

Typically, implementation decisions span the vast range of optimization problems that arise in the management of physical and financial resources. However, it is important to realize that while our measurement and implementation decisions up to now for ranking and selection and bandit problems have been the same, this is only a special case. In Chapter 15, we present some problems where measurement and implementation decisions are very different.

7.3 EXOGENOUS INFORMATION

Exogenous information clearly comes in a wide variety depending on the application. We briefly survey a few common types:

- Stationary processes - This is the simplest process, where observations come from a stationary distribution (with unknown mean, and potentially unknown variance), and where observations of different alternatives are uncorrelated. Increments of the process (that is, new information) may be completely independent of past history, or may exhibit some sort of history dependency.

- Nonstationary processes - We may make an observation of the level of disease in the population, but this can clearly change over time, but it changes in an exogenous way that we do not control.

- Correlated measurements - We might wish to simulate multiple configurations using what are known as common random variables, which means that we use the same sample realizations to help control statistical error. It is important to separate correlated measurements (where W_x^n is correlated with $W_{x'}^n$) and correlated beliefs (where $Cov(\mu_x, \mu_{x'}) > 0$).

- Learning processes - We might have a nonstationary process, but one that improves (or potentially deteriorates) as we sample it, rather than through some exogenous processes. For example, imagine observing how well a baseball player can hit; as we observe the player, his hitting may actually improve.

- State and/or action dependent processes - We generally view exogenous information as dependent on an unknown truth, but exogenous information may be a function of the state (this happens with learning processes described above) and possibly even the action (if we decide to sell stock or expensive equipment, this can depress the market price for these assets).

All of these generalizations represent interesting extensions of our learning model.

7.4 TRANSITION FUNCTIONS

If there is a physical state R^n, we are going to assume we are given a function that describes how it evolves over time. We write it as

$$R^{n+1} = R^M(R^n, x^n, W^{n+1}).$$

The notation $R^M(\cdot)$ represents the "resource transition function." For example, in an inventory problem where R^n represents the amount of inventory (such as water in a reservoir or product on a store shelf) at time n, our resource transition function would be

$$R^{n+1} = R^n + x^n + W^{n+1},$$

where W^{n+1} would represent water from rainfall between n and $n+1$, where W^{n+1} would be positive, or the demand for the product, in which case W^{n+1} would be negative.

We assume that with each decision x^n, we learn something that allows us to update our state of knowledge. We represent this generically using

$$K^{n+1} = K^M(K^n, x^n, W^{n+1}).$$

For example, K^M could represent the Bayesian updating equations (4.1) and (4.2) in a ranking and selection problem. When we want to be really compact, we write the state variable as $S^n = (R^n, K^n)$ and represent its transition function using

$$S^{n+1} = S^M(S^n, x^n, W^{n+1}).$$

Elsewhere in this volume, we have been using the function $S^M(\cdot)$ to represent the updating of the knowledge state when there is no physical state because this notation is more familiar and there is no ambiguity. We suggest using the notation $K^M(\cdot)$ in situations where there is both a physical and a knowledge state, as well as when we want to refer specifically to the updating of the knowledge state.

Most of this volume focuses on problems without a physical state. Chapter 17 addresses learning in the presence of a physical state.

7.5 OBJECTIVE FUNCTIONS

There are several dimensions to the design of an objective function. The first dimension addresses the question, What are we trying to achieve? We need some metric that determines when we are doing a better job of learning. The second concerns the details of measurement costs. The third dimension addresses how we are managing the economics of measurement and the evaluation of the solution.

7.5.1 Designing Versus Controlling

We have already compared two broad settings in which optimal learning arises: *offline learning*, where we conduct a series of measurements to design a process or system under some sort of measurement budget, and *online learning*, where we collect information in the process of managing the system. In the offline case, the economics of measurement are separated from the economics of using the system. On the other hand, in online learning, we are operating the system as we are observing it. The ranking and selection problem is an example of offline learning, whereas multi-armed bandits are an online problem. Below, we briefly sketch how offline and online objective functions might be structured.

Offline Learning

There are many problems where we have a certain budget to find the best decision. There is a measurement cost $C^m(x)$ for a decision x. We let $X^\pi(S)$ represent our policy for measuring x when we are in state S. After we have chosen our measurements $x^1, x^2, \ldots, x^N$, we have to make a final implementation decision that we represent as y which has to fall in a feasible region $\mathcal{Y}$. We formulate the optimization problem as

$$\max_\pi \mathbb{E}^\pi \max_{y \in \mathcal{Y}} C(S, y). \tag{7.1}$$

Our measurements have to be made subject to a measurement budget which we might state as

$$\sum_{n=1}^\infty C^m(x^n) \le B. \tag{7.2}$$

Of course, we assume that we stop measuring after our budget has been consumed.

For many problems, $C^m(x) = 1$, and the budget B represents a limit on the number of iterations (or time periods) that we are allowed to use for measurement. $C(S^\pi, x)$ represents the optimization problem we will solve after we have completed our measurements, given our "state of knowledge" S^π which results from our policy for collecting information.

We note that the total measurement cost on the left-hand side of equation (7.2) is a random variable, so (7.2) means we are saying that a random variable (on the left) has to be bounded by a fixed budget (on the right). If we are using measurement policy π, we might let $C^\pi(\omega)$ be the total measurement costs while following policy π on the sample path ω. For our bound to be true, it means it must be true for every ω for which $P(\omega) > 0$. When this is the case, then we say that the constraint (7.2) holds *almost surely* (which is typically abbreviated "a.s.").

Online Learning

Let $C(S, x)$ be a contribution (or reward, or utility) that we earn if we are in state S and choose action x. Remember that S captures our state of knowledge (in other applications, S would also include the physical state of the system). We make a decision x^n based on our state of knowledge (S^n), where we receive a reward $C(S^n, x^n)$. Let $X^\pi(S)$ be the policy we use to choose the action x. We would like to find a policy to solve

$$\max_{\pi \in \Pi} \mathbb{E} \left\{ \sum_{n=0}^{N} C(S^n, X^\pi(S^n)) \right\}. \tag{7.3}$$

Because we are solving the problem over multiple time periods, we have an incentive to explore certain decisions now so that we may make better decisions in the future. However, we have to pay for those decisions as they are made in the form of reduced rewards.

For online problems, it will typically be the case that we observe as we implement, so we do not have to introduce a separate implementation variable. We also would not have an information collection budget.

7.5.2 Measurement Costs

There are several variations that distinguish problems in terms of measurement costs. These include

- Startup costs - There may be a certain amount of time or money required to start measuring an alternative, making the first observation much more expensive than subsequent observations. For example, perhaps we have to invite a baseball player to training camp, or we have to purchase a device to test it, or we have to create a molecular in a lab to determine its properties.

- Switchover costs - Startup costs address the first measurement of an alternative, but there may be switchover costs, where even after the first measurement, there may be a cost from making an observation of x to an observation of x'. For example, perhaps we are simulating a policy, but it takes time to switch from testing one set of parameters to another because we have to reload a software package or restart a model that has been stopped.

- Alternative specific costs - We often represent the budget for measurements as the number of measurements, which also implies that each alternative "costs" the same. But this may not be true at all. For example, a measurement might involve drilling an exploration well (costs depend on geology) or testing a population for disease (cost might reflect distance to travel, or the type of diagnostic test).

7.5.3 Objectives

Below we provide an indication of some objectives we might use. Perhaps not surprisingly, there is a vast range of possible objectives that we might use. This discussion simply hints at the diversity of perspectives.

Expected Value

Up to now, we have been choosing a policy to maximize our performance. We can write this as

$$V^\pi \quad = \quad \mathbb{E}^\pi \mu_{x^\pi} \tag{7.4}$$

$$= \quad \mathbb{E}^\pi \left\{ \max_{x \in \mathcal{X}} \theta_x^N \right\}, \tag{7.5}$$

where $x^\pi = \arg\max \theta_x^N$. We remind the reader that the expectation is over all beliefs μ and all measurements W. We note that in (7.4), we are using our belief μ_x to evaluate the result of our policy x^π. In (7.5), we are using our estimates θ_x^N. Both objectives produce the same policy, and in expectation they are the same, as we showed in Section 4.4.2. In both versions of the objective function, we are taking an expectation over truths, so we are not changing our fundamental model. However, in practice it will be statistically more reliable to use (7.4).

Expected Opportunity Cost

It is very common to minimize the *opportunity cost*, which measures how much worse we are doing than the optimal (whether we are minimizing or maximizing). This is written as

$$V_{EOC}^\pi \quad = \quad \mathbb{E}^\pi \left\{ \max_x \mu_x - \mu_{x^\pi} \right\}. \tag{7.6}$$

As we saw in Section 5.6, this objective function is also known as "linear loss," because it represents the suboptimality of the alternative selected by policy π. The best possible loss value is zero, indicating that we have found the true best alternative.

In online problems, it sometimes makes sense to consider the average opportunity cost per time step. Thus, if we have N measurements total, we can compute

$$V_{AEOC}^\pi \quad = \quad \frac{1}{N} \mathbb{E}^\pi \left\{ \max_x \mu_x - \mu_{x^\pi} \right\}, \tag{7.7}$$

which represents the average loss incurred by the policy in each time step.

Robust Optimization

In some settings, it makes more sense to focus on the worst case rather than an expectation. This is becoming known in the optimization community as *robust optimization*. Assume there is a bounded set of outcomes Ω. Rather than integrate over this set as we have done up to now, we might propose an objective function of the form

$$V_{robust}^\pi \quad = \quad \min_{\omega \in \Omega} \mu_{x^\pi}(\omega). \tag{7.8}$$

Here, ω could refer to a truth and/or the set of outcomes, although it probably makes most sense to focus on the truth, while continuing to take expectations over observational outcomes. Rather than maximize the expected value, it may make more sense to maximize the minimum value. Robust optimization is attracting increasing attention in the stochastic optimization community (Beyer & Sendhoff (2007), Ben-Tal et al. (2009)), but robust thinking in a learning setting appears to be in its infancy.

Yet a further generalization is to use the concept of *quantile optimization*. If Q_α is the α quantile of a random variable, we may wish to solve

$$V_\alpha^\pi = Q_\alpha \mu_{x^\pi}. \tag{7.9}$$

Quantile optimization is related to robust optimization, because we might choose to maximize the 10th quantile. Quantile optimization is also useful in the presence of heavy-tailed random variables.

Probability of Correct Selection

A different perspective is to focus on the probability that we have selected the best out of a set $\mathcal{X}$ alternatives. In this setting, it is typically the case that the number of alternatives is not too large, say 10 or 20, and certainly not 100,000. Assume that

$$x^* = \arg\max_{x \in \mathcal{X}} \sum_{x \in \mathcal{X}} \mu_x$$

is the best decision (for simplicity, we are going to ignore the presence of ties). If we are using a frequentist perspective, we would make the choice

$$x^n = \arg\max_{x \in \mathcal{X}} \bar{\theta}_x^n.$$

In a Bayesian framework, we would use

$$x^n = \arg\max_{x \in \mathcal{X}} \theta_x^n.$$

Either way, we have made the correct selection if $x^n = x^*$, but even the best policy cannot guarantee that we will make the best selection every time. Let $1_{\{\mathcal{E}\}} = 1$ if the event $\mathcal{E}$ is true, 0 otherwise. We write the probability of correct selection as

$$
\begin{aligned}
P^{CS,\pi} &= \text{probability we choose the best alternative} \\
&= \mathbb{E}^\pi 1_{\{x^n = x^*\}},
\end{aligned}
$$

where the underlying probability distribution depends on our measurement policy π. The probability is computed using the appropriate distribution, depending on whether we are using Bayesian or frequentist perspectives. This may be written in the language of loss functions. We would define the loss function as

$$L^{CS,\pi} = 1_{\{x^n \neq x^*\}}.$$

Although we use $L^{CS,\pi}$ to be consistent with our other notation, this is more commonly represented as L_{0-1} for "0–1 loss."

Note that we write this in terms of the negative outcome so that we wish to minimize the loss, which means that we have not found the best selection. In this case, we would write the probability of correct selection as

$$P^{CS,\pi} = 1 - \mathbb{E}^{\pi} L^{CS,\pi}.$$

Indifference Zone Selection

A variant of the goal of choosing the best is to maximize the likelihood that we make a choice that is almost as good as the best. Assume we are equally happy with any outcome within δ of the best. This is referred to as the *indifference zone*. Let $V^{n,\pi}$ be the value of our solution after n measurements. We require $\mathbb{P}^{\pi}\{\mu_{d^*} = \mu^* | \mu\} > 1 - \alpha$ for all μ where $\mu_{[1]} - \mu_{[2]} > \delta$ and where $\mu_{[1]}$ and $\mu_{[2]}$ represent, respectively, the best and second best choices.

We might like to maximize the likelihood that we fall within the indifference zone, which we can express using

$$P^{IZ,\pi} = \mathbb{P}^{\pi}(V^{n,\pi} > \mu^* - \delta).$$

As before, the probability has to be computed with the appropriate Bayesian or frequentist distribution. We will discuss indifference-zone selection in more detail in Chapter 14.

Least Squared Error

A different form of loss function arises when we want to fit a function to a set of data. In this setting, we think of "x" as a set of independent variables which we choose directly or indirectly. For example, we may be able to choose x directly when fitting a linear regression of the form

$$\begin{aligned} Y(x) &= \theta_0 x_0 + \theta_1 x_1 + \theta_2 x_2 + \cdots + \theta_I x_I + \epsilon \\ &= \theta x + \epsilon, \end{aligned}$$

where Y is the observed response and ϵ is the random error explaining differences between the linear model and the responses. We choose it indirectly when our regression is in the form of basis functions, as in

$$Y(x) = \sum_{f \in \mathcal{F}} \theta_f \phi_f(x) + \epsilon.$$

Classical linear regression assumes that we are given a set of observations which we use to fit a model by choosing θ. Let

$$Y^{n+1} = \theta x^n + \epsilon^{n+1},$$

where θ is the true set of parameters. Our indexing reflects our requirement that x^n be chosen before we observe ϵ^{n+1}. Our measure of performance is given by

$$F(Y^{(N+1)}, x^{(N)}|\bar{\theta}) = \sum_{n=1}^{N} (Y^{n+1} - \bar{\theta}x^n)^2,$$

which is the sample sum of squares given measurements $x^{(N)} = (x^0, \ldots, x^N)$ and observations $Y^{(N+1)} = (Y^1, \ldots, Y^{N+1})$. Ordinary least squares regression fits a model by finding

$$\bar{\theta}^{N+1} = \arg\min_{\bar{\theta}} F(Y^{(N+1)}, x^{(N)}|\bar{\theta}).$$

Let $F^*(Y^{(N+1)}, x^{(N)}) = F(Y^{(N+1)}, x^{(N)}|\bar{\theta}^{N+1})$ be the optimal solution given $Y^{(N+1)}$ and $x^{(N)}$. Sequential estimation starts with $\bar{\theta}^n$, then measures x^n, and finally observes Y^{n+1} from which we compute $\bar{\theta}^{n+1}$. This can be done easily using recursive least mean squares, given by

$$\bar{\theta}^{n+1} = \bar{\theta}^n - H^n x^n (\bar{\theta}^n x^n - Y^{n+1})$$

where H^n is an $I \times I$ scaling matrix that is computed recursively (we cover this in Section 8.2.2).

Our focus is on choosing the measurements x^n. Classical experimental design assumes that we choose $(x^n)_{n=0}^{N}$ first and then fit the model. This is sometimes referred to as batch design because the entire sample is chosen first. This is equivalent to solving

$$\min_{x^{(N)}} \mathbb{E} F^*(Y^{(N+1)}, x^{(N)})$$

where the expectation is over the random variables in $Y^{(N+1)}$.

We focus on sequential design, where we choose x^n given our state S^n, which includes $\bar{\theta}^n$ and the information we need to update $\bar{\theta}^n$. In a sequential learning problem, we have to use some basis for determining how well we have done. In our optimization problems, we want to maximize our expected contribution. This optimization problem determines the values of x that are most interesting. In the area of adaptive estimation, we have to specify the values of x that are most likely to be interesting to us, which we designate by a density $h(x)$ which has to be specified. In an offline learning environment, we want to choose $x^1, \ldots, x^n, \ldots, x^N$ according to a policy π to solve

$$\min_{\pi} \mathbb{E} \int_x \left(Y(x) - \bar{\theta}^\pi x\right)^2 h(x)\, dx,$$

where $Y(x)$ is the random variable we observe given x, and where $\bar{\theta}^\pi$ is the value of $\bar{\theta}$ produced when we select x^n according to π, and when we estimate $\bar{\theta}$ optimally.

This formulation requires that we specify the domain that interests us most through the density $h(x)$. An illustration of the density function arises when we are trying to sample nuclear material over a border or in a region. For such cases, $h(x)$ might be the uniform density over the region in question. When we solve online and offline optimization problems, we do not have to specify $h(x)$ explicitly. The optimization problem (e.g., equation (7.3)), determines the region within $\mathcal{X}$ that is of greatest interest.

Entropy Minimization

Entropy is a measure of uncertainty that can be used for numeric and nonnumeric data. Imagine that we are trying to estimate a parameter μ that we know with uncertainty. If our distribution of belief about μ is continuous with density $f(u)$, a measure of the uncertainty with which we know μ is given by the entropy of $f(u)$, given by

$$H(\mu) = - \int_u f_u \log(f_u) \, du.$$

The logarithm is typically taken with base 2, but for our purposes, the natural log is fine. The entropy is largest when the density is closest to the uniform distribution. If the entropy is zero, then we know μ perfectly. Thus, we can try to take measurements that reduce the entropy of the distribution that describes our knowledge about a parameter.

7.6 EVALUATING POLICIES

The process of evaluating policies is subtle, because it can bias the identification of good policies toward problems with specific properties that may not be clearly identified. Perhaps one of the most difficult issues arises in the interaction between how problems are generated (that is, the "truth") and the relationship between the truth and any prior information that may be used.

Truth from Prior

From the Bayesian point of view, arguably the most natural way to evaluate a learning policy is to use the method described in Chapter 4, where we:

1) Generate a truth $\mu(\psi)$ from a prior.

2) Sample observations $W(\omega)$ from the truth $\mu(\psi)$.

3) Evaluate how well we discover the truth using one of the objectives listed above.

4) Repeat many times for different truths and different sample observations, to evaluate how well we discover the truth on average, and how reliably.

This strategy represents a very controlled experiment, which eliminates potential biases which a particular policy might be able to exploit. However, this approach may bias the evaluation process toward policies that work well with good priors.

Truth from an Alternative Prior

A way of evaluating the robustness of a policy is to generate truths from distributions other than the prior. For example, our prior may be normally distributed, but we might generate truths from a uniform or exponential distribution. Such experiments should focus on the ability of a policy to work with truths that come from distributions with potentially heavy tails, but the expected truth should match the expectation of the priors, which is to say that the prior should be unbiased.

Truth from a Fixed Prior

In the global optimization literature, where we collect noisy observations in an attempt to discover the maximum of an unknown continuous function, it is common to evaluate policies using a few "test functions." In other words, the truth is set to a fixed value that, for some reason, is believed to be a particularly interesting or difficult test case. For example, the Branin function

$$f(x_1, x_2) = \left(x_2 - \frac{5.1}{4\pi^2}x_1^2 + \frac{5}{\pi}x_1 - 6 \right)^2 + 10\left(1 - \frac{1}{8\pi} \right)\cos(x_1) + 10$$

is a commonly used test function in two dimensions. Test functions may be highly nonlinear with multiple local maxima, and so we may view them as posing an especial challenge for an optimization algorithm. If we are able to successfully find the point (x_1^*, x_2^*) that maximizes this difficult function, we may reasonably believe that our policy will be able to learn other functions.

This is a frequentist approach to policy evaluation, where we consider a few fixed test cases rather than sampling truths from a distribution. However, the strength of this approach is that it presents a very clear and well-defined test suite of problems. In Chapter 16, we mention some of these test functions, such as the Branin function, the Ackley function, and others.

Truth from a Biased Prior

In any particular learning situation, the sample realization of a specific truth is going to be higher or lower than the mean of any prior distribution. However, a subtle issue arises if the expected truth is higher or lower than the expectation of the prior, an issue that we touched on in Section 5.7. In that section, we discussed the problem with the prior is consistently higher or lower than the truth. But this discussion did not recognize that it may be possible to identify and correct biases. Even more important to the discussion here is the fact that if there is a bias, then certain types of policies will tend to exploit this bias. For example, we might use epsilon-greedy or interval estimation, each of which has a tunable parameter. In the presence of a bias, we only have to tune the parameters to emphasize more exploration, and we are likely to get a better policy.

Expectations Versus Risks

Most of our presentation focuses on the expected performance of a measurement policy. The probability of correct selection and indifference zone criterion can be viewed as risk-based measures, since they focus on finding alternatives that are within a specified tolerance with a specified probability.

For many information collection problems, risk is a major issue. Finding a workable drug that will help to extend lives with a high probability may be much more attractive than finding the best drug, but with a significant risk of not finding anything. The workable drug may not be ideal, but it may be good enough to gain federal approval which can then be marketed.

In real implementations, we are not allowed to run many sample realizations and take an average. We will use a policy on a single sample path. If we are not successful in finding, for example, a workable drug, we may be left asking whether this was just bad luck, or did we have a bad policy. We would like to minimize the possibility that it was bad luck.

Regret Bounds

In Section 6.3, we described a policy known as upper confidence bounding. For example, for the case of normally distributed priors and rewards, the UCB policy was given by

$$I_x^{UCB1-Normal,n} = \theta_x^n + 4\sigma_W \sqrt{\frac{\log n}{N_x^n}},$$

where N_x^n is the number of times we have observed alternative x after n trials. UCB policies have attracted considerable attention in the community that works on multi-armed bandit problems after it was found that it was possible to bound the number of times that incorrect arms would be tested by $C \log N$, for an appropriately chosen constant C. In fact, it was shown that this was the best possible bound, which is often interpreted to mean that this may be the best possible policy.

In Section 6.4, comparisons were made between a tuned UCB policy and the knowledge gradient policy for problems with normally distributed rewards. Although the knowledge gradient policy outperformed UCB, it is hard to know if this reflects a subtle bias in how the experiments were run. As of this writing, there is insufficient empirical evidence to form any judgments regarding the value of regret bounds in terms of predicting the empirical performance of a UCB policy.

7.7   DISCUSSION

The purpose of this chapter was to provide a sense of the dimensions of learning problems, and a hint of the rich diversity of this emerging class of problems. The remainder of this volume will attempt to chip away at some of these issues, but our treatment is hardly comprehensive.

One of the challenges of optimal learning is that important contributions have been made in specific subcommunities with a style that reflects both the characteristics of specific problem classes as well as the culture of different research communities. For example, the bandit community, which has attracted the attention of computer science, tends to emphasize online problems with relatively small action spaces, no switchover costs, a Bayesian belief model, and a desire to derive provable regret bounds. The simulation optimization community, which tends to focus on finding the best of a small set of designs using discrete event simulation, also deals with small choice sets but uses offline learning of a model with significant switching costs, a frequentist belief model, and a desire to show asymptotic optimality.

7.8   BIBLIOGRAPHIC NOTES

Sections 7.1–7.5 - This representation of learning problems is new, but is based on the modeling framework for dynamic programs presented in Powell (2011). The notion of the "hyperstate" to solve the optimal learning problem was put forth by Bellman & Kalaba (1959); further efforts in this direction were undertaken by Cozzolino et al. (1965), Martin (1967) and Satia & Lave (1973).

Section 7.5 - The different objective functions have been proposed by different authors in different communities. Expected opportunity cost lends itself better to the Bayesian approach (Chick & Inoue 2001), but can also be analyzed in a frequentist setting (Chick 2003, Chick & Wu 2005). A review of early research on indifference zone procedures is given in Bechhofer et al. (1968), with a review of somewhat more recent papers given by Bechhofer et al. (1995).

Section 7.6 - The truth-from-prior approach is used in Bayesian problems; see, for example, Ryzhov et al. (2011). Vermorel & Mohri (2005) and Chhabra & Das (2011) use other evaluation strategies such sampling from an alternative prior or from an empirical dataset. Lai & Robbins (1985) provides the original paper on regret bounds for upper confidence bound policies. Regret bounds continue to be popular, usually in connection with upper confidence bound methods; see, for example, Agrawal (1995), Auer et al. (2002), Auer et al. (2008), Bartlett et al. (2008), Kleinberg et al. (2010), or Srinivas et al. (2010).

PROBLEMS

In the exercises below, you are given a situation that involves learning. For each situation, describe the five fundamental dimensions of a learning model, including:

- Carefully define the state variable, giving variable names to each component. Clearly distinguish the belief state (what you think about unknown parameters whose beliefs are evolving over time) and any physical state variables.

- Define the measurement decision (how you are collecting information) and the implementation decision (what you are doing with the information). Note that these may be the same.

- Define the exogenous information. What are you observing? What is the source of the information? Is the information possibly changing your belief about a parameter?

- Describe the equations that make up the transition function. Distinguish between the equations used to update your knowledge state from those that update any physical or informational state variables that may be present.

- Define your objective function. Here is where you are going to distinguish between offline and online learning. Be sure to differentiate costs of measuring from implementation costs.

Note that in our modeling, we are ignoring the dimension of designing policies. This is addressed in all the remaining chapters of this volume. You may feel that you need information not specified in the problem description, so you should just highlight any missing elements, or make up elements to round out your model.

7.1 An entrepreneur would like to market the use of portable solar panels for recharging cell phones in Africa. The idea is to purchase a number of these solar panels and then to let individuals try to start businesses in different regions of Africa. Each region faces own unique characteristics in terms of need, competition for alternative sources of energy and the ability of the local population to pay. For example, the market responds badly to decisions to raise prices. Focus on the problem faced by a single individual who has to figure out a pricing strategy as quickly as possible.

7.2 A pharmaceutical company is faced with the problem of performing a series of market research studies to determine the best pricing for a drug. The market can be expected to respond to recent pricing behavior, and the performance of the drug. The company would like to strike a balance between maximizing revenue and minimizing costs related to the market research. Market research studies may be run in local regions while the drug is still being marketed nationally.

7.3 The Centers for Disease Control wants to collect information so that it can best understand the scope of an outbreak of a virulent new virus in the northeast of the United States. On a limited budget, the CDC would like to manage a single team of technicians who will set up a tent to test people as they walk by. The tent will typically be set up for 1–3 days before the team moves to another location.

7.4 An analyst is using an expensive computer simulation to model the dynamics of wind and its effect on generating electricity. It is important to understand the impact of very low periods of wind, but obtaining these observations can require simulating years, which can take weeks of time on the computer. The analyst is using the simulator to design the location of wind farms and investments in the power grid. She might run shorter simulations to do quick evaluations to eliminate poor designs, but much longer simulations are needed to obtain accurate estimates of the likelihood that a particular design will be susceptible to blackouts.

CHAPTER 8

LINEAR BELIEF MODELS

In the ranking and selection problem and the multiarmed bandit problem, we assumed we had a finite set of alternatives $x \in \mathcal{X} = \{1, 2, \ldots, M\}$, with a belief μ_x about each of the finite alternatives. Known as a lookup table representation, this model is very flexible in that it does not require any assumed structure among the alternatives. However, such models become very clumsy when the number of alternatives is large.

In this chapter, we make the transition from a lookup table belief model to one that uses a parametric model that is linear in the parameters. We assume that a measurement x can be written as the vector $x = (x_1, x_2, \ldots, x_K)$ where x_k may be discrete or continuous. Given x, we observe a value y where we assume a linear relationship of the form

$$y = \theta_0 + \theta_1 x_1 + \theta_2 x_2 + \cdots + \theta_K x_K + \varepsilon,$$

where ε is typically assumed to be a normally distributed error term with mean 0 and variance σ^2.

The most common motivation for using a parametric model is that the number of alternatives M may be extremely large (say, 100,000 or more), or even infinite (if x is continuous). However, there are applications where we know something about the structure of the function that we can capture with a parametric model. For example, we may be trying to estimate the maximum of a concave function. A quadratic approximation may capture this quite nicely, whereas a discretized approximation may easily lose the natural concavity of the function.

In classical regression model applications, we are given a series of observations $x^1, x^2, \ldots, x^n$, where for each set of explanatory variables x^m there is an observation y^{m+1}. (In traditional batch statistics, we would write y^m as the observation corresponding to measurement x^m, but in our sequential setting, it makes more sense to choose x^m and then observe y^{m+1}.) This data is then used to find a parameter vector θ that minimizes the mean squared error between the predicted model and the actual observations. We often do not have any choice in how the measurements $x^1, x^2, \ldots, x^n$ are chosen.

The problem of determining how to choose a set of measurements is a popular topic in the literature on statistical design of experiments. This literature, however, typically focuses on problems where a design (that is, a sequence of measurements) is chosen before any observations are made (for more on this topic, see Chapter 13). This strategy reflects the nature of the objective function that is used in this work. If the goal is to minimize variance, we can exploit the property that the variance of an estimate is a function of the measurements x^n and not the observations y^n.

In this chapter, we show how the knowledge gradient (for offline or online learning) can be extended to problems where the belief is captured by a model that is linear in the parameters. This result will allow us to tackle problems with thousands of alternatives, capturing correlations in the beliefs between these alternatives but without having to store a covariance matrix with thousands of rows and columns. The complexity of most of the steps in our adaptation of the knowledge gradient will be determined by the number of parameters rather than the number of alternatives.

8.1 APPLICATIONS

It is useful to start with some real applications to provide context to the discussion that follows.

8.1.1 Maximizing Ad Clicks

Imagine that you are an Internet company. You might be selling a set of products, or perhaps you are Google selling ad space. Either way, you may have a set of ads (and possibly thousands of ads) that you could post on the Internet, and you want to find the ad that generates the most ad clicks (the number of times that someone clicks on an ad, indicating interest).

How do you choose which ads to post? Let D_i be the data describing the ith ad. D_i might include all the text, along with descriptors of any graphics. The hard part of this project requires identifying important *features*, which are often represented using a device called *basis functions*. We let $\phi_f(D)$ be a particular basis function in a set $f \in \mathcal{F}$. Examples of basis functions might include

$$
\begin{aligned}
\phi_1(D) &= \text{The number of words in the ad,} \\
\phi_2(D) &= \text{The number of times the word "iphone," "itouch" or} \\
&\quad\text{"ipad" appears,} \\
\phi_3(D) &= \text{1 if the ad involves wireless communication, 0 otherwise,} \\
\phi_4(D) &= \text{1 if the ad involves food, 0 otherwise,} \\
\phi_5(D) &= \text{1 if the ad has color graphics, 0 otherwise,} \\
\phi_6(D) &= \text{1 if the ad claims to save money, 0 otherwise,} \\
\phi_7(D) &= \text{1 if the ad involves travel or vacations, 0 otherwise.}
\end{aligned}
$$

We first quickly realize that a "basis function" is the same as an independent variable in our linear regression model (different words for the same thing). Also, we can capture both individual features as well as combinations of features. For example, some people might be attracted to ads that involve food, while others enjoy travel, but what about ads that do both? We can reflect the combined contribution by adding the marginal effect of each, but we can also introduce a basis function that capture whether an ad covers food in exotic locations.

Using the notation of basis functions, we would write our linear model as

$$ Y = \theta_0 + \sum_{f \in \mathcal{F}} \theta_f \phi_f(D) + \varepsilon. $$

We can make this a bit more compact by introducing the basis function $\phi_0(D) = 1$. Now let $\phi(D)$ be a column vector of basis functions, and θ be a column vector of the coefficients. We can then write

$$ Y = \theta^T \phi + \varepsilon. $$

There are numerous variations of this basic problem which involve identifying websites (or documents) that achieve a particular purpose. For example:

- Researching information about a company that might predict a change in the stock price - The observable information might be an actual change in stock price (which can be automated) or it might involve showing the website to a domain expert who can assess its importance. Since the domain expert is a limited resource, we have to be careful in our selection of websites to be evaluated.

- Finding documents that provide information on terrorist activity - Similar to the previous item, we can scan millions of websites, but we have to choose which ones we should show to a domain expert for evaluation. This allows us to build up a dataset that can be used to calibrate the importance of different features.

8.1.2 Dynamic Pricing

Now consider the same Internet company from the point of view of pricing. Perhaps we are selling textbooks, DVDs, or music downloads. In any case, the Internet offers us a great deal of freedom in setting prices for our product. We can choose a price, wait for a period of time, and observe the resulting revenue, then adjust the price in response. Of course, this is an example of online learning, because our objective is to maximize total revenue.

To keep the modeling simple, we are going to assume that demand is a linear function of price which can be written

$$D(p) = \theta_0 + \theta_1 p,$$

where presumably $\theta_1 < 0$ to create a downward sloping demand function. We do not know θ_0 and θ_1, but we assume that we can charge a price p and then make a noisy observation of the demand $D(p) = \theta_0 + \theta_1 p + \varepsilon$. When we charge a price p, the revenue we earn is given by

$$R(p) = pD(p).$$

The pricing problem also arises outside the e-commerce setting. For example, it could apply to a small business that has a lot of room to change its pricing decisions, but is not prominent enough on the market to be able to change the underlying revenue curve. The business then has to adapt to an existing, unknown true curve. An interesting example of such an enterprise is a recharging station for mobile phones in a developing country. For example, cell phone use in rural Africa is currently experiencing prodigious growth, because it is much easier to build a single cell phone tower in a rural area than to stretch a land line out to every home. Mobile phones can serve as versatile and affordable tools even for many who are not connected to the electric power grid. An entrepreneur with a small power generator, such as a car battery or a solar panel, can serve many users; the question is how the service should be priced.

8.1.3 Housing Loans

The Small Business Administration (SBA) plays the role of granting loans to people whose homes have been destroyed by hurricanes and other natural events. As with any bank, the SBA has to identify people who are good loan prospects. By granting a loan, the SBA is able to observe which people eventually replay the loan. This information can then be used in a regression model to improve future predictions of whether someone might default on a loan. Explanatory variables might include

- The term (length) of the loan.

- The credit score of the applicant.

- Number of months employed.

- Income.

- Did the applicant own the home that was damaged or destroyed?

- Amount of the loan relative to the size of the loss from the storm.

- Amount of collateral.

Normally the SBA would grant loans to people whose probability of repayment is over some amount. However, it is possible for the SBA to grant some loans to applicants whose probability of repayment might be lower simply for the information that might improve their predictive ability for loan defaults.

8.1.4 Optimizing Dose Response

It is often the case that people respond differently to a particular dosage of a medication, forcing physicians to experiment with different dosages. It is natural to try to predict a patient's response (lowering blood sugar, controlling high blood pressure, raising the red blood cell count) using a statistical model. However, sometimes it is necessary to experiment with different dosages to help improve the model (which may involve a term unique to an individual patient).

We might envision a statistical model that predicts response as a function of a patient's body mass index, gender, age, ethnicity and other elements of a patient's history. However, it is often the case that we want to fit a nonlinear function that relates patient response to a nonlinear function which might look like

$$ P_i = \frac{e^{U(x_i|\theta)}}{1 + e^{U(x_i|\theta)}}, $$

where $U(x|\theta)$ is a linear function of independent variables with the general form

$$ U(x|\theta) = \theta_0 + \theta_1 x_1 + \theta_2 x_2 + \dots. $$

In this model, P_i gives the proportion of people in a group i with a specific set of attributes $x_i = (x_{i1}, x_{i2}, \dots)$ who respond to a particular dosage. This is an example of a parametric model that is *nonlinear* in the parameter vector θ. Belief models that are nonlinear in the parameters cause problems with methods such as the knowledge gradient because we lose conjugacy (see Chapter 11 for an in depth analysis of this particular belief model). However, we can overcome this problem by introducing the transformation

$$ \bar{P}_i = \ln\left(\frac{P_i}{1 - P_i}\right). $$

Using this transformation, we obtain the linear regression

$$ \bar{P}_i = \theta_0 + \theta_1 x_1 + \theta_2 x_2 + \cdots. $$

Now we have the response relationship expressed as a linear regression. We can use this model given estimates of θ to optimize a dosage strategy. This strategy, however, depends on estimates of the parameter vector θ. We may wish to attempt dosages simply to learn more about this relationship.

8.2 A BRIEF REVIEW OF LINEAR REGRESSION

Assume we have n measurements of a vector $x = (x_1, \ldots, x_K)$. If x^m is the mth measurement, using the indexing convention we have used throughout this volume, we let y^{m+1} be the observation corresponding to x^m. Recall that our indexing system reflects the property that x^m depends on the observations $y^1, \ldots, y^m$, and we observe y^{m+1} *after* we have chosen x^m. Let

$$
x^n = \begin{pmatrix} x_1^n \\ x_2^n \\ \vdots \\ x_K^n \end{pmatrix}
$$

be a K-dimensional column vector of observations. Often we will let $x_1 = 1$ to represent a constant term. Letting θ be the column vector of parameters, we can write our model as

$$
y = \theta^T x + \varepsilon,
$$

where we assume that the errors $(\varepsilon^1, \ldots, \varepsilon^n)$ are independent and identically distributed. Since θ^n is our estimate of θ after n observations, our best estimate of y is given by

$$
y^n = (\theta^n)^T x^n.
$$

8.2.1 The Normal Equations

We wish to find a parameter vector θ that solves

$$
\min_{\theta} \sum_{m=0}^{n-1} \left(y^{m+1} - \left(\sum_{k=1}^{K} \theta_k x_k^m \right) \right)^2. \tag{8.1}
$$

Let θ^n be the optimal solution for this problem. We can solve this problem very simply. Let X^n be the n by K matrix

$$
X^n = \begin{pmatrix} x_1^1 & x_2^1 & & x_K^1 \\ x_1^2 & x_2^2 & & x_K^2 \\ \vdots & \vdots & \cdots & \vdots \\ x_1^n & x_2^n & & x_K^n \end{pmatrix}.
$$

The vector of observations of the dependent variable is given by

$$
Y^n = \begin{pmatrix} y^1 \\ y^2 \\ \vdots \\ y^n \end{pmatrix}.
$$

The optimal parameter vector θ^n (after n observations) is then given by

$$\theta^n = [(X^n)^T X^n]^{-1} (X^n)^T Y^n. \tag{8.2}$$

From the normal equations, we can compute the covariance matrix for θ^n using a simple matrix identity. If u and w are scalar random variables where $u = Aw$, then we know that $Var(u) = A^2 Var(w)$. If u and v are vectors, and A is a suitably dimensioned matrix, then we can write $Var(u)$ (the covariance matrix for the vector u) as

$$Cov(u) = A Cov(w) A^T,$$

where $Cov(w)$ is the covariance matrix of w. Recall that for matrices A and B, $AB^T = (BA^T)^T$. Also keep in mind that $[(X^n)^T X^n]^{-1}$ is symmetric. Applying this identity to (13.2), where $A = [(X^n)^T X^n]^{-1} (X^n)^T$, we obtain

$$\begin{aligned} Var(\theta^n) &= [(X^n)^T X^n]^{-1} (X^n)^T Cov(Y^n) \left([(X^n)^T X^n]^{-1} (X^n)^T \right)^T \\ &= [(X^n)^T X^n]^{-1} (X^n)^T Cov(Y^n)(X^n)[(X^n)^T X^n]^{-1}. \end{aligned}$$

Since the elements of Y^n are independent, $Cov(Y^n) = \sigma_\epsilon^2 I$ where I is the identity matrix and σ_ϵ^2 is the variance of our measurement error. This allows us to write

$$\begin{aligned} \Sigma^{\theta,n} &= [(X^n)^T X^n]^{-1} (X^n)^T X^n [(X^n)^T X^n]^{-1} \sigma_\epsilon^2 \\ &= [(X^n)^T X^n]^{-1} \sigma_\epsilon^2. \end{aligned}$$

It is important to realize that the matrix X^n is n by K, so computing $\Sigma^{\theta,n}$ is not too difficult.

8.2.2 Recursive Least Squares

There is a shortcut that we can use to do this recursively. The updating equation for θ^n can be computed using

$$\theta^n = \theta^{n-1} + \frac{1}{\gamma^n} B^{n-1} x^n \epsilon^n, \tag{8.3}$$

where ϵ^n is the error given by

$$\epsilon^n = y^n - \theta^{n-1} x^{n-1}. \tag{8.4}$$

The matrix $B^n = [(X^n)^T X^n]^{-1}$. This can be updated recursively without computing an explicit inverse using

$$B^n = B^{n-1} - \frac{1}{\gamma^n} (B^{n-1} x^n (x^n)^T B^{n-1}). \tag{8.5}$$

The scalar γ^n is computed using

$$\gamma^n = 1 + (x^n)^T B^{n-1} x^n. \tag{8.6}$$

Note that if we multiply (8.5) through by σ_ϵ^2, we obtain

$$\Sigma^{\theta,n} = \Sigma^{\theta,n-1} - \frac{1}{\gamma^n}(\Sigma^{\theta,n-1}x^n(x^n)^T\Sigma^{\theta,n-1}),$$

where we scale γ^n by σ_ϵ^2, giving us

$$\gamma^n = \sigma_\epsilon^2 + (x^n)^T\Sigma^{\theta,n-1}x^n.$$

Note that we had to multiply each B^{n-1} in the second term on the right of (8.5) by σ_ϵ^2 so we also divided the second term by σ_ϵ^2, which we did by scaling γ^n.

Thus, we have compact updating equations dimensioned only by the number of parameters, rather than the number of alternatives.

The recursive updating formulas, aside from allowing us to avoid an expensive matrix inversion, allows us to handle an issue that we have not yet considered. There are problems where the observations are nonstationary, which means they are coming from a process that is changing over time. In such settings, we may not want to give all the observations equal weight. We can do this by replacing the objective function (8.1) with

$$\min_\theta \sum_{m=0}^{n-1} \lambda^{n-m}\left(y^{m+1} - \left(\sum_{k=1}^{K}\theta_k x_k^m\right)\right)^2. \tag{8.7}$$

Here, λ is a discount factor that we use to discount older observations. If we use this objective function, our recursive updating equations change only slightly to

$$\gamma^n = \lambda + (x^n)^T B^{n-1}x^n, \tag{8.8}$$

instead of (7.7), while we replace (8.5) with

$$B^n = \frac{1}{\lambda}\left(B^{n-1} - \frac{1}{\gamma^n}(B^{n-1}x^n(x^n)^T B^{n-1})\right).$$

Setting $\lambda = 1$ gives us the original updating equations. Using smaller values of λ reduces the weight on older observations, but also increases the variance of the estimates.

8.2.3   A Bayesian Interpretation

Classical linear regression is, of course, a frequentist method, but we can put linear regression into our standard Bayesian vocabulary. Let μ^i be the true value of alternative i which, in our Bayesian setting, is a random variable given by

$$\mu^i = \theta x^i.$$

Just as μ is a random variable (our truth), so is θ, which is the truth about the effect of each feature. We learned that we get tremendous benefits from exploiting the covariance between two alternatives. The covariance between μ^i and μ^j is given by

$$
\begin{aligned}
Cov(\mu^i, \mu^j) &= Cov\left(\sum_k \theta_k X_k^i, \sum_k \theta_k X_k^j\right) \\
&= \sum_{k,k'} X_k^i X_{k'}^j Cov(\theta_k, \theta_{k'}) = \sum_{k,k'} X_k^i X_{k'}^j \Sigma_{k,k'}^\theta \\
&= e_i^T X \Sigma^\theta X^T e_j,
\end{aligned}
$$

where $\Sigma_{k,k'}^\theta = Cov(\theta_k, \theta_{k'})$ is the covariance between the regression coefficients θ_k and $\theta_{k'}$. If Σ is our original covariance matrix between our beliefs of different alternatives, Σ and Σ^θ are related by

$$
\Sigma = X \Sigma^\theta X^T.
$$

This means that we can write our vector of truths about the value of each alternative as

$$
\mu \sim N(\theta^T X, X \Sigma^\theta X^T).
$$

Here, X is a matrix with a row and column for *each* alternative, which means that it has M rows (one for every alternative which may be an extremely large number) and K columns (one for every feature, which is generally not too large). We then let

$$
\Sigma^n = X \Sigma^{\theta,n} X^T
$$

be the covariance matrix among all M alternatives (hence Σ^n is $M \times M$, while $\Sigma^{\theta,n}$ is $K \times K$). This means that we are getting an estimate of the covariance between every pair of alternatives, including those which we have not tested yet, from the much more compact matrix Σ^θ. Our goal is to avoid having to explicitly compute and store the complete matrix Σ^n.

The important result is that we can compute a covariance matrix among all *possible* alternatives, using only the covariance matrix $\Sigma^{\theta,n}$ among the parameters, and the matrix X of attributes of each possible alternative. Updating $\Sigma^{\theta,n}$ from data is relatively easy using our recursive formulas. Computing Σ^n, given the potentially large number of rows, is still quite manageable given that it involves fairly simple calculations.

8.2.4 Generating a Prior

To perform learning in our Bayesian setting, we have to generate a truth, and then apply a learning algorithm (such as the knowledge gradient) to try to discover the truth. When we had a discrete set of alternatives with normally distributed beliefs, we could simply sample from the normal distribution we used as a prior.

There are several ways to generate a prior for these more complex settings. One, of course, is to collect a small sample of observations from an initial training set $x^0, \ldots, x^{L-1}$, producing observations $y^1, \ldots, y^L$. If L is larger than the number of parameters to be estimated, we can create an initial estimate of the parameter vector θ^0 using the normal equations in equation (13.2). We can then use this initial estimate to obtain an estimate of the variance σ^2 (if this is not known already). In the context of Bayesian learning, this becomes the same empirical Bayes strategy that we have already discussed.

Using an estimate of σ_ϵ^2 obtained from this model, we can estimate an initial covariance matrix for θ from

$$\Sigma^{\theta,0} = [(X^L)^T X^L]^{-1} \sigma_\epsilon^2, \tag{8.9}$$

where X^L is our initial set of observations from the first L observations.

We cannot use equation (8.9) if we have fewer than L observations, and we need at least $L+1$ observations if we are going to obtain a valid estimate of σ_ϵ^2. However, we can still get initial estimates of θ and Σ^θ with fewer than L observations, as long as we have some sort of prior. If we have some starting point, we can use the recursive equations (8.3)–(8.7) for an initial set of training points which may be chosen at random.

A useful strategy for generating an initial set of estimates is to write the regression equation in the form

$$y = \theta_0 + \theta_1(x_1 - \bar{x}_1) + \theta_2(x_2 - \bar{x}_2) + \cdots + \epsilon,$$

where $\bar{x}_i$ is viewed as an average or typical value of the independent variable x_i. In this model, θ_0 should be an average (or typical) value of the observations y, while θ_i captures the marginal impact of x_i on y. For example, consider our pricing model above where the demand $D(p)$ is given by

$$D(p) \approx \theta_0 + \theta_1(p - \bar{p}).$$

Here, we have written the independent variable as the deviation from what we might call a typical price give by $\bar{p}$. In this model, θ_0 would be a prior estimate of the typical demand, while θ_1 captures the effect of changes in the price on demand.

Estimating $\Sigma^{\theta,0}$ is somewhat trickier. An estimate of the diagonal terms can be found by trying to estimate a confidence interval for each coefficient θ_i. So, we might think the right coefficient for our demand curve is $\theta_1 \approx -0.07$, and we might then be willing to say that the right coefficient is probably in the range $(-0.03, -0.10)$. From this, we could infer a standard deviation of, say, 0.015, and use this to create the diagonal elements of $\Sigma^{\theta,0}$.

Estimating the off-diagonal elements of $\Sigma^{\theta,0}$ is inherently more complicated. A first order approximation is to just set them to zero, but this can be dangerous. Consider the set of potential demand functions shown in Figure 8.1. We may be uncertain about the precise line that describes the demand-price tradeoff, but we may feel that they all go through a certain region that might reflect the current demand and current price.

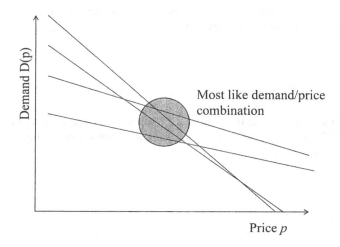

Figure 8.1 Illustration of range of potential demand functions.

The implication is that smaller values of θ_1 correspond to smaller values of θ_0, so our prior on these coefficients would not be independent.

Once we have an initial estimate of a mean vector θ^0 and covariance matrix Σ^0, we can generate a truth μ from a normal distribution using the techniques described in Section 2.4.

8.3 THE KNOWLEDGE GRADIENT FOR A LINEAR MODEL

Now we are ready to derive the knowledge gradient. Recall that the calculation of the knowledge gradient requires computing the function $h(a, b(j))$ where a is a vector with element $a_i = \theta_i^n$ giving the estimate of the value of the ith alternative, and $b(j)$ is a vector with element $b_i(j) = \tilde{\sigma}_i^n(j)$, which is the conditional variance of the change in θ^{n+1} from measuring alternative j. The function $h(a, b)$ is given by

$$h(a, b(j)) = \sum_{i=1}^{M-1} (b_{i+1}(j) - b_i(j)) f(-|c_i(j)|), \qquad (8.10)$$

where

$$c_i(j) = \frac{a_i - a_{i+1}}{b_{i+1}(j) - b_i(j)}.$$

Refer to Section 5.3 for the details of computing the knowledge gradient for correlated beliefs.

Keeping in mind that M may be a very large number, we have to use care in how this is computed. We can use $\theta_i^n = \theta^n x^i$ to compute the expected value, which does not depend on the alternative j that we are measuring.

The harder part is computing $\tilde{\sigma}^n(j)$, which is a vector giving the change in the variance of each alternative i assuming that we have chosen to measure alternative j. Recall that

$$\tilde{\Sigma}^n(j) \;=\; \frac{\Sigma^n e_j (e_j)^T}{\sqrt{\Sigma_{jj}^n + \sigma_\epsilon^2}},$$

which gives us the change in the covariance matrix from measuring an alternative j. The matrix $\tilde{\Sigma}^n(j)$ is $M \times M$, which is quite large, but we only need the jth row, which we compute using

$$\tilde{\sigma}^n(j) \;=\; \frac{\Sigma^n e_j}{\sqrt{\Sigma_{jj}^n + \sigma_\epsilon^2}}. \tag{8.11}$$

Since $\Sigma^n = X \Sigma^{\theta^n} (X)^T$, let X_j be the jth row of X. Then

$$\tilde{\sigma}^n(j) = X_j \Sigma^{\theta^n} X^T.$$

We still have to multiply the $K \times K$-dimensional matrix Σ^{θ^n} times the $K \times M$-dimensional matrix X^T, after which we have to compute equation (8.10) to find the knowledge gradient for each alternative. Even for problems with tens of thousands of alternatives, this can be executed in a few seconds, since K is much smaller than M.

8.4   APPLICATION TO DRUG DISCOVERY

A nice area of application of this logic is the design of molecules to accomplish some purpose, such as storing energy or curing cancer. We start with a base molecule such as that shown in Figure 8.2, where there are five locations $(R_1, R_2, \ldots, R_5)$. At these locations, we can add small molecular components called *substituents*, such as those listed to the right of the molecule in the figure. While we may not be able to put every substituent at every location (a chemist would understand which combinations make sense), the number of permutations and combinations can be quite large. For example, this base molecule and associated set of substituents produced 87,000 possible combinations.

The challenge with testing these molecules is that they are fairly difficult to create. It is necessary to design actual chemical processes to produce these compounds. A knowledgeable chemist can create a number of potential molecular compounds on a white board, but once we decide which compound to test next, the actual testing is a project that can take a day to several days. This is the reason why it is important to sequence the experiments carefully.

The problem of designing molecules arises in many settings. The work in this section was motivated by a problem of finding compounds to cure a form of cancer, but the numerical work was done using data taken from the literature which was

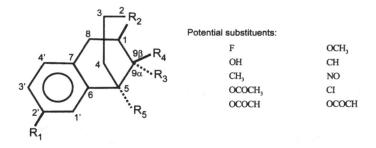

Figure 8.2   Illustration of base molecule with five substituent locations (from Katz & Ionescu 1977).

motivated by a variety of applications. All we require is that we are trying to choose a compound that maximizes (or perhaps minimizes) some metric, such as killing cancer cells.

In practice, the knowledge gradient can be used to produce a ranked list of potential molecules to be tested next. An attraction of this strategy is that it allows a scientist to use expert knowledge to capture dimensions other than the performance of a molecule in terms of a specific metric such as killing cancer cells. For example, a molecule might be toxic, hard to manufacture or insoluble in water.

To put this problem in the context of our model with linear beliefs, let $i = 1, \ldots, I$ be the set of sites (five in our example) and let $j = 1, \ldots, J$ be the set of potential substituents (10 in our example). Let

$$X_{ij} = \begin{cases} 1 & \text{if we put the } j\text{th substituent at the } i\text{th site,} \\ 0 & \text{otherwise.} \end{cases}$$

Now let Y be a measurement of the performance of a particular molecule, which might be the percentage of cancer cells that are killed. We might model the performance of a molecule using

$$Y = \theta_0 + \sum_{i=1}^{I} \sum_{j=1}^{J} \theta_{ij} X_{ij}. \tag{8.12}$$

With this simple model, if we have five sites and 10 substituents, we have 51 parameters to estimate (including the constant intercept θ_0).

To determine the effectiveness of the knowledge gradient, it is necessary to create a truth (following our standard practice) and then try to discover the truth using different methods. The truth was created by using data from an initial set of experiments that was then used to perform an initial fit of the regression model in equation (8.12). This model was then assumed to give us an initial parameter vector θ^0 that we used as our

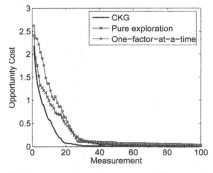

(a) Average opportunity cost over 100 runs using a dataset of 2,640 compounds and a noise standard deviation of 0.1.

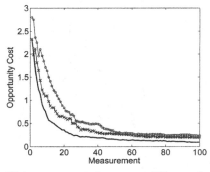

(b) Average opportunity cost over 100 runs using a dataset of 2,640 compounds and a noise standard deviation of 0.5.

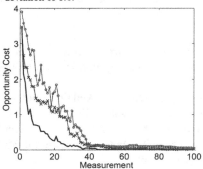

(c) Average opportunity cost over 10 runs using a dataset of 87,120 compounds and a noise standard deviation of 0.1.

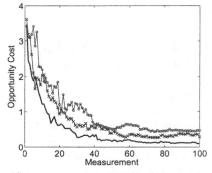

(d) Average opportunity cost over 10 runs using a dataset of 87,120 compounds and a noise standard deviation of 0.5.

Figure 8.3   Comparison of the knowledge gradient to pure exploration and a standard experimental design on a medium and large molecule with different levels of noise (from Negoescu et al. 2011).

prior, and an initial covariance matrix $\Sigma^{\theta,0}$. We made our standard assumption that the true parameter vector $\theta \sim N(\theta^0, \Sigma^{\theta,0})$ follows a normal distribution. We would generate a sample realization of a truth $\theta(\omega)$ by sampling from this distribution.

Once we have a truth (which means we now have a "true" parameter vector θ), we run our experiments by choosing a molecule to test (using some policy), and then sampling its performance using equation (8.12). It is important to recognize that this method of testing a learning policy assumes that the linear *model* is accurate, even if we do not know the true parameter vector. But, there is nothing stopping us from generating observations from some other model, and then using linear regression as an approximation.

The knowledge gradient was compared to a pure exploration policy (choosing molecules at random) and a simple experimental design policy which tests one factor

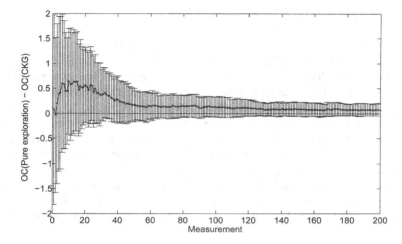

Figure 8.4 Mean and standard deviation of difference in opportunity cost between pure exploration and KGCB using 75 sample paths of 10,000 compounds each and a noise standard deviation of 0.38 (from Negoescu et al. 2011).

(substituent) at a time, cycling through all the substituents. Each of these strategies are used in an offline setting (since this is laboratory experimentation) to collect data to fit our regression model. We then use this regression model to decide which of a much larger set of molecules is best.

Figure 8.3 shows the expected opportunity cost for the knowledge gradient, pure exploration and the one-at-a-time factor design on molecules with 2,640 and 87,120 variations, and with measurement noise of 0.1 (lower than that observed from actual data) and 0.5 (higher than that observed from actual data). These experiments suggest that the knowledge gradient consistently outperforms the competition, with a fairly significant improvement if we wanted to stop at around 20 experiments. Note that we were able to find a near-optimal molecule within approximately 30 experiments.

Care has to be used when evaluating the performance of a measurement policy based on averages. Figure 8.4 shows the actual spread in the results from 75 sample paths, comparing knowledge gradient against pure exploration (positive means that KG is outperforming pure exploration). After 20 measurements, the knowledge gradient policy is almost always outperforming pure exploration. The difference is much more mixed during the first 20 observations, which should be expected. When there is a lot of uncertainty and not enough information to make sensible choices, random exploration can work just as well as the knowledge gradient on specific sample paths. There is no evidence that the knowledge gradient ever underperforms pure exploration; but it is certainly possible that pure exploration can outperform knowledge gradient in the early iterations. Interestingly, the knowledge gradient significantly outperforms pure exploration *on average* in the early iterations, but the spread is quite high.

8.5 APPLICATION TO DYNAMIC PRICING

Let us briefly return to the dynamic pricing application in Section 8.1.2. We are interested in the pricing decision of the African entrepreneur running a cell phone recharging station. This is a small but profitable operation. For example, the entrepreneur can purchase a car battery, charge it in town, and then travel to a village to charge phones for a fee that can be anywhere between $0.10 and $0.50. An alternative with a higher startup cost may be to purchase a small solar system (e.g. a set of solar panels) and then sell the generated power. With a good pricing strategy, the recharging station may see 50–70 customers every day. Assuming that the entrepreneur works five days a week, the resulting revenue can be as high as $50–$100 per week.

This problem provides the basic parameters for our example. Every week, the price per use of the recharging station is chosen from the range $[0.10, 0.50]$. The weekly revenue $R(p)$ can be as high as $50. In this example, we use a simple quadratic belief model,

$$R(p) = p(\theta_0 + \theta_1 p).$$

where we capture the fact that $R(0) = 0$. Our belief about the revenue for the nth week can be written as $R^n(p) = p(\theta_0^n + \theta_1^n p)$. We would typically start with a prior where $\theta_1^0 < 0$, because revenue is likely to be concave (we do not expect it to keep increasing with price forever).

The pricing problem is inherently online: every time we set a price, we collect a revenue. As always, we need to balance the need to maximize revenue with the need to fit a good curve. In this case, we would use the knowledge gradient for online learning, given by

$$X^{KG,n}(s^n) = \arg\max_p R^n(p) + (N - n)\nu_p^{KG,n},$$

where ν_p^{KG} is the offline knowledge gradient corresponding to the value of observing demand at price p.

Let's consider how different learning policies should behave. Figure 8.5 depicts a possible prior on the demand function and resulting revenue curve. Let p^{max} be the price that drives the demand to 0, which means that we should limit our search to prices in the interval $[0, p^{max}]$. Also let p^n be the price that maximizes revenue given our current belief after n measurements, which is to say

$$p^n = \arg\max R^n(p).$$

If we were to stop after n measurements, p^n is the price that we would charge. Otherwise, we would characterize the decision to charge p^n as a pure exploitation policy.

What would we do if we wanted to focus on estimating the demand function? In Figure 8.6(a), we see that if we focus our energies on observations near the middle, we may obtain a wide range of possible functions as a result of our measurement noise. By contrast, if we focus our observations near the endpoints as in Figure 8.6(b), we

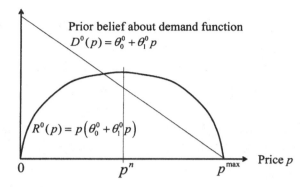

Figure 8.5 Prior demand function and revenue curve.

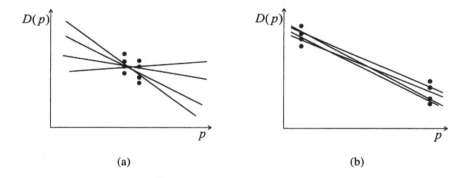

Figure 8.6 Estimating the demand function using (a) observations near the middle and (b) observations near the endpoints.

obtain much more reliable estimates of the demand function. This behavior is well known in the statistics community.

Measuring near the middle offers the potential for maximizing revenue, but we end up learning almost nothing from the observations. By contrast, if we measure near the endpoints, we learn a lot but earn almost nothing. Ideally, we would like to make observations that strike a balance between these two extremes, a property that might be called "sampling the shoulders." This, in fact, is exactly what the knowledge gradient does, in a way that adapts to the number of observations remaining in our budget.

Figure 8.7 illustrates the behavior of a pure exploitation policy (first row), the knowledge gradient policy for offline learning (second row) and the knowledge gradient policy for online learning (third row). These policies were tested for datasets with a moderate level of observation noise (the left column) and a relatively high level of observation noise (right column). The line with solid dots is the initial prior,

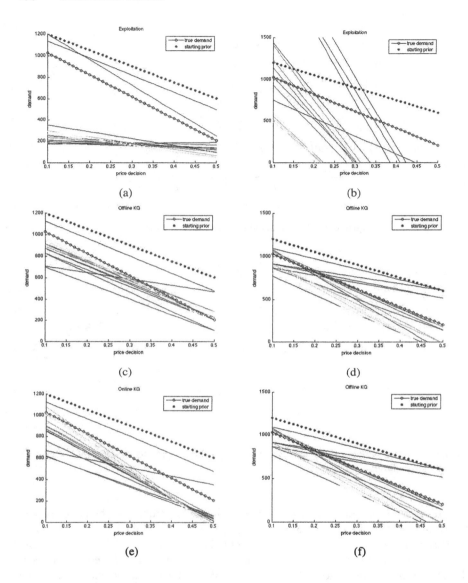

Figure 8.7   The behavior of pure exploitation (first row), offline learning (second row) and online learning (third row) for observations with a moderate level of noise (left column) and a high level of noise (right column).

while the line with open circles is the true demand. The remaining lines represent the estimates of the demand functions for each of the first 20 observations obtained under each policy. The darker lines represent the early iterations, while the lighter lines (which are more clustered) represent the later iterations.

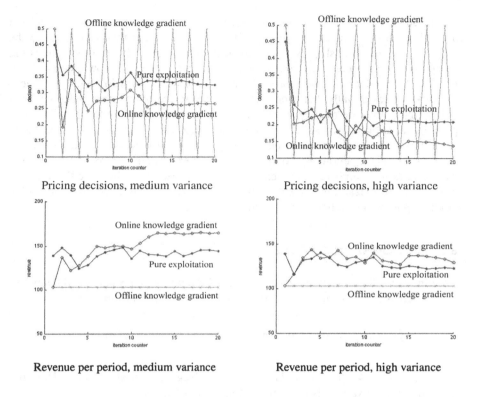

Figure 8.8   The pricing decisions selected by each policy (first row), and revenue earned per period (second row) for moderate and high variance in the measurement noise.

The results show that using a pure exploitation policy does a terrible job identifying the correct demand curve. By contrast, using an offline policy produces the best results, producing an estimate of the demand curve that is quite close to the truth. The online learning policy strikes a balance between the two, producing an estimate of the demand curve that is clearly not as good as what is obtained using an offline learning policy, but much better than the pure exploitation policy.

The first row of Figure 8.8 shows the actual pricing decisions produced by each policy. The offline knowledge gradient behaves exactly as we would expect given the illustration in Figure 8.6(b). The policy alternates between the two endpoints. This behavior is consistent for both the medium and high variance cases. For the medium variance case, the online knowledge gradient does more exploration in the early iterations before settling in to a single point compared to the pure exploitation policy. Note that the online knowledge gradient approaches a pure exploitation policy toward the later iterations. The only reason that it is measuring different points is that it is converging on a different estimate of the demand curve than pure exploitation, as a result of the exploration in the early iterations.

The second row of Figure 8.7 shows the revenue earned per period. Note that the offline knowledge gradient earns a fixed revenue greater than zero because the endpoints were defined slightly interior to the interval $(0, p^{max})$. The more interesting comparison is between the pure exploitation policy and the online knowledge gradient. For both the medium and high variance cases, the pure exploitation policy produces higher revenues initially, but eventually loses out to the online knowledge gradient policy. Note that the relative improvement of the online knowledge gradient over pure exploitation is largest for the moderate noise case. Again, we think this is to be expected. As the measurement noise increases, it is harder for any policy to learn the correct function. By contrast, we would expect the difference between the two policies to diminish as the measurement noise decreases, because the value of information from learning will be minimal, which again pushes the online knowledge gradient policy toward the pure exploitation policy.

8.6   BIBLIOGRAPHIC NOTES

Section 8.2 - This is classic material that can be found in many statistical textbooks such as Hastie et al. (2005), which can be downloaded from

    http://www-stat.stanford.edu/~tibs/ElemStatLearn/

Section 8.3 - 8.4 - This material is based on Negoescu et al. (2011). The drug discovery example is taken from Katz & Ionescu (1977).

Section 8.5 - The experimental results presented here come from work by two undergraduate students at Princeton University. The numerical work was performed by Xiaoyang Long as part of her research for the Program in Applied and Computational Mathematics. The application of pricing cell phone charges was studied by Megan Wong as part of her senior thesis.

PROBLEMS

The exercises below require the knowledge gradient algorithm where the belief model is linear in the parameters. This can be downloaded from

    http://optimallearning.princeton.edu/exercises/KGCBLinReg.m

An example implementation of the algorithm is given in

    http://optimallearning.princeton.edu/exercises/KGCBLinRegEx.m

8.1   You are going to replicate the experiments in Section 8.5 where we try to learn the demand as a function of price, while simultaneously trying to maximize revenue.

We start by assuming that the demand as a function of price is given by

$$D(p) = \theta_0 + \theta_1 p.$$

Our revenue is given by $R(p) = pD(p)$, and prices are assumed to range between .1 and .5. Assume the true value of $\theta = (\theta_0, \theta_1) = (1233, -2055)$. Normally we would sample this truth from a normal distribution, but we are going to assume a single truth to illustrate how well we discover this truth.

Now assume we start with a high prior $\theta^0 = (1350, -1500)$, with a starting covariance matrix of

$$\Sigma^{\theta,0} = \begin{pmatrix} 62{,}500 & 0 \\ 0 & 2{,}500{,}000 \end{pmatrix}.$$

a) Set up and run the knowledge gradient with a linear belief model for $N = 20$ iterations assuming a low measurement noise of $\sigma_\epsilon^2 = 62{,}500$, limiting your search of prices to the range $[.1, .5]$. Do this for both online and offline learning (recall that Section 6.4 describes how to compute the online knowledge gradient from the offline version). Plot the prices chosen and the updated estimate of the demand curve after each iteration.

b) Now sample prices using a pure exploitation policy which maximizes the revenue at each iteration, and compare your results to those obtained using the online and offline knowledge gradient.

c) Repeat (a) and (b) using $\sigma_\epsilon^2 = 625{,}000$.

d) Repeat (a) and (b) using $\sigma_\epsilon^2 = 6{,}250{,}000$.

e) Compare the performance of the three algorithmic strategies under the different noise levels.

8.2 Repeat exercise 8.1, but this time start with a low prior of $\theta^0 = (1350, -4500)$.

8.3 Repeat exercise 8.1, but now assume that the truth is normally distributed around the prior with mean $\theta^0 = (\theta_0^0, \theta_1^0) = (1233, -2055)$ and variance

$$\Sigma^{\theta,0} = \begin{pmatrix} 62{,}500 & 0 \\ 0 & 2{,}500{,}000 \end{pmatrix}.$$

CHAPTER 9

SUBSET SELECTION PROBLEMS

In the ranking and selection problem, we have to pick the best out of a set of discrete choices. Now assume that we have to pick the four best rowers to man a four-person shell for rowing competitions. Perhaps the coach has identified seven people who, based on the performance on rowing machines, appear to be the best. But it is well known that people perform differently in a boat, and people interact in ways that affect their performance.

Our challenge is to find the best subset of a discrete choice set. This problem, known as *subset selection*, arises in numerous settings and has a classical formulation as a mathematical program. For example, we might have M items, where $x_i = 1$ if we have chosen i to be in our subset. Normally we have some sort of budget constraint, so we have to choose the vector x to satisfy this constraint. If we knew that item i made a contribution c_i, we could choose our elements by solving

$$\max_x \sum_{i=1}^{M} c_i x_i,$$

Optimal Learning. By Warren B. Powell and Ilya O. Ryzhov
Copyright © 2012 John Wiley & Sons, Inc.

subject to the constraints

$$\sum_{i=1}^{M} a_i x_i \leq B,$$
$$x_i \leq 1,$$
$$x_i \geq 0.$$

If $a_i = 1$, then B is simply a limit on how many items we can have (for example, four rowers). The coefficient a_i might be the cost of using i, in which case B is a monetary budget constraint. In this case, we have an instance of the well-known knapsack problem.

Subset selection problems include forming a team, identifying investment portfolios, prescribing a course of treatment for an illness, and choosing tools to solve a problem. In all of these problems, we have to choose a subset of elements that interact with each other in a relatively complex way. Section 9.1 discusses some examples of such problems in more detail. Our specific interest is in problems where we do not have a simple deterministic, linear objective function. Because of the interactions between elements, the problem is harder than simply not knowing the coefficients c_i. However, we may have a learning budget allowing us to experiment with several subsets before we have to choose one. For example, a basketball coach may hold some practice games to learn how well different players work together, before committing to a lineup for the season.

We can apply optimal learning concepts to subset selection. In theory, the tools developed in Chapters 4 and 5 already allow us to handle this problem. We can view subset selection as an instance of a ranking and selection problem with correlated beliefs. Each alternative in the ranking and selection problem corresponds to one possible subset. In the problem of choosing four rowers out of seven possible candidates, we thus have a total of $M = \binom{7}{4} = 35$ alternatives. We model the values of different subsets separately. If we choose rowers {Anne, Mary, Cathy, Tara} and {Anne, Mary, Susan, Tara}, these two subsets would be considered as two distinct alternatives, each with its own value. However, our beliefs about these two values are heavily correlated, because these two subsets have three elements in common. Learning about the effectiveness of one subset should also provide information about the other subset.

As long as we do a good job modeling our prior beliefs (particularly our beliefs about the correlations between alternatives), we can then apply one of the learning techniques from Chapter 4, or the knowledge gradient method from Chapter 5, to decide which subset we want to learn about. Section 9.2.2 discusses two strategies for setting the prior covariances. Conceptually, we can approach the problem of subset selection entirely within the framework of ranking and selection. For smaller problems such as choosing four rowers out of seven, this approach will work fine.

However, one characteristic specific to subset selection problems is that they tend to be quite large. Suppose that we need to choose five items from a list of ten. Five and ten are small numbers, but the number of possible subsets is $\binom{10}{5} = 252$. If we slightly increase the size of the problem, and now choose six items from a list of twelve, the number of subsets is $\binom{12}{6} = 924$. The number of alternatives grows combinatorially as we add more items to our list. Choosing ten items from a list of twenty results in 184,756 alternatives!

In Chapter 4, we did not really specify a problem size. Rather, we analyzed a generic problem with M alternatives. In theory, the techniques we developed can handle any M. In practice, they may be slow when M is very large. This issue is especially severe in subset selection, when even fairly small problems produce large numbers of subsets. We will need to develop further modifications to our methods that not only obtain good performance, but do so in a reasonable amount of time in larger problems.

9.1 APPLICATIONS

Subset selection problems come in many varieties. We are primarily interested in applications where there is an interaction between the subsets.

- Designing a diabetes treatment - Type 2 diabetes is an illness where the body is unable to properly react to insulin, a hormone used to remove sugar from the bloodstream. Most diabetes drugs fall into one of several categories; for example, sensitizers increase the body's sensitivity to insulin, whereas secreta-gogues stimulate the body to secrete more insulin. A course of treatment for a single patient may contain drugs from multiple categories, as well as multiple complementary drugs within a single category. Clinical trials can help us learn how well two drugs complement each other.

- Choosing a starting basketball lineup - The coach wants to find five players who provide the best performance over the first quarter of a basketball game. He knows how many points each player scores from past games, and has other statistics such as rebounds, assists and blocks, but the simple reality is that there is no easy formula that predicts how people will interact with each other. Each time he tries out a team, he observes the total score (which might be viewed as a correction to a simpler formula). The coach has 12 players, implying that he has to pick the best set of 5 out of a team of 12, which means evaluating 792 different teams. Fortunately, the observations should be correlated. We would not know the correlations exactly, but we might start with a covariance matrix that is proportional to the number of players in common between two teams.

- Collecting information for a venture capitalist - The VC is looking at a list of 100 companies, divided into sectors such as web applications, energy, finance, and health. Within these, there are subfields such as companies making contributions to solar energy (new materials, new collection devices, new storage devices) and finance (risk management software, portfolio balancing, tools for pricing credit derivatives, and forecasting). The VC has a rough idea of the prospects of each company from their literature and a phone call. He has time to visit about 20 of these companies. After a visit, his opinion may go up or down on the prospects of the company. After he makes the visits, he will allocate $100 million in funding. He does not want to concentrate too much money in a narrow area (solar panels), or even in a broader area (energy), which might be dramatically affected by a single random event (such as a carbon tax).

- Installing energy saving technologies - A company installs a variety of technologies to save energy costs in industrial buildings. These include computer-controlled thermostats, solar plating on windows, fluorescent light bulbs, insulation for air ducts, and solar panels on the roof. The company has ways of estimating the energy savings from each technology based on the configuration of the building and the weather patterns. But the company has found that the combined contribution of all the technologies does not always match what it expects if it were to install each technology in isolation. Six months after each project, the company does an energy audit and can use the information to guide the choice of projects. Given the expense of each technology, it has to carefully choose how to experiment with different technologies so that it will better understand for the future how component technologies interact.

- R&D portfolio optimization - The Department of Energy is trying to decide which proposals, if funded, would advance our knowledge of hydrogen fuel cells. The proposals each address different components of the fuel cell. Some of these proposals address competing technologies (what is the best material for the cathode), while others are complementary (the total throughput of a fuel cell may be restricted by three separate components). Funding a proposal means that a component gets better by some random amount, but the amount by which the overall costs drops depends on changes in all the other technologies. We need to find the best set of proposals to fund to maximize the likelihood of a low cost fuel cell.

- Product assortment planning - A retail company has to choose a set of products to offer at its outlet store, or to display in a storefront window. The profitability of a product, or its effectiveness at drawing customers to a store, may also depend on the presence of other products in the assortment (e.g., accessories for an iPod). It is also important to display a diverse set of products that may attract buyers from different demographic groups. We have a short period of time to experiment with different assortments at a particular store in an attempt to discover the best one.

- Shortest path problems - Finding a path through a network with uncertain costs (or travel times) involves finding a subset of links out of the full set of links in the network. Of course, these links have to form a path that joins origin to destination.

A common feature of many subset selection problems is that there are interactions between elements within a set. If the contributions of different elements in a subset were purely additive, then these problems would be best solved using the linear belief model of Chapter 8.

9.2 CHOOSING A SUBSET USING RANKING AND SELECTION

We show how ranking and selection can be used to find a subset in the context of designing a diabetes treatment. In the latter stages of clinical trials, the search has typically been narrowed to a relatively small set of the most promising drugs. We can create a combination treatment consisting of multiple drugs from this set.

The effectiveness of a diabetes treatment can be measured in terms of blood sugar reduction. Blood sugar level is measured in millimoles per liter, after the patient has not eaten for eight hours; this is known as the "fasting plasma glucose" or FPG level. For a healthy person, the FPG level is about 4-6 mmol/L. A diabetic can show FPG levels of up to 10-15 mmol/L. A single drug can reduce FPG level by an amount between 0.5 and 2 mmol/L.

We design a combination treatment for two reasons. First, two drugs yield a bigger FPG reduction than one, and second, combination treatments tend to have lower risks of side effects. Prescribing three drugs instead of one may give us an FPG reduction between 2 and 5 mmol/L. Due to interactions between drugs, FPG reduction obtained from the combination is not a straightforward sum of the reductions we might observe from individual drugs.

9.2.1 Setting Prior Means and Variances

Let us consider an example with six drugs. Our list may include certain basic elements such as conventional treatment (recommending diet and exercise to the patient), metformin (the first drug of choice in most cases), and insulin injections, as well as newer compounds such as rosiglitazone, glibenclamide, and chlorpropamide. Our goal is to find the best possible combination of three of these drugs. The total number of subsets is still quite small, $\binom{6}{3} = 20$.

Let $\mathcal{S} = \{1, ..., 6\}$ be our set of $S = 6$ drugs, from which we have to choose a subset of size s. Let μ_x be the true value of a subset $x \subseteq S$, that is, the true FPG reduction achieved by treatment x on average. Now suppose that we have a large population of patients in roughly the same poor health condition. We divide them into N groups and then we prescribe a treatment for each group. When we assign the $(n+1)$st group to treatment x, our observation W_x^{n+1} is the average FPG reduction across all the patients in the group. Sample averages are approximately normal, so we make our usual modeling assumption $W_x^{n+1} \sim \mathcal{N}(\mu_x, \sigma_W^2)$, where the variance σ_W^2 is assumed known (but somehow estimated or guessed in practice).

All we need now in order to apply optimal learning is a prior distribution on the values μ_x. We will use a multivariate normal distribution (accounting for correlations between subsets). The prior mean θ_x^0 on the true value of treatment x tends to be easier to choose, because it comes directly from our domain knowledge about the problem. In light of what we know about blood sugar levels, we might let θ_x^0 be a number between 2 and 5 mmol/L. The prior variance Σ_{xx}^0 represents a rough guess about the range of our uncertainty about the true value. If we are reasonably sure that the true value is between 2 and 5, we might let the variance be 0.5, meaning that we believe that μ_x falls in the range $\theta_x^0 \pm 2 \cdot \sqrt{0.5}$.

9.2.2 Two Strategies for Setting Prior Covariances

While prior means and variances can be chosen with the help of knowledge about the problem, covariances between the true values are trickier to estimate. In actual clinical trials, we can observe the effectiveness of a treatment directly, but covariances between treatments can only be inferred from the results of individual trials. Fortunately, we can still obtain good performance with a simple, heuristic covariance structure. Our prior mainly needs to capture the fact that subsets are more heavily correlated if they have more elements in common.

Our first rule of thumb is based purely on the number of such elements. In the problem of choosing three drugs from a list of six, let $x = \{1, 2, 3\}$ and $y = \{2, 3, 5\}$ be two possible treatments. These subsets have $2/3$ elements in common, so we can simply choose $2/3$ as the correlation coefficient of these two subsets in our prior distribution. That is, the prior correlation coefficient is set using

$$\rho^0_{xy} = \frac{1}{S} \left| x \cap y \right|.$$

The prior covariance is simply

$$\Sigma^0_{xy} = \rho^0_{xy} \sigma^0_x \sigma^0_y. \tag{9.1}$$

We calculate these covariances for every possible x and y. This simple heuristic produces high correlations between subsets with more common elements.

In some problems, we may have more domain knowledge about individual items on our list than about subsets. For example, we may have access to clinical data about the drug rosiglitazone, tested independently of other drugs. Or, we may have some data on the travel time on a particular street or region, but no travel time data for a complex travel route that takes us through multiple regions. In this case, it may be easier to construct a prior on the value of a subset using our information about individual components.

Let μ^{ind}_i be the true value of the individual component $i \in S$ (e.g. the FPG reduction achieved by the single diabetes drug i). Using our domain knowledge, we construct a prior $\mu^{ind}_i \sim \mathcal{N}\left(\theta^{ind,0}_i, \left(\sigma^{ind,0}_i\right)^2\right)$. For the diabetes setting, we may believe that μ^{ind}_i falls in the range 1.25 ± 0.75 mmol/L. Then, our belief about a treatment x is given by

$$\theta^0_x = \sum_{i \in x} \theta^{ind,0}_i, \tag{9.2}$$

and the covariance of our beliefs about treatments x and y is given by

$$\Sigma^0_{xy} = \sum_{i \in x \cap y} \left(\sigma^{ind,0}_i\right)^2. \tag{9.3}$$

This heuristic implicitly assumes that the values of individual diabetes drugs are independent, which may not be the case. For example, the FPG reductions for multiple drugs in the same class will be correlated. If a patient reacts adversely to sensitizers, all drugs of this type will tend to perform poorly. In any case, however, our prior is not able to capture all the nuances of the problem. Instead, the prior is meant to provide some rough guidelines about the values of different alternatives and the interactions between them. Even if we assume that individual items on our list are independent, subsets will still be correlated if they have common elements. The correlation structure in (9.3) once again captures the fact that subsets with many common elements should have similar values. As we make decisions, the information we collect will allow us to obtain more accurate beliefs. To put it another way, we start by assuming a simple linear objective function, then leave it to our learning algorithm to refine these initial beliefs.

We now proceed as usual. Once we have a prior (μ^0, Σ^0), we rely on our standard Bayesian updating equations to change our beliefs from that point. If we choose treatment x for the trial at time n, our beliefs for subsets y will change according to the equations

$$\theta_y^{n+1} = \theta_y^n - \frac{W_x^{n+1} - \theta_x^n}{\sigma_W^2 + \Sigma_{xx}^n} \Sigma_{xy}^n,$$

$$\Sigma_{y,y'}^{n+1} = \Sigma_{y,y'}^n - \frac{\Sigma_{x,y}^n \Sigma_{x,y'}^n}{\sigma_W^2 + \Sigma_{xx}^n}.$$

This gives us everything we need to run an algorithm such as knowledge gradient. If there are only 20 possible subsets, it is smooth sailing from here.

9.3 LARGER SETS

Let us switch gears and consider a slightly larger problem, namely the energy port-folio selection example from Section 9.1. A recent study by McKinsey & Company (2007) has identified a number of promising technologies for reducing greenhouse gas emissions. These new technologies have a higher energy efficiency, and also help to reduce energy costs. A partial list of these technologies includes: residential light-ing; energy-efficient water heaters and appliances; improved ventilation systems and air conditioners; heating systems and furnaces; solar technology; shell improvements for residential buildings (e.g., insulation); and fuel economy packages for cars.

Suppose that we have a total of 12 promising technologies, and our goal is to create a portfolio of six. The McKinsey study provides estimates for the abatement potential (the emissions reduction) of these technologies. The estimates represent the total abatement, measured in gigatons of CO_2 per year (ranging from 0.2 to 3.0 for different technologies), that could be achieved by implementing the technology across the entire United States. We can scale down these estimates and use the strategy from (9.2) and (9.3) to create a prior for the energy efficiency of a single portfolio of six technologies, applied to a single building.

The total number of energy portfolios in this example is 924, about 50 times more than we had in the problem of designing a diabetes treatment. We can still run the correlated KG algorithm from Chapter 5, but we will see that it will run much more than 50 times slower on this problem. The reason is because, in a problem with M subsets, the time required to compute correlated KG factors for all the subsets is proportional to $M^2 \log M$. This computational cost grows faster than the problem size. Considering that subset selection problems are already prone to very large sizes, this is a serious issue. We now discuss a way to reduce this cost using Monte Carlo simulation.

9.3.1 Using Simulation to Reduce the Problem Size

The basic idea is straightforward. First, we want to narrow down our large number of subsets to a much smaller number of subsets that appear to be promising. We then calculate knowledge gradients only for this reduced choice set, choosing a single alternative to measure.

Intuitively, when faced with a very large number of choices, we might ignore most of them and focus only on choosing from the top candidates. This behavior is also typical of optimal learning algorithms. Interval estimation, Gittins indices, knowledge gradients, and other techniques all have an initial period of extensive exploration, but quickly start vacillating between a small number of choices that consistently look good. Although we know that, in the limit, KG is guaranteed to measure every alternative infinitely often, in practice most alternatives go for very long periods of time without attracting attention.

Suppose that we are at time n, with beliefs θ^n and Σ^n. (As before, θ^n refers to a vector of estimates for the values of possible subsets, which are enumerated and arranged into a list.) We can use Monte Carlo simulation to generate K samples from our distribution of belief $\mathcal{N}(\theta^n, \Sigma^n)$. Let $\bar{\mu}^n(\omega_k)$ denote the kth sample, for $k = 1, ..., K$. Since we are generating from a multivariate normal distribution, each $\bar{\mu}^n(\omega_k)$ is a vector of numbers. The notation $\bar{\mu}_x^n(\omega_k)$ represents the simulated value of the individual subset x in the kth sample.

From the kth sample, we can find $\bar{x}^n(\omega_k) = \arg\max_x \bar{\mu}_x^n(\omega_k)$. This is the "winner" of the kth sample, or the subset that had the highest simulated value on the sample path ω_k. Now, if we group together all of the winners, we will have a reduced set of at most K alternatives (at most, because there may be some duplicates). We will make our measurement decision based purely on these K subsets, ignoring all other possible subsets.

To see how this works, consider a numerical example. Suppose that we have five alternatives with $\theta^n = (10, 13, 12, 16, 8)^T$ and

$$\Sigma^n = \begin{bmatrix} 22 & 13 & 14 & 15 & 14 \\ 13 & 16 & 18 & 13 & 12 \\ 14 & 18 & 24 & 12 & 14 \\ 15 & 13 & 12 & 14 & 11 \\ 14 & 12 & 14 & 11 & 16 \end{bmatrix}.$$

We decide to reduce the number of alternatives. To that end, we generate $K = 4$ samples from the distribution $\mathcal{N}(\theta^n, \Sigma^n)$. Table 9.1 gives the simulated values for all five alternatives on all four sample paths. Observe that, according to our prior, alternative 4 is believed to be the best. This same alternative is also the winner on $2/4$ sample paths. However, alternatives 2 and 3 each win on a single sample path, even though these alternatives are believed to be the second and third best according to the prior. The set of winners is thus $\{2, 3, 4\}$. In this way, Monte Carlo sampling can give us a reasonably diverse choice set, which includes many promising alternatives, but does not include alternatives that seem to be hopelessly bad.

It can be shown that the marginal distribution of (μ_2, μ_3, μ_4) is multivariate normal with parameters

$$\theta^{MC,n} = \begin{bmatrix} 13 \\ 14 \\ 15 \end{bmatrix}, \quad \Sigma^{MC,n} = \begin{bmatrix} 16 & 18 & 13 \\ 18 & 24 & 12 \\ 13 & 12 & 14 \end{bmatrix}.$$

That is, if we throw out alternatives 1 and 5, the distribution of our belief about the remaining alternatives is the same as before. We just need to throw out the rows and columns of Σ^n, and the elements of θ^n, having to do with alternatives 1 and 5. We obtain the reduced prior $\mathcal{N}(\theta^{MC,n}, \Sigma^{MC,n})$, where the notation MC refers to the use of Monte Carlo sampling to reduce the prior.

We can formalize this procedure as follows. Let K_0 be the number of unique winners obtained from Monte Carlo sampling (with duplicates removed; this number would be 3 in the preceding example). We can enumerate the winners as $\bar{x}_1^{MC,n}, \ldots,$ $\bar{x}_{K_0}^{MC,n}$. We can define a matrix A^n of size $M \times K_0$ by

$$A^n = \left[e_{\bar{x}_1^{MC,n}}, \ldots, e_{\bar{x}_{K_0}^{MC,n}} \right].$$

Recall that e_x is a vector of zeroes with only the xth component set to 1. In the preceding example, we would write

$$A^n = \begin{bmatrix} 0 & 0 & 0 \\ 1 & 0 & 0 \\ 0 & 1 & 0 \\ 0 & 0 & 1 \\ 0 & 0 & 0 \end{bmatrix}.$$

Table 9.1 Simulated values of five alternatives based on the prior distribution.

x	$\bar{\mu}_x^n(\omega_1)$	$\bar{\mu}_x^n(\omega_2)$	$\bar{\mu}_x^n(\omega_3)$	$\bar{\mu}_x^n(\omega_4)$
1	15.59	11.38	3.73	10.48
2	17.24	18.49	7.30	**18.98**
3	18.06	**20.77**	2.61	18.88
4	**19.32**	17.97	**12.53**	18.73
5	12.37	13.41	4.08	11.55

Then, we can calculate the reduced prior using the equations

$$\theta^{MC,n} = (A^n)^T \theta^n,$$
$$\Sigma^{MC,n} = (A^n)^T \Sigma^n A^n.$$

We can now run the knowledge gradient algorithm using $\theta^{MC,n}$ and $\Sigma^{MC,n}$ as the prior instead of the original parameters θ^n and Σ^n. The computational cost is thus at most $K^2 \log K$, and likely less if there are many duplicates. We choose K to be much smaller than M. In a problem with 924 alternatives, we can obtain good results with $K = 30$.

9.3.2 Computational Issues

Monte Carlo simulation allows us to greatly reduce the cost of calculating knowledge gradients. However, it carries a new cost for running the simulations. Recall from Section 2.4 that a single sample from a multivariate normal distribution can be created using the equation

$$\bar{\mu}^n (\omega) = \theta^n + C^n Z,$$

where $Z = (Z_1, ..., Z_M)$ is a vector of standard normals, and C^n is a "square-root matrix" satisfying $C^n (C^n)^T = \Sigma^n$. The standard normals are easy enough to generate, but calculating the square root matrix can be costly. In the worst case, the time required to compute C^n is proportional to M^3, which is actually greater than the $M^2 \log M$ needed to compute the full set of knowledge gradients in the first place.

Fortunately, there is an elegant work-around for this issue. The square-root matrix can be computed recursively using the formula

$$C^{n+1} = C^n - \frac{\Sigma^n e_{x^n} e_{x^n}^T C^n}{(\sigma_W^2 + \Sigma_{x^n x^n}^n) \left(1 + \sqrt{\frac{\sigma_W^2}{\sigma_W^2 + \Sigma_{x^n x^n}^n}}\right)}. \tag{9.4}$$

This equation is entirely analogous to (2.23). Given our decision x^n at time n, C^{n+1} can be calculated directly from the beliefs at time n. The updating equation (9.4) even has a similar form to (2.23), and requires the same computational effort.

Thus, we only need to perform the expensive calculations required to compute the square root of a matrix once, to obtain C^0 from Σ^0. In MATLAB, this can be done using the routine chol (for Cholesky factorization, the name for the square root procedure). After that, we keep track of the square root matrix at each time step, and the next square root is computed recursively at a low cost. We are saving a lot of computing time by extending our knowledge state to include C^n as well as Σ^n.

It may also sometimes happen that, due to numerical rounding errors, the prior covariance matrix Σ^0 that we build using the methods from Section 9.2.2 may be reported by MATLAB as not being positive semidefinite (a necessary condition for computing square root matrices). This issue can often be resolved by simply adding a diagonal matrix $\varepsilon \cdot I$ to Σ^0. Here, I is the identity matrix and ε is a small positive number (e.g., 0.001). This quick fix is often enough for computing C^0, while still preserving the general correlation structure of subset selection.

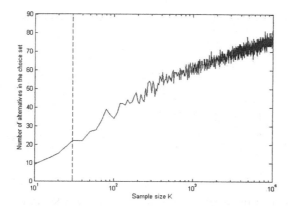

Figure 9.1 Growth of the size K_0 of the reduced choice set as a function of the sample size K.

9.3.3 Experiments

Earlier, we mentioned that the number K of Monte Carlo samples can be chosen to be much smaller than the total number M of subsets. One important question is exactly how many samples we need to take. Figure 9.1 shows how the size K_0 of the reduced choice set (recall that this is the number of winners obtained from Monte Carlo sampling) increases with the sample size K on a variant of the energy portfolio selection problem. Due to the random sampling, the curve is noisy, but we can see that K_0 grows fairly slowly. With $K = 30$ samples, we obtain a reduced choice set of $K_0 \approx 20$ alternatives. This number is roughly doubled when K increases to 300. Thus, as long as our prior presents a roughly accurate picture of the values of the alternatives (in Chapter 7 we referred to this as a "truth-from-prior" problem), we can safely ignore the vast majority of the possible subsets and focus only on several dozen of the most promising choices.

Figure 9.2(a) illustrates the performance, expressed using the average opportunity cost metric from (7.7), of the Monte Carlo KG (MCKG) policy as a function of the sample size K. We see that, once $K \geq 30$, performance begins to level off, and adding more samples does not yield substantially better results. Figure 9.2(b) shows the number of unique alternatives (energy portfolios) measured by MCKG for different sample sizes K. Here, we see that small sample sizes lead to more erratic behavior. In each iteration, our reduced choice set is randomly missing some important, promising portfolios, and it also randomly includes some portfolios that have previously been under-represented. Our policy thus measures more portfolios in an effort to gain enough information about the most relevant alternatives. On the other hand, once the sample size is large enough that all of these important portfolios are consistently included in the reduced choice set, the policy is able to more accurately discern which alternatives need to be measured.

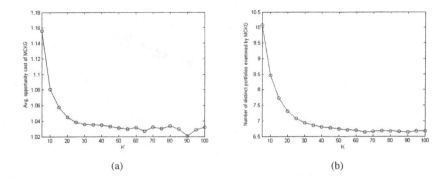

(a)   (b)

Figure 9.2   Illustration of (a) performance of the Monte Carlo KG (MCKG) policy, and (b) the number of alternatives measured by the policy as the sample size K increases. Results are taken from Ryzhov & Powell (2009*a*).

Finally, Figure 9.3 compares the performance of MCKG for a fixed sample size $K = 25$ to several other policies. In addition to our usual mainstays – approximate Gittins indices, interval estimation, and pure exploitation – we also apply the independent KG policy, in which we apply the formula from (5.10) to make a decision, thus temporarily ignoring the correlations between subsets when we make decisions, but incorporating them into our learning model. In Section 5.3, this approach was also called "hybrid KG." Although this idea is clearly not as good as correlated KG, it makes sense to use it in this setting because of its much lower computational cost: the time to calculate the formula for each alternative is proportional to M, as opposed to $M^2 \log M$ for correlated KG.

We find that Monte Carlo KG still outperforms the other policies. These results are consistent with what we know about correlated KG from Section 5.3. The Monte Carlo component does not seem to harm performance, as long as K is around 20 or 30. Note that Figure 9.3 does not include a comparison with the full correlated KG policy (computing correlated KG factors for the full choice set). Even in our version of energy portfolio selection, which is still not unreasonably large, it simply takes too much time to run correlated KG. The Monte Carlo modification allows us to work around this computational limitation.

9.4 VERY LARGE SETS

Eventually, subset selection problems become so large that the very framework of ranking and selection breaks down. For example, assume the Department of Energy is considering 100 proposals, from which they can fund 20. In this case, the number of potential subsets is 100 choose 20, which is approximately 5×10^{20}. We will simply not have enough memory to store a covariance matrix with 10^{20} rows and columns.

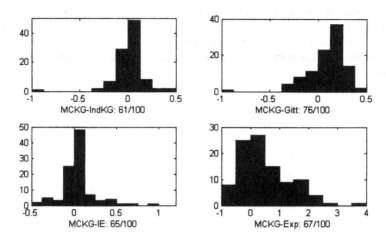

Figure 9.3 Performance of the MCKG policy relative to other policies for $K = 25$. The numbers indicate the number of problems (e.g., 61/100) where MCKG outperformed the competing policy. Results taken from Ryzhov & Powell (2009a).

Let $C(\rho)$ be a cost function where $\rho = (\rho_1, \ldots, \rho_M)$ is a set of technology parameters. Let $x = (x_1, \ldots, x_M)$ be a portfolio, where $x_i = 1$ means we are funding a proposal to improve technology i.

One strategy starts by replacing our objective function with a linear approximation of the form

$$F(x) = \sum_{i=1}^{m} x_i v_i.$$

Here, v_i is an approximation of the marginal value of adding the ith project to the portfolio. We approximate these values using $\bar{v}_i$, which we estimate using Monte Carlo methods. For example, assume we have a portfolio x^n at the nth iteration. The updated technology parameters for the perturbed portfolio are

$$\rho_1^{n+1,i}(W, x) = \rho_0 + \hat{x}_j^{n,i} W_j^{n+1,k}(\omega^{n+1}).$$

where $W_j^{n+1,k} \sim U[0, \tilde{\rho}_j^k]$ is a random variable that describes the technology change of the kth technology of the jth project if that project were funded. We would like to choose x to solve

$$\max_x \mathbb{E} C(\rho_1(W, x).$$

We are not able to compute the expectation, so instead we compute a sample realization

$$v^{n+1} = C\left(x^n, W(\omega^{n+1})\right). \tag{9.5}$$

We now compute a sample estimate of the marginal value of each project. We do this by looping over all the projects. Each project i is forced either in or out of the base portfolio (depending on whether it was out of or in the base portfolio, respectively), and then find the new portfolio given this constraint. The new portfolio is found by solving

$$\hat{x}^{n,i} = \arg\max_y \sum_{j=1}^{m} \bar{v}_j^n y_j,$$

subject to

$$\sum_{j=1}^{m} c_j y_j \leq b,$$

$$y_i = 1 - x_i^n,$$

$$y_j \in \{0,1\}, \; j \neq i.$$

A sample estimate of the marginal value of the new portfolio is given by

$$\hat{v}_i^{n+1} = \begin{cases} C(\rho_1^{n+1,i}) - v^{n+1} & \text{if } x_i^n = 1, \\ v^{n+1} - C_1(\rho_1^{n+1,i}) & \text{if } x_i^n = 0. \end{cases} \qquad (9.6)$$

Finally, we smooth these sample estimates to find an averaged estimate using

$$\bar{v}_i^{n+1} = \alpha_n \hat{v}_i^{n+1} + (1 - \alpha_n)\bar{v}_i^n, \qquad (9.7)$$

where $0 < \alpha_n \leq 1$ is a stepsize.

We have to remind ourselves that we are still solving a problem where the elements of the subset interact in a fairly complex way. We capture these interactions in the manner by which we calculate the marginal values $\hat{v}_i^{n+1}$.

9.5 BIBLIOGRAPHIC NOTES

Section 9.1 - See also Miller (2002) for additional applications of subset selection in statistics, and Horrace et al. (2008) for applications in economics.

Section 9.3 - The energy portfolio application is also considered by Ryzhov & Powell (2009*a*) and Ryzhov & Powell (2009*b*). The first of these references also proposes the Monte Carlo KG policy. The recursive updating equation for the square root matrix can be found, for example, in Kaminski et al. (1971).

PROBLEMS

9.1   There are six primary drugs that can be used to treat diabetes, and these are often prescribed in groups of three at a time. There are 20 ways of choosing three drugs from a set of six (known as a "cocktail"), assuming all possible combinations make sense. Enumerate these 20 cocktails, and create a covariance matrix Σ where entry $\Sigma_{ij} = \sigma^2 N_{ij}$, where N_{ij} is the number of drugs in common between drug cocktail i and drug cocktail j (so, $0 \leq N_{ij} \leq 3$). Let $\sigma^2 = .35^2$ be the variance in our prior distribution of belief about the effect of a drug cocktail. Finally, let $\sigma_\epsilon^2 = .55^2$ be the variance of an observation.

Assume that our prior on the blood sugar level produced by each cocktail is the same and is equal to 5.0.

a) Randomly generate a truth from this prior for each cocktail. Create the covariance matrix and use the knowledge gradient for correlated beliefs to search for the best cocktail using $N = 20$ measurements. You may use the MATLAB code (first introduced in Chapter 5) that can be downloaded from

   http://optimallearning.princeton.edu/exercises/KGCorrBeliefs.m

   An example illustration of the KGCB algorithm is given in

   http://optimallearning.princeton.edu/exercises/KGCorrBeliefsEx.m

b) Now repeat (a) for 100 different truths, and summarize how well you discover each truth using 20 observations.

c) Next we are going to use the methods described in this chapter to solve a much larger problem, but we are going to use them on this small test problem. Repeat (a), but this time you are going to randomly sample 5 out of the 20 combinations, and limit your calculation of the KG factor to these three (the easiest way to modify the code is to compute the knowledge gradient for all 20 as you are doing, but then choose a subset of 5, and finally choose the drug cocktail with the best knowledge gradient out of these five). Perform $N = 100$ samples and compare the performance to the results you obtained when you computed the knowledge gradient for all the alternatives.

d) Finally, repeat (d) for all 100 truths and report on the average performance.

9.2 Verify that 9.4 produces a valid square root matrix, that is, $C^{n+1} \left(C^{n+1} \right)^T = \Sigma^{n+1}$.

CHAPTER 10

OPTIMIZING A SCALAR FUNCTION

Optimizing scalar functions arises in a variety of settings, such as choosing the price of a product, the amount of memory in a computer, the temperature of a chemical process, or the diameter of the tube in an aerosol gun. Important special cases are functions which are unimodular, as depicted in Figure 10.1, where there is a single local maximum, or concave. In this chapter, we review some specialized algorithms designed for this particular problem class.

We begin our presentation using an unusual problem setting for this volume, which is an unknown but deterministic function which can be measured perfectly. We then transition back to our more familiar setting of noisy functions.

10.1 DETERMINISTIC MEASUREMENTS

We begin by assuming that $f(x)$ is differentiable and unimodular (that is, it has a single local maximum). We assume the region we are searching is bounded, so we can scale the search region to be between 0 and 1. Initially, we assume the optimum x^* can be anywhere in this interval with equal likelihood. If we measure the derivative $f'(x)$ at $x = 0.5$, we can observe whether $f'(.5) > 0$ or $f'(.5) < 0$. If the derivative is negative (as shown in Figure 10.1), then we know that $0 \leq x^* \leq .5$. We can

Optimal Learning. By Warren B. Powell and Ilya O. Ryzhov
Copyright © 2012 John Wiley & Sons, Inc.

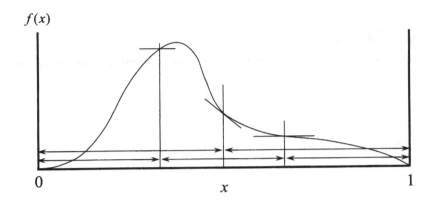

$f(x)$

0

x

1

Figure 10.1   A unimodular function, showing the slope at the midpoint, and the height of the function at 1/3 and 2/3.

eliminate the portion of the interval greater than .5. This means we can redefine the entire problem on the interval $(0, .5)$ and repeat the process. Let ℓ^n be the length of the interval in which the optimum may lie after n iterations, where $\ell^0 = 1$. It is easy to see that $\ell^n = .5^n$. Under the assumptions of the problem, this is the fastest possible rate of reduction that we can achieve.

Unfortunately, there are many problems where we cannot compute the derivative, but we can compute the function. For example, a transportation company may have a model that evaluates the on-time service when the fleet size is x. The company may vary x, re-running the model each time, but the model may not provide a derivative of the performance measure with respect to x. Instead, we have to resort to a kind of hunt-and-peck, trying different values of x. But, we can use our unimodular structure effectively. For example, we can try $x^1 = .2$ and $x^2 = .8$. If $f(.2) > f(.8)$, then we can eliminate the region $(.8, 1)$ from further consideration. Of course, this only eliminates 20 percent of the interval.

Imagine that we have a measurement budget of $N = 2$ measurements. In this case, the best strategy is to measure $x^1 = .5^-$ and $x^2 = .5^+$, by which we mean slightly less than .5 and slightly more than .5. Comparing $f(.5^-)$ and $f(.5^+)$ allows us to effectively eliminate half the interval, just as we did when we could compute a derivative (we are basically computing a numerical derivative).

Next consider the case where $N = 3$. If we first measure $x^1 = 1/3$ and $x^2 = 2/3$, we eliminate either $(0, 1/3)$ or $(2/3, 1)$. Assume that we eliminate the upper interval (as we did in Figure 10.1). Now we are left with the interval $(0, 2/3)$, but we already have a measurement at the midpoint $x = 1/3$. We are going to use our final measurement at a point slightly above or below $1/3$ to determine the final interval of uncertainty, which will have width $1/3$.

We repeat this exercise one more time for $N = 4$, but now we are going to assume that the optimum is at $x = 0$ (but we have to construct our measurements without knowing this), so that we are always eliminating the upper part of the remaining interval. If we measure $x^1 = 2/5$ and $x^2 = 3/5$, we would eliminate $(3/5, 1)$, and we are left with the interval $(0, 3/5)$ with a measurement at $2/5$. Conveniently, this is at the two-thirds point of the interval $(0, 3/5)$, with two remaining measurements. If we measure $x = 1/5$, we are now in the same situation we were when we started with $N = 3$, but on an interval of width $3/5$. We eliminate the upper interval, leaving us with the interval $(0, 2/5)$ and a measurement at the midpoint $1/5$. We use our final measurement at a point slightly higher or lower than the measurement at $1/5$, giving us a final interval of $1/5$.

Now compare this result to what we would have obtained if we had used the bisection search with numerical derivatives. If $N = 2$, we would have measured just above and below .5, giving us a final interval of width .5. If $N = 4$, we would have done this twice, giving us an interval of width $.5^2 = .25$, which is greater than the interval we obtained of $1/5 = .2$. How did we accomplish this?

There is a pattern in this logic. When we eliminated the interval $(3/5, 1)$, we were left with the interval $(0, 3/5)$ and a measurement at $2/5$. Rescaling all the numbers so that the interval is of length 1, we get an interval of $(0, 1)$ with a measurement at $2/3$. If we eliminated the lower part of the interval $(0, 2/5)$, we would be left with the interval $(2/5, 1)$ (which still has length $3/5$) and a measurement at $3/5$. Rescaling gives us an interval of length 1, and a measurement at $1/3$. Either way, we end up with a measurement we would have made anyway if we only had 3 measurements.

We can formalize the algorithm as follows. Let $f^1 < f^2 < f^3 < \ldots$ be an increasing sequence of integers. If we are allowed three measurements, assume we first measure f^1/f^3 and f^2/f^3. If we eliminate the upper part of the interval, we are left with an interval $(0, f^2/f^3)$. If we eliminate the lower part of the interval, we are left with $(f^1/f^3, 1)$. We would like the width of these intervals to be the same, so we are going to require that

$$\frac{f^2}{f^3} = 1 - \frac{f^1}{f^3},$$

or, rearranging slightly,

$$f^3 = f^1 + f^2.$$

Similarly, if we have four measurements, we first measure f^2/f^4 and f^3/f^4. Repeating the exercise above, rejecting the upper range leaves us with an interval of f^3/f^4, while rejecting the lower range leaves us with the interval $(1 - f^2/f^4)$. Again equating these gives us

$$\frac{f^3}{f^4} = 1 - \frac{f^2}{f^4},$$

or

$$f^4 = f^3 + f^2.$$

Using proof by extrapolation, we see that if we are making N measurements, we want

$$f^N = f^{N-1} + f^{N-2}. \tag{10.1}$$

Furthermore, this has to be true for all $n < N$. Equation (10.2) defines what is known as the Fibonacci sequence, comprising the numbers $(1, 1, 2, 3, 5, 8, \ldots)$, where we initialize the sequence using $f^0 = 1$, and then let $f^1 = 1, f^2 = 2, f^3 = 3, \ldots$.

It is possible to show that the Fibonacci sequence produces an optimal search sequence for a finite measurement budget (under the assumptions of our problem). By this we mean that for a given value of N, no other method will produce a smaller final interval in which the optimal solution may lie. This means that our search is optimal, because we learn the most within our budget.

So what if our budget is unlimited? We are going to start by hypothesizing that in the limit, we are going to measure the interval at two points, which we are going to denote by $1 - r$ and r, where $.5 < r < 1$. If we measure $1 - r$ and r and then eliminate the upper interval, we are left with the interval $(0, r)$. Now assume we are going to measure the same points within this interval, which would occur at $r(1 - r)$ and r^2. We want the larger of these two points to coincide with the smaller of the two measurements in the original interval, so that at each iteration, we are measuring only one additional point (as we did with the Fibonacci search). This means that we require

$$r^2 = 1 - r,$$

or

$$r^2 + r - 1 = 0.$$

We solve this using the quadratic formula which gives the roots for the equation $ar^2 + br + c = 0$ as

$$r = \frac{-b \pm \sqrt{b^2 - 4ac}}{2a}.$$

Using only the positive part gives us the solution

$$
\begin{aligned}
r &= \frac{-1 + \sqrt{1+4}}{2} \\
&= \frac{-1 + \sqrt{5}}{2} \\
&= 0.618.
\end{aligned}
$$

The quantity $r = .618$ is known as the *golden section* or *golden ratio*. We note that we get the same result if we had eliminated the lower portion of the interval, since our measurements are naturally symmetric. We also note that if f^n is the nth Fibonacci number, then

$$\lim_{n \to \infty} \frac{f^{n-1}}{f^n} \to r = 0.618.$$

Thus, the golden section search is the limit of the Fibonacci search.

The Fibonacci search and the golden section search are both examples of optimal learning in that they give the fastest improvement in our knowledge, measured by the length of the interval where we think the optimum might be. The Fibonacci search guarantees the smallest possible interval after a fixed (and known) number of measurements. The golden section search gives the fastest *rate* of convergence for an algorithm that will be run an infinite number of times. To put it another way, we account for the effect that our next measurement will have on the length of the uncertain interval.

10.2 STOCHASTIC MEASUREMENTS

We now return to our more familiar setting where measurements of the function are noisy. We consider the stochastic version of bisection search, where we assume we have access to a noisy indicator that tells us whether we think the optimum is to the left or the right of our measurement. In generalizing the bisection search we make a particular assumption about the measurement noise. We explain this assumption by noting that the bisection search operates by separating two regions of the search space: the region to the left of x^* and the region to the right. Measuring the derivative of the function $f(x)$ at a point x reveals in which part of the search space x belongs.

Even without a function $f(x)$, we can use a bisection search to solve any problem where we want to find the boundary between two regions, and where measuring a point reveals in which region it resides. In the error-free case, this revelation is always correct. In the noisy version of the problem we instead assume that the revelation is incorrect with a probability that is *known* and *constant*. While these conditions are typically not going to be satisfied in practice, they provide an elegant search model and they are certainly more realistic than assuming the measurement is perfect.

This assumption of constancy would tend to be met in applications where the transition between regions is abrupt. As an example, if a city's water supply were contaminated with a dangerous chemical we would want to localize the extent of contamination as quickly as possible, and if the chemical did not dissolve well in water but instead tended to stay concentrated, we would find a situation with this abrupt transition between contaminated and uncontaminated water. In contrast, when we measure a smooth function with additive noise, noise tends to cause incorrect region assignments more frequently near the function's maximum. With this in mind, we should be careful in applying the stochastic bisection algorithm presented here to situations not meeting its assumptions.

10.2.1 The Model

We formulate our problem mathematically by again supposing that we have a point x^* whose location is unknown beyond that it resides in the interval $[0, 1]$. The point x^* corresponds to the boundary between the two regions in $[0, 1]$, so in the water contamination example, the water in region $(x^*, 1]$ would be contaminated and the water in region $[0, x^*]$ would not. We adopt a Bayesian prior density p_0 on the location of this point x^*, where $p_0(x)$ gives the likelihood (density) that $x^* = x$. It could be uniform on $[0, 1]$ if we had little real information about its location, or it could be some more complicated distribution expressing a stronger belief. We then suppose that we are offered the opportunity to take a sequence of measurements $x^0, x^1, \ldots, x^N$ in the interval. With each measurement x^n, we get a noisy response $\hat{y}^{n+1}$, suggesting into which region x^n resides. Given x^n and x^*, this response will be independent of all other responses and will have the distribution

$$\hat{y}^{n+1} = \begin{cases} I_{\{x^* \leq x^n\}} & \text{with probability } q, \\ I_{\{x^* > x^n\}} & \text{with probability } 1 - q. \end{cases}$$

We may also express this as $\mathbb{P}\{\hat{y}^{n+1} = I_{x^* \leq x^n}\} = q$. Here q is the probability our measurement is correct, and we assume this probability is known and constant. Equivalently, $1 - q$ is the error rate in our measurements.

10.2.2 Finding the Posterior Distribution

These measurements alter our prior belief about the location of x^*, giving us a posterior belief, all according to Bayes' rule. We use the notation p^n to denote the posterior density at time n. To write the updating rule for p^n explicitly, we introduce two pieces of notation. Let F^n be the cumulative distribution function of the posterior at time n, by which we mean that, for any dummy variable x,

$$F^n(x) := \mathbb{P}\{x^* \leq x \mid p^n\} = \int_{[0,x]} p^n(x) \, dx.$$

We may also think of $F^n(x)$ as giving the probability that x is in the region to the left of x^*, and in our water contamination example we think of it as giving the probability that the water at x is not contaminated. As our second piece of notation, let g be the function defined by $g(a, 1) = a$ and $g(a, 0) = 1 - a$, where a will be a probability (such as the probability the true value is to the left or the right). Now we are ready to compute our updating rule.

Noting that nature's correct response to the measurement x^n would be $I_{\{x^* \leq x^n\}}$, we write

$$\mathbb{P}\{\hat{y}^{n+1} = y \mid x^*, x^n\} = \begin{cases} q & \text{if } y = I_{\{x^* \leq x^n\}}, \\ 1 - q & \text{if } y \neq I_{\{x^* \leq x^n\}}, \end{cases}$$
$$= g(q, I_{\{y = I_{\{x^* \leq x^n\}}\}}).$$

So, $g(q, 1) = q$ corresponds to two cases: Either $x^n < x*$, and we observe $\hat{y}^{n+1} = 1$ which correctly indicates that the optimum $x*$ is greater than our measured point; or $x^n > x*$ and $\hat{y}^{n+1} = 0$, which correctly indicates that the optimum x^* is less than our measured point. The outcome $g(q, 0) = 1 - q$ corresponds to the opposite of both of these cases.

We may also write

$$
\begin{aligned}
\mathbb{P}\{\hat{y}^{n+1} = y \mid p^n\} &= \mathbb{P}\{x^* \leq x^n \mid p^n\}\mathbb{P}\{\hat{y}^{n+1} = y \mid x^* \leq x^n\} \\
&\quad + \mathbb{P}\{x^* > x^n \mid p^n\}\mathbb{P}\{\hat{y}^{n+1} = y \mid x^* > x^n\} \\
&= F^n(x^n)g(q,y) + (1 - F^n(x^n))g(1-q,y) \\
&= g(qF^n(x^n) + (1-q)(1 - F^n(x^n)), y),
\end{aligned}
$$

where the last line may be seen by considering the cases $y = 0$ and $y = 1$ separately. Fixing some dummy variable y, we may then use Bayes' rule and these two relations to write

$$
\begin{aligned}
p^{n+1}(x)\, dx &= \mathbb{P}\{x^* \in dx \mid p^n, \hat{y}^{n+1} = y\} \\
&= \frac{\mathbb{P}\{\hat{y}^{n+1} = y \mid p^n, x^* = x\}\mathbb{P}\{x^* \in dx \mid p^n\}}{\mathbb{P}\{\hat{y}^{n+1} = y \mid p^n\}} \\
&= \frac{g\left(q, I_{\{y = I_{\{x \leq x^n\}}\}}\right)}{g\left(qF^n(x^n) + (1-q)(1 - F^n(x^n)), y\right)}p^n(x)\, dx.
\end{aligned}
$$

Substituting $\hat{y}^{n+1}$ for y shows that our updating rule is

$$
p^{n+1}(x) = \frac{g\left(q, I_{\{\hat{y}^{n+1} = I_{\{x \leq x^n\}}\}}\right)}{g\left(qF^n(x^n) + (1-q)(1 - F^n(x^n)), \hat{y}^{n+1}\right)}p^n(x). \tag{10.2}
$$

The essential content of this updating rule is as follows. Our observation, if correct, would tell us into what region x^n lies and would tell us whether x^* is to the left or right of this measurement point. Let us give the name "suggested region" to the region in which the observation, if correct, would indicate x^* resides. Since we know that the observation is only correct with probability q, we multiply the density in the suggested region by q, and we also multiply the density in the other region by $1 - q$. This leaves us with a density which does not integrate to 1, so we then finish the update by normalizing.

As a numerical example, assume that $q = .7$, $x^n = .6$, and that we observe $\hat{y}^{n+1} = 1$. Assume that the distribution $p^n(x)$ is uniform, as depicted in Figure 10.2(a). Figures 10.2(b) and 10.2(c) show the conditional distribution given $\hat{y}^{n+1} = 1$ and $\hat{y}^{n+1} = 0$, respectively. Finally, Figure 10.2(d) shows the updated distribution for p^{n+1} given $\hat{y}^{n+1} = 1$, overlaid on top of the original uniform distribution for $p^n(x)$. This distribution is computed using

$$
p^{n+1}(x) = \begin{cases}
\frac{g(.7,1)}{g(.7(.6)+.3(.4),0)}p^n(x) = \frac{0.7}{.54}p^n(x) = 1.296p^n(x) & x < x^n, \\
\frac{g(.7,0)}{g(.7(.6)+.3(.4),0)}p^n(x) = \frac{0.3}{.54}p^n(x) = 0.556p^n(x) & x > x^n.
\end{cases}
$$

Let us briefly consider the case when our measurements are always correct. Then $q = 1$ and we are back in the deterministic case. Suppose for the moment that p^n is uniform on some interval $(a^n, b^n]$. Then our updating rule can be simplified to

$$
p^{n+1}(x) = \frac{g\left(1, I_{\{\hat{y}^{n+1} = I_{\{u \leq x^n\}}\}}\right)}{g(F^n(x^n)), \hat{y}^{n+1})}\frac{I_{\{x \in [a^n, b^n]\}}}{b^n - a^n} = \begin{cases}
\frac{I_{\{x \in (a^n, x^n]\}}}{x^n - a^n} & \text{if } \hat{y}^{n+1} = 1, \\
\frac{I_{\{x \in (x^n, b^n]\}}}{b^n - x^n} & \text{if } \hat{y}^{n+1} = 0.
\end{cases}
$$

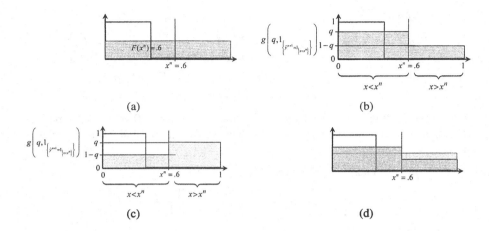

Figure 10.2 Illustration of the effect of an observation of $\hat{y}$ for $x^n = .6$, starting with an initial uniform prior (a). The new beliefs are shown in (b) and (c) for $\hat{y} = 1$ and $\hat{y} = 0$. The updated distribution $p^{n+1}(x)$ is given in (d).

Thus we see that the posterior is now still uniform but on some smaller interval, where either the points to the left or right of x^n have been removed. Thus, if we begin with a uniform prior, i.e., with $p^0(x) = I_{x \in [0,1]}$, then our posterior at each time will again be uniform on a smaller interval. Comparing our knowledge in the deterministic case, where we had an interval in which we knew x^* resided, to our knowledge here suggests that a natural way to express knowledge that x^* lies in an interval is through a uniform probability distribution on that interval.

10.2.3 Choosing the Measurement

Now let us return to the stochastic case where $q < 1$. With our updating rule in hand, we could take any given method for choosing our measurements, and compute the posterior density of the location of x^* after our allotted number of measurements N, but before proceeding to say which measurement methods are best we need a way to evaluate the quality of the knowledge that we arrive to at the final time. In the deterministic case we knew that x^* resided in an interval, and we evaluated how happy we were with our final knowledge according to the length of this interval. In the stochastic case, however, we no longer have an interval but instead a density expressing a continuum of belief about the location of x^*.

The objective function we use should correspond to our notion of length in the deterministic case, and it should punish greater uncertainty. There are many choices, but one possibility is the entropy. Denote the entropy of any particular density p on the location of x^* by $H(p)$,

$$H(p) := -\int_0^1 \log_2(p(x))p(x)\,dx.$$

The entropy corresponds to uncertainty about x^* in several senses. First, measurements always decrease entropy on average, no matter what is measured. Second, the entropy is largest for the uniform prior, which we may understand intuitively as the density we have when we are most uncertain about x^*. Third, the entropy approaches $-\infty$ as our posterior density sharpens to a point at the true location of x^*.

Additionally, the entropy corresponds in a very nice way to our use of interval length as the objective function in the deterministic case. In the deterministic case, we may know at a point in time that x^* is in $[a, b]$, but we have no information about it beyond that. A natural belief to take on the location of x^* in this situation is the uniform density on $[a, b]$, which is $p(x) = I_{x \in [a,b]}/(b-a)$. This density has entropy

$$H(p) = -\int_a^b \log_2(1/(b-a))/(b-a)\,dx = \log_2(b-a),$$

which is a strictly increasing function of the length $b - a$ of the interval. Thus minimizing the length of the interval $[b, a]$ is in some sense equivalent to minimizing the entropy. We characterize this equivalence more concretely later, when we show that the stochastic bisection algorithm is the same as the deterministic bisection algorithm when the probability q of measurement error is 0. Finally, an additional and very important reason for using entropy is that it provides an analytic and easy-to-use solution to the sequential problem.

Now with the transition function worked out and the entropy as our objective function, we have a well-defined sequential information collection problem. This problem can be solved and the optimal solution computed using dynamic programming. Let us define our value function V^n by taking $V^n(p^n)$ to be the smallest value of $\mathbb{E}[H(p^N) \mid p^n]$ than can be achieved starting from the density p^n at time N. Bellman's principle then tells us that

$$V^n(p^n) = \min_x \mathbb{E}\left[V^{n+1}(p^{n+1}) \mid p^n, x^n = x\right].$$

In addition, we know that $V^N(p^N) = H(p^N)$ since there are no measurements left to make at time N.

We can use Bellman's recursion to compute V^{N-1} from V^N. Since $V^N = H$, this recursion is $V^{N-1}(p^{N-1}) = \min_x \mathbb{E}\left[H(p^N) \mid x^{N-1} = x, p^{N-1}\right]$. Rather than computing this here, we simply state the following formula, which can be confirmed by direct computation:

$$\min_x \mathbb{E}[H(p^{n+1}) \mid x^n = x, p^n] = H(p^n) - q\log_2 q - (1-q)\log_2(1-q) - 1, \quad (10.3)$$

and the minimum is achieved by choosing x to be a median of p^n. The median of p^n is defined to be the point where $F^n(x) = 1/2$. If there is more than one median, any point satisfying $F^n(x) = 1/2$ achieves the minimum.

Now, using (10.3) and the Bellman recursion, we see that V^{N-1} is given by

$$V^{N-1}(p^{N-1}) = H(p^{N-1}) - q\log_2 q - (1-q)\log_2(1-q) - 1,$$

and that the optimal decision x^{N-1} is the median of p^{N-1}. Moreover, we see that the form of the value function at time $N-1$ is the same as it is at time N, but with a constant subtracted. This constant does not depend on p^{N-1}, nor does it depend on n. This tells us that, if we repeat the computation of the Bellman recursion, we will find that $V^{N-2}(p^{N-2}) = H(p^{N-2}) - 2\left(q\log_2 q + (1-q)\log_2(1-q) + 1\right)$ and that in general

$$V^n(p^n) = H(p^n) - (N-n)\left(q\log_2 q + (1-q)\log_2(1-q) + 1\right). \qquad (10.4)$$

Furthermore, since the minimizer of the Bellman recursion at each time n is the median of the density p^n at that time, we have discovered that the optimal policy is to always measure at the median. Denoting the optimal measurement at time n by $x^{*,n}$, we summarize this conclusion by saying that $x^{*,n}$ is such that

$$F^n(x^{*,n}) = 1/2.$$

Let us spend a few moments interpreting these results. First, the computation (10.3) tells us that if our goal is to minimize the expected entropy after one measurement, then the best x^n has the expected posterior entropy equal to the original entropy at time n minus a deterministic factor $q\log_2 q + (1-q)\log_2(1-q) + 1$. This factor is actually the mutual information between the measurement $\hat{y}^{n+1}$ and x^*, given that we measure at the median. This fact can be confirmed by computing the mutual information directly from its definition, although we do not perform this computation here.

We can view this reduction another way: The expected reduction in entropy about x^* is equal to the information about x^* contained in the measurement, and this mutual information is maximized if we measure at the median of x^*. The mutual information is largest at the median because at this point we are "maximally uncertain" about to which region, $[0, x^*]$ or $(x^*, 1]$, it belongs since our belief assigns an equal probability of $1/2$ to each possibility. This conveys a general principle of information collection: Often the measurement that is most valuable is the one whose result is least predictable. Put another way, if we already knew the result of a measurement before we took it, there would be no point in actually taking that measurement.

Then, the formula (10.4) shows that this general principle applies not just to single but to multiple measurements. That is, the best we can do is to measure each time at the median of our belief, which is the measurement whose outcome is most uncertain, and the resulting decrease in expected entropy of x^* is equal to the sum of the mutual information in all the measurement outcomes combined. From measurement n onward, this decrease is $(N-n)\left(q\log_2 q + (1-q)\log_2(1-q) + 1\right)$ since there are $N-n$ measurements left to make and each contributes $q\log_2 q + (1-q)\log_2(1-q) + 1$.

We may also gain insight by taking the special case $q = 1$ and comparing to the deterministic case. As previously noted, when $q = 1$ and when we begin with a uniform prior on $[0, 1]$, the posterior remains uniform on a smaller interval. Then, the median of any such uniform posterior is simply the middle of the interval, and so we again always measure in the middle of the current interval, just as we did with deterministic bisection. Thus we see that the stochastic bisection algorithm reduces in the error-free case to exactly the classic bisection algorithm.

10.2.4 Discussion

We make one additional note about the usefulness of the stochastic bisection algorithm when our objective function is something different than the entropy objective we have assumed here. Although we have not shown it here, the decrease in entropy one obtains from measuring at the median is *deterministic*, even though the final density p^N itself is certainly random. This is similar to the situation in the deterministic case, where the location of the final interval containing x^* is unknown a priori, but the length of that interval is deterministic as long as we use the bisection measurement rule.

This is a very nice property because it means that the optimal policy for the entropy objective function is also optimal for a broader class of objective functions. In particular, if our objective function is $\mathbb{E}[L(H(p^N))]$, where L is some concave increasing function, then again one can show that the same stochastic bisection rule is optimal. We can think of L as inducing some kind of risk aversion, in that using it would indicate we fear uncertainty about x^* more than than we hope for certainty. For example, earlier we saw that the length of an interval is equal to the logarithm of the entropy of a uniform distribution on this interval, and so if we wanted to minimize the length of the final interval in the deterministic case, then perhaps we should minimize the logarithm of the entropy rather than just the entropy. But we have already minimized using this criterion, since the logarithm is concave and increasing and the stochastic bisection algorithm is optimal for this objective function as well.

10.3 BIBLIOGRAPHIC NOTES

Section 10.1 - For a proof of the optimality of the Fibonacci search sequence for unimodular functions, see Avriel & Wilde (1966).

Section 10.2 - The stochastic bisection algorithm originates from Horstein (1963). The optimality properties of this algorithm with respect to entropy minimization are discussed by Jedynak et al. (2011).

PROBLEMS

10.1 How many iterations of a Fibonacci search sequence are needed to ensure that we can find the optimum of a unimodular function within 1 percent?

10.2 Consider the function $f(x|\alpha, \beta) = x^{\alpha-1}(1-x)^{\beta-1}$ for $0 \le x \le 1.0$. Perform the Fibonacci search to find the optimum of this function for the following values of (α, β).

a) $\alpha = 8, \beta = 2$.

b) $\alpha = 3, \beta = 12$.

c) $\alpha = 6, \beta = 8$.

10.3 Repeat exercise 10.2 using the golden section search.

10.4   Assume that we are trying to find the maximum of the function $f(x) = x^6(1 - x)^2$, but now we are going to assume that we can compute the derivative.

a) Using the derivative to indicate whether the optimum is to the left or the right, perform eight iterations of deterministic bisection search and report your best estimate of the optimal solution. Plot your distribution of belief describing where the optimum lies after eight observations.

b) Now assume that we can estimate the sign of the derivative correction with probability $q = .70$ (even though we are computing it perfectly). Use the method described in Section 10.2 to find the optimum. Again plot your distribution of belief describing the location of the optimum.

c) How would your answer to (b) change if $q = .5$? You should be able to answer this without repeating any of the calculations.

CHAPTER 11

OPTIMAL BIDDING

Imagine that industrial customers come to you requesting price quotes on contracts to provide a product of some type. This might be laptops for a large company, a component for a product that the customer is building (such as disk drives for the laptops), or materials used in a manufacturing process. You have to quote a price, recognizing that you have to match or beat competing prices. If you win the contract, you might be left wondering if you could have asked for a little more. If you lose the contact, you are kicking yourself if you feel that a slightly lower price would have helped you win the contract.

Figure 11.1 illustrates the economics of different prices. There is a breakeven price p^b, below which you will lose money on the contract. Then there is a price point $\bar{p}_c$ for each customer c. If you quote a price above this number, you will lose the contract. The problem, of course, is that you do not know $\bar{p}_c$. As the figure illustrates, increasing the price above p^b produces dramatic increases in profits, especially for competitive, commodity products.

We are going to assume that we have several opportunities to quote a price and observe a buy/not-buy decision from the customer. We are not allowed to observe $\bar{p}_c$ directly, even after the fact. After each iteration of quoting a price and observing a decision of whether or not to accept the bid, we have an opportunity to use what we

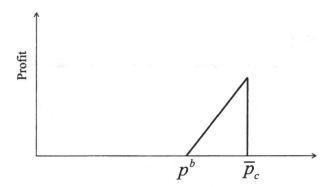

Figure 11.1   Illustration of the value of quoting a price p that is greater than the break-even price p^b but lower than the cuttoff price $\bar{p}_c$ for customer c.

have learned before deciding on our next bid. We know that we have to bid prices above p^b, and we know what prices have been accepted in the past. If we discover that we can get higher prices, the impact on our profits can be significant. Our challenge is trying to learn $\bar{p}_c$ while balancing the profits we realize against the information gained while trying higher prices (and potentially losing some contracts).

We can also consider a version of this problem from the customer's point of view. Suppose that there are now multiple sellers offering a single product. A customer makes a bid for the product at a price of his or her choosing. The bid is successful if there is a seller willing to accept it, but the customer does not get to observe the sellers' willingness to sell before making the bid. In this case, higher bids are more likely to be successful, but the customer can save money by finding the optimal price. Bidding too low and failing to secure an offer carries an opportunity cost (perhaps we are forced to wait a period of time before bidding again). We might call this the "Priceline problem," because of its similarities to the business model of the well-known online travel agency.

Overall, we can see a clear potential for optimal learning in the bidding problem. In fact, variations of this problem have attracted a great deal of attention in the revenue management community, precisely from an optimal learning perspective. However, it is less clear precisely how learning should be applied. The problem has a number of features that bring us outside the scope of the clean, fundamental models presented in Chapters 4 and 6. For example, in the bidding problem, we are not able to observe an unbiased sample of the truth. That is, in the industrial setting, the company does not get to observe the exact amount that the client was willing to pay for the contract, only whether or not the quoted price was accepted. As we will see in this chapter, this creates an additional challenge for our analysis. We are no longer able to make use of an elegant conjugate prior, as in Chapter 2.

New challenges call for new methods. In contrast with our earlier focus on knowledge gradient methods, we look at a simpler greedy policy for the bidding problem.

However, this policy is still in line with our general approach to optimal learning. We will use a Bayesian model to represent our uncertainty about customer demand. Our policy incorporates this uncertainty into the decision-making. As a result, we will tend to quote higher prices to the customers than we would without the learning dimension. How much higher will depend on the amount of uncertainty we have. By doing so, we will take on more risk with the first few contracts, but the information we collect will help us to make money in the long run. The bidding problem shows that optimal learning provides value in a slightly messier setting that goes beyond the standard models we have discussed up to this point.

11.1 MODELING CUSTOMER DEMAND

Suppose that we are working with a sequence of bids. The customers' behavior is modeled using the concept of *valuation* or *willingness to pay*. The nth customer values a product or service at W^n. We make the sale as long as this valuation is at least as much as our quoted price p, that is, $W^n \geq p$. Our revenue, in this case, is the price p. If $W^n < p$, our revenue is zero. Thus, our *expected revenue*, for a fixed price p, is given by

$$R(p) = p \cdot P(W^n \geq p). \tag{11.1}$$

In most applications, we will work with the *expected profit*

$$P(p) = (p - c) \cdot P(W^n \geq p), \tag{11.2}$$

where c is the cost of the product to the seller.

In real life, many customers may not have an exact number for the most they are willing to pay. In any case, we will never observe this quantity, only whether it is larger or smaller than p. However, the idea of the valuation gives us a way to think about the bidding problem formally and put a number on the probability that a price will be accepted. Initially, let us assume that the valuations $W^0, W^1, \dots$ are independent and identically distributed. Thus, while different customers can have different valuations, all customers come from the same population. This may be a reasonable assumption if a type of industrial contract serves a particular market (such as semiconductor manufacturers). Later, in Section 11.1.2, we begin to push the boundaries of this assumption.

11.1.1 Some Valuation Models

Nearly any standard distribution might be used to model customer valuation. We list several distributions that have been used in the literature on bidding. We will make use of some of these models to illustrate some of the issues inherent in the problem, although our main focus will be the logistic model of Section 11.1.2. Many of these same distributions also appeared in Chapter 2 as sampling models.

Uniform Valuation The simplest model assumes that the valuation follows a uniform distribution, $W^n \sim U[a, b]$. It follows that the probability of making the sale is

$$P(W^n \geq p) = \begin{cases} 1, & p < a, \\ \frac{b}{b-a} - \frac{p}{b-a}, & p \in [a, b], \\ 0, & p > b, \end{cases}$$

which is a linear function of p. This is known as *linear demand*. If we assume that $a = 0$, the only parameter in this model is the maximum valuation b. If this parameter is unknown, a natural choice for a distribution of belief would be a Pareto prior (see Section 2.3.3).

The uniform valuation is clean, but involves several strong assumptions. We are assuming that there is a fair amount of variation; customers are equally likely to have low or high valuations. Additionally, we assume that there is a cutoff point. A high enough price is guaranteed to lose the sale.

Exponential Valuation An exponential valuation assumes that $W^n \sim Exp(\lambda)$. The demand curve then has the simple form $P(W^n \geq p) = e^{-\lambda p}$. As in the linear model, higher prices are less likely to be successful. However, now any price will have a non-zero probability of being successful. We are assuming that there will be a small proportion of customers willing to pay very large prices.

Lognormal Valuation In the lognormal model, we assume that

$$P(W^n \geq p) = 1 - \Phi\left(\frac{\log p - \mu}{\sigma}\right),$$

where μ and σ are the parameters of the demand curve. This model implies that $\log W^n \sim \mathcal{N}(\mu, \sigma^2)$. If we had a way to observe the exact values W^n, we could put a normal prior on μ and treat $\log W^n$ as a normal observation, enabling the use of the normal-normal learning model.

11.1.2 The Logit Model

The logit model is a particularly attractive method of approximating the probability that a customer will accept a bid. The logic model expresses the probability of making a sale as

$$P(W^n \geq p) = \frac{1}{1 + e^{-(\mu_1 - \mu_2 p)}}. \tag{11.3}$$

When $\mu_2 > 0$, plotting (11.3) as a function of p yields a logistic curve, a well-known mathematical model for predicting demand, population growth, technology adoption, and other cyclical phenomena. Figure 11.2(a) gives an example for a particular choice of μ_1 and μ_2. We see a classic S-curve, flipped around so that higher prices lead to lower success probabilities. Balancing the decreasing success probability with the increasing potential revenue, the expected revenue function (11.1) has a maximum at approximately $p^* \approx 3.9$.

The parameters μ_1 and μ_2 determine the customer's reaction to different price offers. We can view μ_1 as the *market share* of the seller. Even if the seller were to

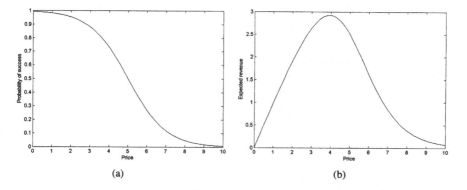

Figure 11.2 Example of (a) probability of making a sale and (b) expected revenue under a logistic model with $\mu_1 = 5, \mu_2 = 1$.

give away the product for free, the probability of success would only be $\frac{1}{1+e^{-\mu_1}}$; that is, some customers would still prefer to buy from a competitor. Essentially, these customers have a negative valuation of our product, and there is no way we could convince them to buy it. The higher the value of μ_1, the higher the market share and the closer $P\left(W^n \geq 0\right)$ is to 1.

The second parameter μ_2 can be viewed as the *price sensitivity* of the customers. Large values of μ_2 tend to make the curve in Figure 11.3 steeper, causing the probability of success to decrease faster as the price goes up. We typically require that $\mu_2 > 0$, to preserve the S-curve shape of the success probability. Negative values of μ_2 would imply that higher prices are more likely to be successful, which does not make sense for our problem.

One important advantage of the logistic model is that it allows us to move beyond the assumption that the customers come from the same population. We could, for example, allow the demand curve to depend on attributes of the customer, as given by

$$P(W^n \geq p) = \frac{1}{1 + e^{-(\mu^T x^n - p\mu_{P+1})}}. \tag{11.4}$$

Here, $x^n = [x_1^n, x_2^n, ..., x_P^n]^T$ is a vector representing P attributes of the customer. For example, the attributes could reflect the location of the customer and its size. The vector μ contains the parameters assigned to the attributes. We also have a single price sensitivity parameter $\mu_{P+1} > 0$. This model is an example of the *logistic regression* technique in statistics, which fits a set of parameters to a success probability. Throughout this chapter, we mostly focus on the simple two-parameter model of (11.3), but it is important to remember that our analysis can easily be extended to the multi-parameter case.

Optimal learning comes into play when we, as the seller, do not know the exact values of the parameters μ_1, μ_2. We may have some prior estimates of these parameters, based on past sales figures. However, what makes this problem especially difficult is that even small changes to the parameters will greatly change the expected revenue

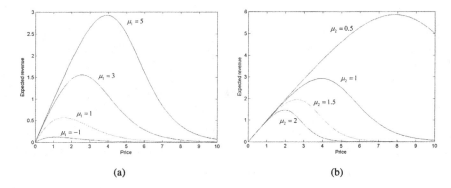

(a)   (b)

Figure 11.3 Sensitivity of the revenue curve to changes in (a) market share and (b) price parameter.

function. Figure 11.3(a) shows the expected revenue curve for different values of μ_1, with μ_2 fixed at 1. Figure 11.3(b) shows the same curve for different values of μ_2, with μ_1 fixed at 5.

Higher values of μ_1 move the optimal price to the right, but they also expand the magnitude of the entire revenue curve. Increasing μ_1 from 1 to 3 moves the optimal price roughly from 1.5 to 2.5, but triples the expected revenue collected in the process. Smaller values of μ_2 (that is, closer to zero) have the same two effects, but move the optimal price more than they increase optimal revenue. When we allow both parameters to change at the same time, as in Figure 11.4, even small changes will allow for a wide range of possible optimal prices and revenues.

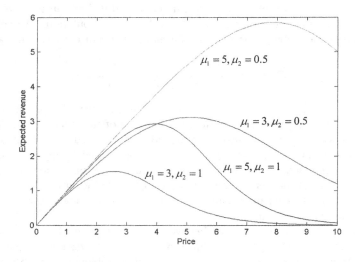

Figure 11.4 Sensitivity of the revenue curve to changes in both parameters.

11.2 BAYESIAN MODELING FOR DYNAMIC PRICING

We will use a Bayesian model to represent our uncertainty about the parameters. In this setting, it becomes especially important to construct our prior in such a way as to realistically cover our range of uncertainty about the optimal price. Furthermore, the nature of the observations in this problem creates additional challenges. We are not able to observe the customer valuations W^n. Rather, we only observe whether or not $W^n \geq p$. The clean, conjugate Bayesian models introduced in Chapter 2 cannot be directly applied, and we need to do additional work to be able to use them.

11.2.1 A Conjugate Prior for Choosing Between Two Demand Curves

Before we delve into the intricacies of nonconjugate Bayesian learning, let us first begin with a simple, stylized model that allows for a conjugate prior. In dynamic pricing, our observations are binary: either the customer accepts the offer (that is, $W^n \geq p$), or not. One way to obtain a conjugate prior is if the truth is binary, as well. Suppose that there are only two possible demand curves. Each curve has known, fixed parameters, but we do not know which curve is the right one for describing customer valuations. Figure 11.5 gives an example where we are trying to choose between a uniform valuation on the interval $[3, 7]$, and an exponential valuation with parameter 0.2.

This model is somewhat stylized, but may be useful in some cases. It may be that we, as the seller, have a large amount of historical data on sales figures, enough to fit any particular type of demand curve. We could conduct a statistical analysis to fit a uniform valuation model, or an exponential model, or perhaps a logistic model. For each model, we would fit a different set of parameters. However, we are not sure which type of demand curve is most appropriate. A logistic model may be fundamentally better-suited to the data than a uniform model. In that case, the binary-truth problem may help us to distinguish between two competing types of demand models.

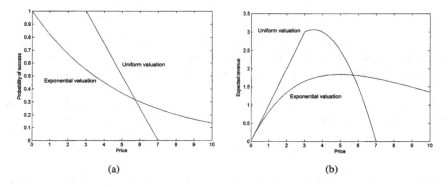

(a)   (b)

Figure 11.5 Examples of (a) demand curves and (b) revenue curves for a pricing problem with two truths.

Let f_1 and f_2 be the two demand curves under consideration. That is, $f_i(p) = P(W^n \geq p)$ under two different models $i = 1, 2$. Let A be the event that f_1 is the correct demand curve; then, A^c is the event that f_2 is correct. We begin with a prior probability $q^0 = P^0(A)$. For a fixed price p, the revenue function is given by

$$R(p) = p[q^0 f_1(p) + (1 - q^0) f_2(p)]. \tag{11.5}$$

We then make a pricing decision p^0 and observe either $W^1 \geq p^0$ or $W^1 < p^0$. First, let us consider the case where $W^1 \geq p^0$, that is, we make the sale with price p^0. Using Bayes' rule, we can derive

$$P(A \mid W^1 \geq p^0) = \frac{P(W^1 \geq p^0 \mid A)P(A)}{P(W^1 \geq p^0 \mid A)P(A) + P(W^1 \geq p^0 \mid A^c)P(A^c)}. \tag{11.6}$$

Observe that

$$
\begin{aligned}
P(W^1 \geq p^0 \mid A) &= f_1(p), \\
P(W^1 \geq p^0 \mid A^c) &= f_2(p), \\
P(A) &= q^0.
\end{aligned}
$$

We can let $q^1 = P(A \mid W^1 \geq p^0)$ denote our posterior probability of the event A. Then, (11.6) becomes

$$q^1 = \frac{q^0 f_1(p)}{q^0 f_1(p) + (1 - q^0) f_2(p)}.$$

Repeating the same analysis for the event that $W^1 < p^0$ produces the updating equation

$$q^1 = \frac{q^0 (1 - f_1(p))}{q^0 (1 - f_1(p)) + (1 - q^0)(1 - f_2(p))}.$$

Let $X^n = I_{\{W^n \geq p^n\}}$. That is, $X^n = 1$ if our price p^n is successful, and $X^n = 0$ otherwise. Then, we obtain a clean updating formula

$$q^{n+1} = \frac{q^n f_1(p^n)^{X^{n+1}} (1 - f_1(p^n))^{1-X^{n+1}}}{q^n f_1(p^n)^{X^{n+1}} (1 - f_1(p^n))^{1-X^{n+1}} + (1 - q^n) f_2(p^n)^{X^{n+1}} (1 - f_2(p^n))^{1-X^{n+1}}}. \tag{11.7}$$

This may be the simplest possible conjugate model. We put a simple discrete (actually binary) distribution of belief on the probability that the truth is given by a particular demand curve. When we make a binary observation, the posterior distribution is also discrete.

We might then apply a variety of techniques to make our pricing decision. A common approach in the literature is to use a simple myopic policy,

$$p^n = \arg \max_p \mathbb{E} R(p) = \arg \max_p p[q^0 f_1(p) + (1 - q^0) f_2(p)].$$

We could also apply some of the ideas from Section 5.5.4, where the observation was also binary, to this problem. There is one important issue to keep in mind, however. Observe that Figure 11.5(a) has a point where the two demand curves intersect. That is, there is a price $\bar{p}$ for which $f_1(\bar{p}) = f_2(\bar{p}) = \bar{f}$. Substituting this price into (11.7) gives us

$$
\begin{aligned}
q^{n+1} &= \frac{q^n \bar{f}^{X^{n+1}} (1 - \bar{f})^{1-X^{n+1}}}{q^n \bar{f}^{X^{n+1}} (1 - \bar{f})^{1-X^{n+1}} + (1 - q^n) \bar{f}^{X^{n+1}} (1 - \bar{f})^{1-X^{n+1}}} \\
&= \frac{q^n \bar{f}^{X^{n+1}} (1 - \bar{f})^{1-X^{n+1}}}{\bar{f}^{X^{n+1}} (1 - \bar{f})^{1-X^{n+1}}} \\
&= q^n.
\end{aligned}
$$

If our policy chooses the price $\bar{p}$, our beliefs about $P(A)$ will remain unchanged. Furthermore, since the policy determines what to do based on our beliefs, it will continue to choose the same price $\bar{p}$ thereafter. We would thus get stuck with the same beliefs forever. For that reason, the price $\bar{p}$ is known as the *noninformative price* or the *confounding price*.

Fortunately, since we have both demand curves specified exactly (we just don't know which is the right one), we can calculate the confounding price before we set out to solve the problem. The literature suggests a simple fix for the problem of confounding. We simply fix another price $\tilde{p} \neq \bar{p}$ beforehand. If our policy tells us to choose the price $\bar{p}$, we choose $\tilde{p}$ instead. Otherwise, we follow the policy.

11.2.2 Moment Matching for Nonconjugate Problems

Although the simple model of Section 11.2.1 has its uses, we are really interested in the case where we do not know the parameters of the demand curve. In any case, before we settle on two demand curves to choose from, we first need to find good parameters for those curves. So, we might focus on one particular type of curve, and then try to learn the right parameters for that type. This will lead us to the problem of non-conjugacy.

Let us illustrate this issue using a simple example. Suppose that the customers' valuations are drawn from a uniform distribution on the interval $[0, B]$, where B is unknown. Recalling Chapter 2, a natural choice of prior distribution on B is the Pareto distribution. If $B \sim Pareto(\alpha^0, b^0)$, and if we were able to observe the exact valuation $W^1 \sim U[0, B]$, we could apply the conjugate update

$$
\begin{aligned}
\alpha^1 &= \alpha^0 + 1, \\
b^1 &= \max(b^0, W^1).
\end{aligned}
$$

Unfortunately, we are never able to see the exact valuation. However, we can still derive a posterior distribution of belief, given the incomplete information that we do observe. For simplicity, let us assume that we set a price $p^0 < b^0$. We will first consider the case where $W^1 < p^0$, that is, we lost the sale. We apply Bayes' rule and write

$$
g(u \mid W^1 < p^0) = \frac{P(W^1 < p^0 \mid B = u)g(u)}{P(W^1 < p^0)}. \tag{11.8}
$$

Conditionally given $B = u$, the valuation W^1 has a uniform distribution on $[0, u]$. Thus,

$$P(W^1 < p^0 \mid B = u) = \begin{cases} \frac{p^0}{u}, & p^0 < u, \\ 1, & p^0 \geq u. \end{cases}$$

Next, the likelihood that $B = u$ is given by the Pareto density,

$$g(u) = \frac{\alpha^0 \left(b^0\right)^{\alpha^0}}{u^{\alpha^0+1}} \qquad \text{for } u > b^0.$$

Thus, in our calculations, we are implicitly assuming that $u > b^0$, since $B > b^0$ by the definition of a Pareto distribution. Since we are also assuming $p^0 < b^0$, it follows that $p^0 < u$ for all possible u. The numerator of the right-hand side of (11.8) thus becomes

$$P(W^1 < p^0 \mid B = u)g(u) = \frac{\alpha^0 p^0 \left(b^0\right)^{\alpha^0}}{u^{\alpha^0+2}}. \tag{11.9}$$

Integrating this expression from b^0 to infinity, we obtain the denominator of the right-hand side of (11.8),

$$P(W^1 < p^0) = \int_{b^0}^{\infty} \frac{\alpha^0 p^0 \left(b^0\right)^{\alpha^0}}{u^{\alpha^0+2}} du = \frac{\alpha^0}{\alpha^0+1} \frac{p^0}{b^0}. \tag{11.10}$$

Dividing (11.9) by (11.10) yields

$$g(u \mid W^1 < p^0) = \frac{\left(\alpha^0 + 1\right) \left(b^0\right)^{\alpha^0+1}}{u^{\alpha^0+2}}.$$

This is a Pareto density with parameters $\alpha^1 = \alpha^0 + 1$ and $b^1 = b^0$. That is, if we lose the sale, conjugacy is maintained. The conjugate updating equations hint at this. From (11.8), we see that $b^1 = \max(b^0, W^1)$, if we could observe the exact valuation W^1. However, we are assuming that $p^0 < b^0$. Thus, if we observe $W^1 < p^0$, it follows automatically that $\max(b^0, W^1) = b^0$. We do not really need to know the exact value of W^1 in this case. For any value of W^1 less than b^0, we will not change the scale parameter of our distribution of belief.

In the other case, however, we run into trouble. Suppose that $W^1 \geq p^0$, with $p^0 < b^0$ as before. Then,

$$P\left(W^1 \geq p^0 \mid B = u\right) = \begin{cases} 1 - \frac{p^0}{u}, & p^0 < u, \\ 0, & p^0 \geq u, \end{cases}$$

and

$$P(W^1 \geq p^0 \mid B = u)g(u) = \frac{\alpha^0 \left(b^0\right)^{\alpha^0}}{u^{\alpha^0+1}} - \frac{\alpha^0 p^0 \left(b^0\right)^{\alpha^0}}{u^{\alpha^0+2}}, \qquad u > b^0.$$

Integrating this quantity from b^0 to infinity yields

$$P(W^1 \geq p^0) = 1 - \frac{\alpha^0}{\alpha^0+1} \frac{p^0}{b^0},$$

and the posterior density turns out to be

$$g(u \mid W^1 < p^0) = \frac{\frac{\alpha^0 \left(b^0\right)^{\alpha^0}}{u^{\alpha^0+1}} - \frac{\alpha^0 p^0 \left(b^0\right)^{\alpha^0}}{u^{\alpha^0+2}}}{1 - \frac{\alpha^0}{\alpha^0+1}\frac{p^0}{b^0}}, \qquad u > b^0. \tag{11.11}$$

What are we to do with this strange expression? It does not correspond to any standard density. (The difference of two Pareto densities is *not* the density of the difference of two Pareto random variables!) It is certainly not a Pareto density.

We get around this problem by using a technique called *moment-matching*. Define a random variable U that has the density given by (11.11). As long as we have the density, we can compute the first two moments of U as follows. First,

$$\begin{aligned}
\mathbb{E}U &= \int_{b^0}^{\infty} u \frac{\frac{\alpha^0 \left(b^0\right)^{\alpha^0}}{u^{\alpha^0+1}} - \frac{\alpha^0 p^0 \left(b^0\right)^{\alpha^0}}{u^{\alpha^0+2}}}{1 - \frac{\alpha^0}{\alpha^0+1}\frac{p^0}{b^0}} \\
&= \frac{\frac{\alpha^0 b^0}{\alpha^0-1} - p^0}{1 - \frac{\alpha^0}{\alpha^0+1}\frac{p^0}{b^0}}.
\end{aligned} \tag{11.12}$$

Similarly,

$$\begin{aligned}
\mathbb{E}(U^2) &= \int_{b^0}^{\infty} u^2 \frac{\frac{\alpha^0 \left(b^0\right)^{\alpha^0}}{u^{\alpha^0+1}} - \frac{\alpha^0 p^0 \left(b^0\right)^{\alpha^0}}{u^{\alpha^0+2}}}{1 - \frac{\alpha^0}{\alpha^0+1}\frac{p^0}{b^0}} \\
&= \frac{\frac{\alpha^0 \left(b^0\right)^2}{\alpha^0-2} - \frac{\alpha^0 b^0 p^0}{\alpha^0-1}}{1 - \frac{\alpha^0}{\alpha^0+1}\frac{p^0}{b^0}}.
\end{aligned} \tag{11.13}$$

Let Y be a random variable following a Pareto distribution with parameters α^1 and b^1. The first and second moments of Y are given by

$$\mathbb{E}Y = \frac{\alpha^1 b^1}{\alpha^1 - 1}, \qquad \mathbb{E}(Y^2) = \frac{\alpha^1 \left(b^1\right)^2}{\alpha^1 - 2}.$$

Moment matching works by setting the moments of Y equal to the moments of U and solving for α^1 and b^1 in terms of α^0 and b^0. Essentially, we are forcing the posterior distribution to be Pareto, and choosing α^1 and b^1 to make this Pareto distribution resemble the actual posterior distribution, at least up to the first two moments. To do this, we have to solve two sets of nonlinear equations

$$\frac{\alpha^1 b^1}{\alpha^1 - 1} = \frac{\frac{\alpha^0 b^0}{\alpha^0-1} - p^0}{1 - \frac{\alpha^0}{\alpha^0+1}\frac{p^0}{b^0}},$$

$$\frac{\alpha^1 \left(b^1\right)^2}{\alpha^1 - 2} = \frac{\frac{\alpha^0 \left(b^0\right)^2}{\alpha^0-2} - \frac{\alpha^0 b^0 p^0}{\alpha^0-1}}{1 - \frac{\alpha^0}{\alpha^0+1}\frac{p^0}{b^0}}.$$

The solution gives us the nonconjugate updating equations,

$$\alpha^1 = 1 + \sqrt{\frac{\mathbb{E}\left(U^2\right)}{\mathbb{E}\left(U^2\right) - \left(\mathbb{E}U\right)^2}},$$

$$b^1 = \frac{\alpha^1 - 2}{\alpha^1 - 1}\frac{\mathbb{E}\left(U^2\right)}{\mathbb{E}U},$$

where $\mathbb{E}U$ and $\mathbb{E}\left(U^2\right)$ are given in (11.12) and (11.13) in terms of α^0 and b^0. Notice that, for compactness, the updating equation for b^1 is written in terms of α^1. To use these equations, we should first compute α^1, then use that value to find b^1.

Keep in mind that we have not computed updating equations for the case where $p^0 \geq b^0$. In this case, neither success nor failure will give us a Pareto posterior, and we have to repeat the above analysis. Choosing a different demand curve and prior distribution would also require a new round of moment-matching, and most likely new sets of nonlinear equations to solve. Moment-matching is no easy task. However, it is the simplest way to obtain clean updating equations in a problem where the observations are incomplete; that is, we can only obtain imperfect information about the customer valuations.

11.2.3 An Approximation for the Logit Model

The issue of nonconjugacy is also present in the logit model. There is no natural choice of prior for the logistic regression parameters μ_1, μ_2 in (11.3). That is, there is no clean prior distribution that is conjugate with observations sampled from the logistic distribution. The simplest choice of prior, from the decision-maker's point of view, is a multivariate normal distribution on (μ_1, μ_2). Especially if we extend our model to incorporate customer attributes, as in (11.4), a multivariate normal prior would allow us to include correlations in our beliefs about the parameters of different attributes. We followed this same approach in Chapter 8 when we fit a linear regression model to our observations. As we saw earlier, the linear regression model admits an elegant conjugate normal–normal learning model.

Unfortunately, there is no direct analog in the case of logistic regression, because our observations are now binary, of the form

$$X^n = \begin{cases} 1 & \text{if } W^n \geq p^n, \\ 0 & \text{otherwise.} \end{cases}$$

Recall that $X^n = 1$ if we make the sale, and $X^n = 0$ if we do not. If we start with a normal prior on the logistic parameters, then see an observation of this form, the posterior distribution will not be normal. However, there is an approximation that gives us a set of recursive updating equations. Like we did in (11.2.2) with moment matching, we can use this approximation to force the posterior to be normal. If $\mu = (\mu_1, \mu_2)$ has a multivariate normal distribution with mean vector $\theta^n = (\theta_1^n, \theta_2^n)$

and covariance matrix Σ^n, these approximate recursive updating equations are

$$\Sigma^{n+1} = \left((\Sigma^n)^{-1} + 2\lambda \left(\xi^n \right) \left(x^n \right) \left(x^n \right)^T \right)^{-1}, \tag{11.14}$$

$$\theta^{n+1} = \Sigma^{n+1} \left((\Sigma^n)^{-1} \theta^n + \left(X^{n+1} - \frac{1}{2} \right) x \right). \tag{11.15}$$

The vector $x^n = (1, -p^n)^T$ corresponds to the explanatory variables in (11.4). The function λ is given by

$$\lambda \left(\xi \right) = \frac{\tanh \left(\xi/2 \right)}{4\xi}.$$

The value ξ^n is an artificial parameter used in the approximation of the measurement precision. This parameter is also updated recursively, using the equation

$$\xi^{n+1} = \sqrt{(x^n)^T \Sigma^{n+1} x^n + (x^n)^T \theta^{n+1}}.$$

We can apply the same technique we used for correlated normal beliefs back in Chapter 2 to get a cleaner form for the updating equations

$$\theta^{n+1} = \theta^n + \frac{\frac{X^{n+1} - \frac{1}{2}}{2\lambda(\xi^n)} - (x^n)^T \theta^n}{\frac{1}{2\lambda(\xi^n)} + (x^n)^T \Sigma^n x^n} \Sigma^n x^n, \tag{11.16}$$

$$\Sigma^{n+1} = \Sigma^n - \frac{\Sigma^n x^n (x^n)^T \Sigma^n}{\frac{1}{2\lambda(\xi^n)} + (x^n)^T \Sigma^n x^n}. \tag{11.17}$$

These equations closely resemble the recursive updating rules we used for linear models in Section 8.2.2. Writing the update in this way gives us some insight into the way the approximation works. The quantity $\frac{1}{2\lambda(\xi)}$ is used in two ways. First, in the denominator of the fractional terms in (11.16) and (11.17), it serves as a stand-in for the variance of the observation. In the correlated normal model considered in Section 2.2.3, the same role is played by the measurement noise. Second, in the numerator of the fractional term in (11.16), the same quantity is used to convert the binary observation $X^{n+1} - \frac{1}{2}$ into a continuous quantity. We can rewrite (11.3) as

$$\mu_1 - \mu_2 p^n = \log \left(\frac{P \left(W^{n+1} \geq p^n \right)}{1 - P \left(W^{n+1} \geq p^n \right)} \right). \tag{11.18}$$

The right-hand side of (11.18) is called the *log-odds* of making a sale. The quantity $(x^n)^T \theta^n$ can be viewed as our prediction of the log-odds. Thus, the continuous quantity $\frac{X^{n+1} - \frac{1}{2}}{2\lambda(\xi^n)}$ can be interpreted as an approximate observation of the log-odds.

We are forcing the problem into the framework of Chapter 8, where our regression model is used to predict continuous observations. To do this, we are artificially creating a set of continuous responses from our binary observations. If $X^{n+1} = 1$, that is, we make the sale, the continuous response is positive, and we conclude that the log-odds are more likely to be greater than zero, and adjust our prior if it is under-estimating them. If we lose the sale and $X^{n+1} = 0$, the continuous response is negative, leading us to believe that the log-odds are more likely to be negative.

One advantage of this approach is that it can easily handle customer attributes, as in (11.4). We simply place a multivariate prior on the vector $\mu = (\mu_1, ..., \mu_{P+1})$ on the parameters of the customer attributes, as well as on the price sensitivity. We then make a pricing decision p^n and apply (11.16) and (11.17) using $x^n = (x_1^n, ..., x_P^n, -p^n)$.

It is important to understand that our posterior distributions in this model are not really normal, just as our posterior distributions in Section 11.2.2 were not really Pareto. The updating equations in (11.16) and (11.17) can only serve as an approx-imation of the way we learn in this problem. The approximation may not always be accurate. We do not necessarily need to use this approach. For example, we can simply collect our observations $X^1, X^2, ..., X^{n+1}$ and fit a logistic regression model to obtain estimates of μ_1 and μ_2. This approach may give us better fits for some particular values of μ, but it is frequentist in nature. It does not incorporate any idea of our prior uncertainty about the parameters, as represented by the prior covariance matrix Σ. Thus, on average across many different truths, we may do better with the Bayesian approximation.

There is no perfect model for learning in this setting, making it difficult to create a sophisticated learning policy. For example, if we try to construct a look-ahead policy such as knowledge gradient, we are forced to rely on an approximation for the predictive distribution of the future beliefs, and another approximation for the optimal implementation decision under those future beliefs. Even so, we can still improve our decision-making by considering some concepts of optimal learning, such as the idea of our uncertainty about the problem parameters.

11.3 BIDDING STRATEGIES

Recall that the revenue function for the logit demand model, under the parameters μ_1 and μ_2, is given by

$$R(p; \mu_1, \mu_2) = \frac{p}{1 + e^{-(\mu_1 - \mu_2 p)}}.$$

Suppose that we have some estimates θ_1^n and θ_2^n of the logistic parameters in (11.3). These estimates may be obtained using the approximate Bayesian model from Section 11.2.3, or they may come from a frequentist statistical procedure. Either way, once we have these estimates, a simple and intuitive course of action would be to simply assume that these are the true values, and make a pricing decision by solving

$$p^n = \arg \max_p R\left(p; \theta_1^n, \theta_2^n\right) = \arg \max_p \frac{p}{1 + e^{-\left(\theta_1^n - \theta_2^n p\right)}}. \tag{11.19}$$

For simplicity, let us assume that the set of possible prices is finite. Then, it is very easy to compute (11.19). We need only plug our estimated values into the revenue function and solve. This is known as a *point-estimate* policy, or a *certainty-equivalent* policy. We are making the decision that would be optimal if the true values were exactly equal to our estimates.

11.3.1   An Idea From Multi-Armed Bandits

The main insight of optimal learning, however, is that the true values are *not* exactly equal to our estimates. If we assume that they are, we may under-perform. Consider a very simple bandit problem with a single arm. The one-period reward of the arm follows an exponential distribution with unknown parameter λ. We use a gamma-exponential learning model from Section 2.3.1, and assume that $\lambda \sim Gamma\,(a, b)$.

The average one-period reward is $\frac{1}{\lambda}$, a very simple function of λ. Our estimate of λ is $\mathbb{E}\lambda = \frac{a}{b}$. If we assume that the true value is exactly equal to our estimate, then our resulting estimate of the average one-period reward becomes $\frac{1}{\mathbb{E}\lambda} = \frac{b}{a}$. This reasoning seems straightforward, but it ignores the uncertainty in our beliefs about λ.

Instead of merely plugging in our estimate of λ into the average reward, let us view that reward as a function of a random variable λ. We can then take an expected value of that reward over our distribution of belief, arriving at

$$\mathbb{E}\left(\frac{1}{\lambda}\right) = \frac{b}{a-1}, \tag{11.20}$$

a different estimate of the average reward. This approach accounts for the fact that λ is unknown. Not only do we have an estimate of this parameter, we also have some amount of uncertainty about that estimate. That uncertainty is encoded in the gamma distribution that we use to represent our beliefs.

Suppose now that we have M independent arms. The one-period reward obtained by playing arm x is exponential with parameter λ_x, and we assume that $\lambda_x \sim Gamma\,(a_x, b_x)$. Suppose also that our goal is simply to use a greedy, pure-exploitation strategy for pulling arms. On average over many truth values, the policy that plays $\arg\max_x \frac{b_x}{a_x-1}$ will do better than the policy that plays $\arg\max_x \frac{b_x}{a_x}$. We do better by incorporating the uncertainty in our beliefs into our decision-making.

11.3.2   Bayes–Greedy Bidding

We apply the same idea to the bidding problem. Rather than using (11.19) to make decisions, we take an expectation of the revenue function over our distribution of belief,

$$p^n = \arg\max_p \mathbb{E}R\,(p; \mu_1, \mu_2) = \arg\max_p \mathbb{E}\frac{p}{1 + e^{-(\mu_1 - \mu_2 p)}}. \tag{11.21}$$

Of course, this requires us to have a distribution of belief in the first place. Thus, this policy only really makes sense if we use the Bayesian model from Section 11.2.3 to learn about the unknown parameters. Under this model, we can assume that, at time n, $\mu \sim \mathcal{N}\,(\theta^n, \Sigma^n)$. After we make a decision p^n and observe X^{n+1}, we use (11.16) and (11.17) to change our beliefs. We refer to this policy as a *Bayes–greedy* policy, to distinguish it from the point-estimate greedy policy.

A well-known property of multivariate normal distributions is that a linear function of a multivariate normal vector is also normal. That is, if $\mu \sim \mathcal{N}\,(\theta, \Sigma)$ and c is a vector, then $c^T x \sim \mathcal{N}\left(c^T \theta, c^T \Sigma c\right)$. In our case, $c = (1, -p)^T$ and

$$
\begin{aligned}
c^T \theta &= \theta_1 - \theta_2 p, \\
c^T \Sigma c &= \Sigma_{11} - 2p\Sigma_{12} + p^2 \Sigma_{22}.
\end{aligned}
$$

Here we are using the fact that $\Sigma_{12} = \Sigma_{21}$ due to the symmetry of the covariance matrix. Consequently, we can rewrite (11.21) as

$$
p^n = \arg\max_p p \cdot \mathbb{E}\left(\frac{1}{1 + e^{-Y}}\right), \tag{11.22}
$$

where $Y \sim \mathcal{N}\left(\theta_1 - \theta_2 p, \Sigma_{11} - 2p\Sigma_{12} + p^2 \Sigma_{22}\right)$. Note that Y is a one-dimensional normal random variable. The right-hand side of (11.22) now seems straightforward: we need to compute an expectation over a normal density. We can view $\mathbb{E}\left(\frac{1}{1+e^{-Y}}\right)$ as the Bayesian probability of success.

Unfortunately, this expectation is impossible to compute analytically by integration. We have to resort to yet another approximation. One approach is to generate a large number of Monte Carlo samples of Y from a normal distribution with the appropriate mean and variance. We can then take a sample average

$$
\mathbb{E}\left(\frac{1}{1 + e^{-Y}}\right) \approx \frac{1}{K} \sum_{k=1}^{K} \frac{1}{1 + e^{-Y(\omega_k)}}, \tag{11.23}
$$

where $Y(\omega_k)$ is the kth sample realization. Figure 11.6(a) plots the right-hand side of (11.23), as a function of the mean and standard deviation of Y. For very large values of K, we will get more accurate estimates. However, generating enough samples may be fairly expensive computationally.

From this figure, we can make an interesting observation. If we fix a value of the standard deviation, and view the left-hand side of (11.23) as a function of the mean

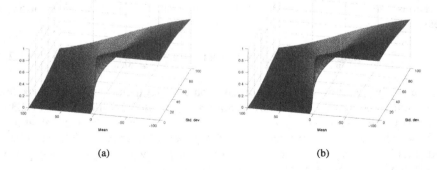

(a)   (b)

Figure 11.6 Bayesian success probability, as a function of the mean and standard deviation of Y using (a) Monte Carlo simulation and (b) closed-form approximation.

of Y, this function also appears to be a logistic curve. In fact, this is borne out by observing that, if $Var\,(Y) = 0$, then

$$\mathbb{E}\left(\frac{1}{1 + e^{-Y}}\right) = \frac{1}{1 + e^{-\mathbb{E}Y}},$$

which is itself a logistic function in $\mathbb{E}Y$. If we fix a larger value of the standard deviation, the expectation continues to look like a logistic function, only with a gentler slope.

This insight has led to a clever closed-form approximation for the difficult expectation. We can write

$$\mathbb{E}\left(\frac{1}{1 + e^{-Y}}\right) \approx \frac{1}{1 + e^{-\frac{\mathbb{E}Y}{\gamma}}}, \tag{11.24}$$

where

$$\gamma = \sqrt{1 + \frac{\pi^2}{8}Var(Y)}.$$

Figure 11.6(b) plots this approximation for different values of $\mathbb{E}Y$ and $Var\,(Y)$. The result does not exactly match the values we got from Monte Carlo sampling (there are errors on the order of 0.01), but one can easily see that the two surfaces are quite close.

Converting this result back into the language of our bidding problem, our Bayes–greedy policy makes pricing decisions according to the rule

$$p^n = \arg\max_p \frac{p}{1 + e^{-\frac{\theta_1^n - \theta_2^n p}{\gamma^n(p)}}}, \tag{11.25}$$

where

$$\gamma^n\,(p) = \sqrt{1 + \frac{\pi^2}{8}\left(\Sigma_{11}^n - 2p\Sigma_{12}^n + p^2\Sigma_{22}^n\right)}.$$

Interestingly, the calculation in (11.25) is very similar to what we use for the point-estimate policy. The only difference is that we divide the point estimate $\theta_1^n - \theta_2^n p$ by an additional factor $\gamma^n\,(p)$ that depends on the uncertainty in our beliefs (the covariance matrix), as well as the price decision p.

11.3.3 Numerical Illustrations

We will examine the performance of Bayes–greedy pricing in an example problem. Suppose that we are running an online bookstore and selling copies of a certain textbook (perhaps this one!) over the Internet. Suppose, furthermore, that the cost of buying the textbook from the publisher wholesale is \$40 per copy. We are thus interested in maximizing the profit function (11.2) with $c = 40$. We will never set a price below \$40, and it is also highly unlikely that anyone will buy the book for more than, say, \$110.

Figure 11.7 shows a realistic starting prior ($\theta_1^0 = 15$, $\theta_2^0 = 0.2$) for a logistic demand curve in this setting. Notice that the prior logistic curve is a bit narrower than the proposed range of \$40–\$110. This prior seems to be saying that we believe customer valuations to be contained roughly in the range $[50, 100]$. However, we also create a prior covariance matrix

$$\Sigma^0 = \begin{bmatrix} 30 & 0 \\ 0 & 0.03^2 \end{bmatrix}, \tag{11.26}$$

which allows for some uncertainty in this belief. Note the difference in magnitude between the variances on the market share and price sensitivity. Figure 11.7 also displays three possible true demand curves, representing scenarios where customer valuations are higher, lower, or about the same relative to the prior. The parameters of these truths are as follows:

High truth:	$\mu_1 = 25,$	$\mu_2 = 0.25$
Medium truth:	$\mu_1 = 10,$	$\mu_2 = 0.125$
Low truth:	$\mu_1 = 20,$	$\mu_2 = 0.3$

The variation in the price sensitivity for these three scenarios is much smaller than the variation in the market share. This is reflected in our choice of Σ^0. By starting with a prior curve roughly in the middle of our price range, then placing some uncertainty on the parameters, we are able to allow for a wide range of true demand curves.

Figure 11.8(a) shows the uncertainty factor $\gamma^0(p)$ as a function of the pricing decision p, for the first time step of this problem. We see that $\gamma^0(p)$ increases with price. We can think of this as expressing the risk involved in choosing higher prices: the penalty for losing the sale is greater, but so is the potential reward if the truth is

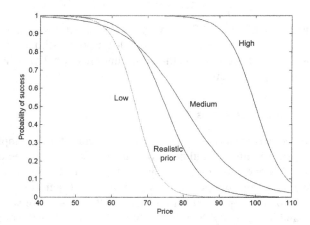

Figure 11.7   Comparison of our starting prior distribution with several possible truth values in a logistic valuation model, along with a realistic prior.

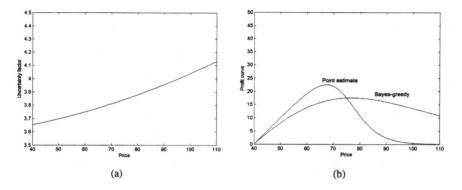

(a)   (b)

Figure 11.8 Graphs of (a) the uncertainty factor $\gamma(p)$ and (b) the modified profit curve as a function of price.

higher than we think. Figure 11.8(b) compares the estimated profit curves

$$P^{PE}\left(p; \theta_1^0, \theta_2^0\right) \;=\; \frac{p - c}{1 + e^{-\left(\theta_1^0 - \theta_2^0 p\right)}}, \tag{11.27}$$

$$P^{BG}\left(p; \theta_1^0, \theta_2^0\right) \;=\; \frac{p - c}{1 + e^{-\frac{\theta_1^0 - \theta_2^0 p}{\gamma^0(p)}}}, \tag{11.28}$$

based on the point estimate and the Bayesian distribution, respectively. These are the functions maximized by the point-estimate and Bayes–greedy policies. We see that the Bayes–greedy curve is wider, representing higher uncertainty or variation in the profit, and the maximum is shifted right. In other words, the Bayes–greedy policy places more value on exploratory behavior such as setting higher prices, and (usually) makes more aggressive pricing decisions than the point-estimate policy.

It remains to show how this aggressive pricing impacts performance. Figure 11.9 compares the performance (profit, averaged over 1000 sample paths) and pricing decisions of the point-estimate and Bayes–greedy policies in the first 20 iterations, on each of the three truths graphed in Figure 11.7. We also compare these policies to a frequentist logistic regression technique, a standard approach for fitting a logistic distribution to binary observations. We have focused on Bayesian models and algorithms in most earlier chapters. In this setting, however, the Bayesian modeling assumptions do not hold, and our updating equations are approximations. Thus, it is relevant to compare to a frequentist technique in order to see whether the inaccuracy of the Bayesian model has any negative impact on performance. We deliberately use the same set of axes for each comparison, to get a better sense of the magnitude of the difference between policies.

The graphs reveal several interesting behaviors. First, the Bayes–greedy policy consistently prices more aggressively than the point-estimate policy, particularly in the very early iterations. If these aggressive decisions are unsuccessful (for example, if the truth is lower than we believe), the policy quickly corrects its behavior and

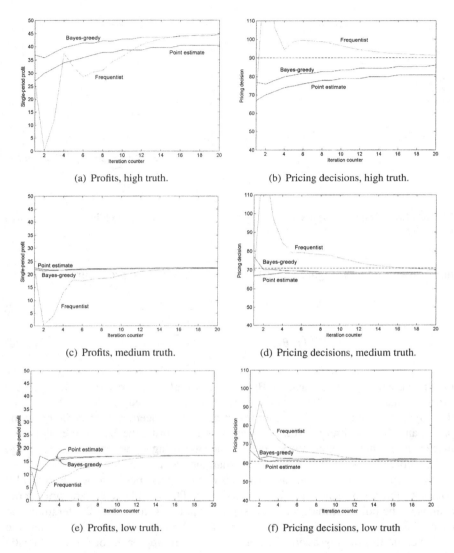

(a) Profits, high truth.

(b) Pricing decisions, high truth.

(c) Profits, medium truth.

(d) Pricing decisions, medium truth.

(e) Profits, low truth.

(f) Pricing decisions, low truth

Figure 11.9 Average profits and pricing decisions for different policies across 20 iterations.

chooses lower prices. If the optimal price (shown as a dashed line) is higher than expected, both point-estimate and Bayes–greedy gradually ramp up their pricing decisions. However, because Bayes–greedy starts out more aggressively, it is able to make a bigger profit more quickly. Thus, the Bayes–greedy policy exhibits a kind of robustness. If the truth is higher than expected, Bayes–greedy adds considerable value (roughly an extra $10 profit per iteration in Figure 11.9(a)), and if the truth is lower than expected, Bayes–greedy tends to lose the first sale, but adjusts almost immediately and obtains comparable profits to the point-estimate policy in subsequent iterations.

The frequentist method is able to fit a good model eventually. After around 20 iterations, it comes closer to the optimal pricing decision than the two policies using the approximate Bayesian model. At the same time, it does not use any prior information, and so it requires some time in order to fit a good model. In the first ten iterations, it exhibits volatile behavior and consistently chooses prices that are unreasonably high, even in the case when the truth itself is higher than we think. It would seem that Bayes–greedy is able to achieve robust performance in the critical early stages of the problem. In some applications, such as the setting of pricing industrial contracts, ten or twenty iterations may be all we ever get.

11.4 WHY DOES IT WORK?*

11.4.1 Moment Matching for Pareto Prior

We now solve the system of nonlinear equations

$$\frac{\alpha^1 b^1}{\alpha^1 - 1} = \frac{\frac{\alpha^0 b^0}{\alpha^0 - 1} - p^0}{1 - \frac{\alpha^0}{\alpha^0 + 1} \frac{p^0}{b^0}} = \mathbb{E}U, \qquad (11.29)$$

$$\frac{\alpha^1 \left(b^1\right)^2}{\alpha^1 - 2} = \frac{\frac{\alpha^0 \left(b^0\right)^2}{\alpha^0 - 2} - \frac{\alpha^0 b^0 p^0}{\alpha^0 - 1}}{1 - \frac{\alpha^0}{\alpha^0 + 1} \frac{p^0}{b^0}} = \mathbb{E}(U^2). \qquad (11.30)$$

To find the solution for b^1, we simply divide (11.30) by (11.29). This yields precisely

$$b^1 = \frac{\alpha^1 - 2}{\alpha^1 - 1} \frac{\mathbb{E}\left(U^2\right)}{\mathbb{E}U}. \qquad (11.31)$$

It remains to solve for a^1. We substitute (11.31) into (11.29) to obtain

$$\frac{\alpha^1 \left(\alpha^1 - 2\right)}{\left(\alpha^1 - 1\right)^2} \frac{\mathbb{E}\left(U^2\right)}{\mathbb{E}U} = \mathbb{E}U,$$

whence

$$\frac{\alpha^1 \left(\alpha^1 - 2\right)}{\left(\alpha^1 - 1\right)^2} = \frac{(\mathbb{E}U)^2}{\mathbb{E}\left(U^2\right)}. \qquad (11.32)$$

Let $A = \frac{(\mathbb{E}U)^2}{\mathbb{E}(U^2)}$. We now rewrite (11.32) as a quadratic equation

$$(A - 1) \left(\alpha^1\right)^2 - 2 \left(A - 1\right) \alpha^1 + A = 0.$$

The solution is found using the quadratic formula,

$$
\begin{aligned}
\alpha^1 &= \frac{2\,(A-1) + \sqrt{4\,(A-1)^2 - 4A\,(A-1)}}{2\,(A-1)} \\
&= \frac{2\,(A-1) + \sqrt{4\,(A-1)^2 - 4\,(A-1)^2\,\frac{A}{A-1}}}{2\,(A-1)} \\
&= 1 + \sqrt{1 + \frac{A}{1-A}} \\
&= 1 + \sqrt{\frac{1}{1-A}}.
\end{aligned}
$$

Observe now that

$$
\frac{1}{1-A} = \frac{1}{1 - \frac{(\mathbb{E}U)^2}{\mathbb{E}(U^2)}} = \frac{\mathbb{E}(U^2)}{\mathbb{E}\,(U^2) - (\mathbb{E}U)^2},
$$

as required. Notice that the quantity inside the square root is greater than 1, so we take the positive quadratic root.

11.4.2 Approximating the Logistic Expectation

The elegant approximation to the logistic expectation in (11.24) is based on two transformations involving the Gauss error function. First, we approximate

$$
\frac{1}{1 + e^{-y/c}} \approx \frac{1}{2} + \frac{1}{2}\mathrm{erf}\!\left(\frac{\pi}{4c}y\right), \tag{11.33}
$$

for fixed values of y and c. The notation erf represents the Gaussian error function,

$$
\mathrm{erf}(y) = \frac{2}{\sqrt{\pi}} \int_0^y e^{-z^2}\,dz.
$$

Letting $Y \sim \mathcal{N}\left(\mu, \sigma^2\right)$, it can be shown that

$$
\begin{aligned}
\mathbb{E}\!\left[\frac{1}{2} + \frac{1}{2}\mathrm{erf}\!\left(\frac{\pi}{4}Y\right)\right] &= \int_{-\infty}^{\infty} \left(\frac{1}{2} + \frac{1}{2}\mathrm{erf}\!\left(\frac{\pi}{4}y\right)\right) \frac{1}{\sqrt{2\pi\sigma^2}} e^{-\frac{(y-\mu)^2}{2\sigma^2}}\,dy \\
&= \frac{1}{2} + \frac{1}{2}\mathrm{erf}\!\left(\frac{\sqrt{\pi}}{4\gamma}\mu\right), \tag{11.34}
\end{aligned}
$$

with

$$
\gamma = \sqrt{1 + \frac{\pi}{8}\sigma^2}.
$$

Applying (11.33) with $c = 1$, calculating (11.34), and using (11.33) again with $c = \gamma$ yields

$$
\mathbb{E}\!\left(\frac{1}{1 + e^{-Y}}\right) \approx \frac{1}{1 + e^{-\mu/\gamma}},
$$

as required.

11.5   BIBLIOGRAPHIC NOTES

Section 11.1 - A seminal work on dynamic pricing with stochastic demand is the paper by Gallego & Van Ryzin (1994), which also considers some of the valuation models described in Section 11.1.1. This work, however, assumes a known demand distribution (but with finite inventory), and uses dynamic programming to characterize the optimal pricing policy. The general problem of optimal learning, that is, gradually discovering the true demand distribution through experimentation, was posed by Aghion et al. (1991). A connection to multi-armed bandits was made by Rothschild (1974); this work shares our modeling assumption of binary observations, but focuses on a simple two-armed bandit model with only two possible pricing decisions. A slightly different connection, looking at the problem from the buyer's point of view, is made in Bergemann & Välimäki (1996). Theoretical properties of general classes of valuation models (mostly suboptimality bounds) are derived in Broder & Rusmevichientong (2010a), while Broder & Rusmevichientong (2010b) focuses specifically on the logistic model. Parametric log-normal valuation models were studied under the name of "log-linear models" by Kalyanam (1996) and Carvalho & Puterman (2003), with the assumption that exact demands could be observed. A logistic valuation model was studied by Carvalho & Puterman (2005). This work also derives a KG-like algorithm based on a one-step look-ahead idea, though without an explicit Bayesian model. Other models for dynamic pricing incorporate customer arrival rates, which may be unknown. Bayesian models for such problems may assume that the set of possible truths is finite (Araman & Caldentey 2009), or they may use gamma priors on the arrival rate (Farias & Van Roy 2010).

Section 11.2 - The simple binary-truth conjugate model in Section 11.2.1, and the insight into confounding prices, come from Harrison et al. (2010). Moment matching is a general technique for approximate Bayesian inference. It is used in multiple communities and known under various names such as "assumed density filtering" (Minka 2001), "moment updating" (Dearden et al. 1998), and "monitoring" (Boyen & Koller 1998). An application of this technique to dynamic pricing with binary observations can be found in Chhabra & Das (2011). The moment-matching approximation for Pareto priors is the subject of ongoing work by Chhabra & Das (2012). The approximate Bayesian updating equations for the specific case of logistic regression were proposed by Jaakkola & Jordan (2000).

Section 11.3 - Point-estimate policies are also known as "certainty equivalent" policies, and are frequently studied in the literature on optimal learning in dynamic pricing (for example, see Aviv & Pazgal 2005). Sometimes, as in Harrison et al. (2010) and Broder & Rusmevichientong (2010a), simple modifications to these policies are made to allow for more exploration (or, for instance, to avoid the confounding prices described in Section 11.2.1). The distinction between point estimates and Bayesian estimates in multi-armed bandit problems with exponential rewards is made in Ryzhov & Powell (2011c). The approximation of the logistic expectation is due to Crooks (2009). The work by Araman & Caldentey (2009) proposes a method that is similar to our Bayes–greedy policy, but uses a model in which the set of possible truths is discrete, and it is possible to obtain exact observations from the true demand distribution.

PROBLEMS

11.1 Imagine that we are using a logistics curve to model the probability that a customer will accept a bid, which gives us the revenue function

$$R(p; \theta_1, \theta_2) = \frac{p}{1 + e^{-(\theta_1 - \theta_2 p)}}. \tag{11.35}$$

Assume the prior is $\theta^0 = (4, 2)$ for prices that are between 0 and 5.

a) Discretize prices into increments of 0.10 and find the optimal price $p^{Exp,1}$ given this prior. This represents the choice you would make using a pure exploitation policy.

b) Assume we choose a price $p^0 = 2$ and we observe that the customer accepts the bid (that is, $W^1 = 1$). Use equations (11.16) and (11.17) to find an updated model. Note that we did not need to assume a variance for the measurement noise. Why is this?

c) Now use the Bayes–greedy policy in equation (11.25) to compute the price $p^{BG,1}$. Try to explain intuitively the difference between $p^{BG,1}$ and $p^{Exp,1}$.

d) Simulate the Bayes–greedy policy for 20 measurements, using (11.16) and (11.17) to update your beliefs. At each iteration, using the beliefs generated by the Bayes–greedy policy, compute $p^{Exp,1}$, and compare the two policies over 20 iterations.

e) Finally, simulate the pure exploitation policy and the Bayes–greedy policy for 20 iterations. Repeat this process 100 times and summarize the difference.

CHAPTER 12

STOPPING PROBLEMS

Stopping problems are a simple but important class of learning problems. In this problem class, information arrives over time, and we have to choose whether to view the information or stop and make a decision. In this setting, "learning" means to continue to continue to receive information. We do not choose which information to observe, as we have done in the past, but we do have to decide whether to continue observing or stop and make a decision.

In this chapter we consider two classical stopping problems. The first is the sequential probability ratio test, where we have to quickly identify when a signal from some source is changing. The second is a classic problem known as the secretary problem, where we have to find the best out of a sequence of offers which arrive one at a time.

12.1 SEQUENTIAL PROBABILITY RATIO TEST

A fundamental problem arises when we need to decide as quickly as possible when something is changing. For example, we may think we are observing data being

Optimal Learning. By Warren B. Powell and Ilya O. Ryzhov
Copyright © 2012 John Wiley & Sons, Inc.

generated by the sequence

$$W^n = \mu^0 + \epsilon^n, \tag{12.1}$$

where we assume that ϵ is normally distributed with mean 0 and variance σ^2. But we are aware that the mean might change to μ^1, which means the observations would come from the model

$$W^n = \mu^1 + \epsilon^n. \tag{12.2}$$

We would like to design a method that allows us to determine when the mean has changed as quickly as possible.

A more general statement of the problem would be that the observations W^n are coming from an initial model that we refer to as the null hypothesis H_0. We then want to determine when the data is coming from a different model that we refer to as the alternative hypothesis, H_1. We start with a prior probability $\rho_0 = \rho_0^0$ that H_0 is true. Similarly, ρ_1^0 is our prior probability that H_1 is true. After we observe W^1, we would update our prior using Bayes' rule to obtain

$$
\begin{aligned}
\rho_0^1 &= P(H_0|W^1) \\
&= \frac{P(W^1|H_0)P(H_0)}{P(W^1)},
\end{aligned}
$$

where $P(W^1) = \rho_0 P(W^1|H_0) + \rho_1^0 P(W^1|H_1)$. The quantity $\rho_1^1 = P(H_1|W^1)$ would be worked out similarly. For example, if our data are normally distributed, we would write

$$P(W^1 = w|H_0) = \frac{1}{\sqrt{2\pi}\sigma} \exp\left[-\frac{1}{2}\left(\frac{(w-\mu^0)}{\sigma}\right)^2\right].$$

Let $p_0(w^n) = P(W^n = w^n|H_0)$. After n observations, we can write the posterior probability as

$$
\begin{aligned}
\rho_0^n &= \frac{\rho_0 \Pi_{k=1}^n p_0(w^k)}{\rho_0 \Pi_{k=1}^n p_0(w^k) + \rho_1^0 \Pi_{k=1}^n p_1(w^k)} \\
&= \frac{\rho_0 \lambda^n(w^1,\dots,w^n)}{\rho_0 + \rho_1^0 \lambda^n(w^1,\dots,w^n)},
\end{aligned}
$$

where

$$\lambda^n(w^1,\dots,w^n) = \Pi_{k=1}^n \frac{p_0(w^k)}{p_1(w^k)}.$$

We can write $S^n = (W^1, W^2, \dots, W^n)$ as being the set of all measurements, or it could be a sufficient statistic (such as the mean and variance of the normal distribution) that captures everything we need to know from previous measurements. Later, we let $\lambda^n = \lambda^n(S^n) = \lambda^n(w^1,\dots,w^n)$.

To solve our problem, we need to make two decisions. The first is whether to continue to observe W^n, or to stop and make a decision. If we decide to stop, we then have to choose between H_0 and H_1. We let

$$X^\pi(S^n) = \begin{cases} 1 & \text{if we decide to stop and make a decision,} \\ 0 & \text{if we continue observing.} \end{cases}$$

$$Y^\pi(S^n) = \begin{cases} 1 & \text{if we decide } H_1 \text{ is correct,} \\ 0 & \text{if we decide } H_0 \text{ is correct.} \end{cases}$$

We let π denote a policy consisting of the two functions (X^π, Y^π). Given a policy π, there are two mistakes we can make. The first is a false alarm, which means we stop and conclude that H_1 is true when in fact H_0 is true, and a miss, which means that H_1 is true, but we did not pick it up and we still think H_0 is true. We define the probability of these two events using

$$\begin{aligned} P_F^\pi &= \quad \text{the probability we conclude } H_1 \text{ is true (the false alarm) given} \\ & \qquad H_0 \text{ when using policy } \pi \\ &= \quad \mathbb{E}[Y^\pi(S^n)|H_0], \\ P_M^\pi &= \quad \text{the probability we conclude } H_0 \text{ is true given } H_1 \text{ when using} \\ & \qquad \text{policy } \pi \\ &= \quad \mathbb{E}[1 - Y^\pi(S^n)|H_1]. \end{aligned}$$

Now let ρ_0 be the prior probability that H_0 is true. We can define the overall probability of an error using

$$P_e = (1 - \rho_0)P_F^\pi + \rho_0 P_M^\pi.$$

Of course, we can minimize this error by taking many measurements. The number of measurements is given by

$$N^\pi = \min\{n|X^\pi(S^n) = 1\}.$$

N^π is a random variable that depends on our policy (the decision function X^π) and the observations $(W^1, W^2, \ldots, W^n)$. We can assign a "cost" c to each measurement, giving us a utility function

$$U^\pi(c) = P_e + c\mathbb{E}N^\pi.$$

Here, c is not a true cost in the sense of being measured in units of dollars per measurement. Rather, it is a scaling coefficient that allows us to combine the probability of being wrong with the number of measurements. Our challenge is to find a policy π that minimizes the utility function $U^\pi(c)$.

The problem can be solved (approximately) by exploiting some nice structural properties. We first write the conditional risk as

$$\begin{aligned} r_0^\pi &= P_F^\pi + c\mathbb{E}[N^\pi|H_0], \\ r_1^\pi &= P_M^\pi + c\mathbb{E}[N^\pi|H_1], \\ r^\pi &= \rho_0 r_0^\pi + (1 - \rho_0)r_1^\pi. \end{aligned}$$

We would like to find a policy π that minimizes the risk, given by

$$R^0(\rho_0) = \min_{\pi} r^{\pi}.$$

We start by observing that if $\rho_0 = 1$ (which means we are positive that H_1 is true), then we can stop and choose $Y = 1$ with no risk of a false positive (and $N = 0$), which means that the risk is $R^0(1) = 0$. The same reasoning tells us that $R^0(0) = 0$. It is also possible to show that $R^0(\rho)$ is concave in ρ.

Assume that we stop after making no measurements. If we choose $Y = 0$ then the risk is $R^0(\rho_0|Y = 0) = \rho_0$ (which is the same thing as saying that if I have to choose now with no information, my probability of being right is my original prior that H_0 is true). Similarly, if we choose $Y = 1$ then the risk is $R^0(\rho_0|Y = 1) = 1 - \rho_0$. Or, we could choose to make a single measurement (sort of like choosing curtain number 3). In this case, we want to choose the measurement policy π that solves

$$R^1(\rho_0^0) = \min_{\{\pi, N > 0\}} \rho_0^0 r_0^{\pi} + (1 - \rho_0^0) r_1^{\pi}.$$

Here, we are solving the same problem as we were at time 0, but we are now forcing ourselves to take at least one measurement. So, our policy will be to take a single measurement (which means $N > 0$), and then we have to do the best we can choosing between $Y = 0$, $Y = 1$ or taking yet another measurement.

So, we want the smallest of ρ_0^0 (corresponding to stopping and choosing H_0), $1 - \rho_0^0$ (corresponding to stopping and choosing H_1) and $R^1(\rho^0)$ (which means take another observation and repeat the process). The problem is depicted in Figure 12.1. Here, we plot the lines ρ_0^0 and $1 - \rho_0^0$, and the concave function $R^1(\rho^0)$, all as a function of ρ_0^0. If c is large enough, it is possible that the midpoint of $R^1(\rho^0)$ is greater than .5, in which case the best choice is to stop right away ($N = 0$) and choose between H_0 and H_1. Now assume that the maximum of $R^1(\rho^0)$ is less than .5. In this case, take one measurement and compute the posterior ρ_0^1. We then divide the horizontal axis into three regions: $\rho_0 < \rho^L$, $\rho^L \le \rho_0 \le \rho^U$, and $\rho_0 > \rho^U$. If $\rho_0 < \rho^L$, then the best choice is to choose $Y = 0$. If $\rho_0 > \rho^U$, then we stop and choose $Y = 1$. If $\rho^L \le \rho_0 \le \rho^U$, then we make another measurement and repeat the process. After each measurement, we face the same problem, where the only change is that we have a new prior.

This seems like a pretty simple rule. The only challenge is finding ρ^L and ρ^U. We begin by computing the likelihood ratio

$$L^n(S^n) = \Pi_{k=1}^n \frac{p_1(W^k)}{p_0(W^k)} = \Pi_{k=1}^n L(W^k),$$

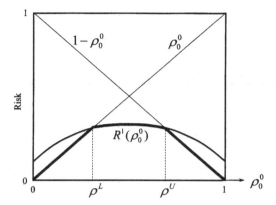

Figure 12.1 The expected risk as a function of the prior probability that H_0 is true.

where $L(W^k) = \frac{p_1(W^k)}{p_0(W^k)}$ is the likelihood ratio for a single observation. We next use Bayes' rule to compute the posterior $\rho_0^{n+1}(W^n)$ as follows:

$$
\begin{aligned}
\rho_1^{n+1}(S^n) &= \frac{p_1(S^n)(1 - \rho_0^n)}{\rho_0^n p_0(S^n) + (1 - \rho_0^n)p_1(S^n)} \\
&= \frac{L^n(S^n)}{L^n(S^n) + \rho_0^n/(1 - \rho_0^n)} \\
&= f(L^n(S^n), \rho_0^n/(1 - \rho_0^n)),
\end{aligned}
$$

where $f(\ell, \beta) = \ell/(\ell + \beta)$. For $\beta \geq 0$, $f(\ell, \beta)$ is strictly increasing for $\ell \geq 0$. This means that if $0 < \rho_0^n < 1$, $\rho_0^{n+1}(S^n)$ is strictly increasing with $L^n(S^n)$. This means that determining if $\rho_0^{n+1}(S^n) \leq \rho^L$ or $\rho_0^{n+1}(S^n) \geq \rho^U$ is the same as testing if $L^n(S^n) \leq A$ or $L^n(S^n) \geq B$, where A and B satisfy

$$
\begin{aligned}
f(A, \rho_0^n/(1 - \rho_0^n)) &= \rho^L, \\
f(B, \rho_0^n/(1 - \rho_0^n)) &= \rho^U.
\end{aligned}
$$

We can solve for A and B, which gives us

$$
A = \frac{\rho_0^n \rho^L}{(1 - \rho_0^n)(1 - \rho^L)},
$$

$$
B = \frac{\rho_0^n \rho^U}{(1 - \rho_0^n)(1 - \rho^U)}.
$$

This means that we can write our policy in the form

$$
L^n(S^n) = \begin{cases} \geq B, & \text{stop and choose } Y^n = 1, \\ \leq A, & \text{stop and choose } Y^n = 0, \\ \text{otherwise} & \text{take an additional observation.} \end{cases}
$$

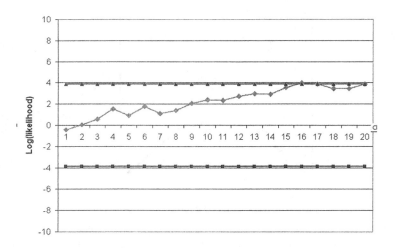

Figure 12.2 Sample path of log of the SPRT likelihood function.

Hence, it is easy to see why this rule is known as the *sequential probability ratio test* (*SPRT*). The *SPRT* is controlled by the parameters A and B, and hence we refer to the rule as $SPRT(A, B)$.

Finding A and B exactly is difficult, but we can find good estimates using *Wald's approximation* which gives us

$$P_F^\pi \approx \frac{(1 - A)}{B - A},$$

$$P_M^\pi \approx \frac{A(B - 1)}{B - A}.$$

We then choose the acceptable probability of a false positive, P_F^π, and the probability of a miss, P_M^π, and then solve for A and B, giving us

$$A = \frac{P_M^\pi}{1 - P_F^\pi},$$

$$B = \frac{1 - P_M^\pi}{P_F^\pi}.$$

This rule is (approximately) optimal in the sense that it meets these goals for the probability of a false positive and the probability of missing a change with the fewest number of measurements. We note that it is customary to work in terms of the logarithm of L.

Figure 12.2 plots the log of the likelihood for a set of sample observations. After 16 observations, we conclude that H_1 is true.

12.2 THE SECRETARY PROBLEM

The so-called secretary problem is one of the first (formally defined) learning problems. The motivation of the problem is determining when to hire a candidate for a job (presumably a secretarial position), but it can also be applied to reviewing a series of offers for an asset (such as selling your house or car). The problem involves the tradeoff between observing candidates (which allows us to collect information) and making a decision (exploiting the information). As with the sequential probability ratio test in the previous section, our decision is when to stop reviewing new offers and accept the most recent offer. In contrast with our work on Bayesian models, at the heart of the secretary problem is that we assume that we know absolutely nothing about the distribution of offers.

12.2.1 Setup

Assume that we have N candidates for a secretarial position (you can also think of these as offers to purchase an asset). Each candidate is interviewed in sequence and assigned a score that allows us to compare him or her to other candidates (if we are trying to sell an asset, these scores are the offers to purchase the asset). While it may be reasonable to try to maximize the expected score that we would receive, in this case, we want to maximize the probability of accepting the highest score out of all that have been, or might be, offered. We need to keep in mind that if we stop at candidate n, then we will not interview candidates $n+1, \ldots, N$. Also, we only have the option of accepting the last candidate or interviewing the next one. Once we have turned down a candidate, we cannot return to that candidate at a later time.

Let

$$W^n \quad = \quad \text{score of the } n\text{th candidate.}$$

$$S^n \quad = \quad \begin{cases} 1 & \text{if the score of the } n\text{th candidate is the best so far,} \\ 0 & \text{if the score of the } n\text{th candidate is not the best so far,} \\ \Delta & \text{if we have stopped already.} \end{cases}$$

$$\mathcal{S} \quad = \quad \text{state space, given by } (0, 1, \Delta), \text{ where the states 0 and 1 mean that we are still searching, and } \Delta \text{ means we have stopped the process.}$$

$$\mathcal{X} \quad = \quad \{0 \text{ (continue)}, 1 \text{ (stop)}\}, \text{ where "stop" means that we hire the last candidate interviewed.}$$

Because the decision function uses the most recent piece of information, we define our history as

$$h^n \quad = \quad \{W^1, \ldots, W^n\}.$$

To describe the system dynamics, it is useful to define an indicator function

$$I^n(h^n) \quad = \quad \begin{cases} 1 & \text{if } W^n = \max_{1 \le m \le n}\{W^m\}, \\ 0 & \text{otherwise.} \end{cases}$$

which tells us if the last observation is the best. Our transition function can now be
given by

$$S^{n+1} = \begin{cases} I^n(h^n) & \text{if } x^n = 0 \text{ and } S^n \neq \Delta, \\ \Delta & \text{if } x^n = 1 \text{ or } S^n = \Delta. \end{cases}$$

To compute the one-step transition matrix, we observe that the event the $(n+1)$st
applicant is the best has nothing to do with whether the nth was the best. As a result,
we can write the conditional probability that $I^{n+1}(h^{n+1}) = 1$ using

$$\mathbb{P}[I^{n+1}(h^{n+1}) = 1 | I^n(h^n)] = \mathbb{P}[I^{n+1}(h^{n+1}) = 1].$$

This simplifies the problem of finding the one-step transition probabilities. By definition we have

$$\mathbb{P}[S^{n+1} = 1 | S^n, x^n = 0] = \mathbb{P}[I^{n+1}(h^{n+1}) = 1].$$

$I^{n+1}(h^{n+1}) = 1$ if the $(n+1)$st candidate is the best out of the first $n+1$, which
clearly occurs with probability $1/(n+1)$. So

$$\mathbb{P}(S^{n+1} = 1 | S^n, x^n = 0) = \frac{1}{n+1},$$
$$\mathbb{P}(S^{n+1} = 0 | S^n, x^n = 0) = \frac{n}{n+1}.$$

Our goal is to maximize the probability of hiring the best candidate. So, if we do not
hire the last candidate, then the probability that we hired the best candidate is zero. If
we hire the nth candidate, and the nth candidate is the best so far, then our reward is
the probability that this candidate is the best out of all N. This probability is simply
the probability that the best candidate out of all N is one of the first n, which is n/N.
So, the conditional reward function is

$$C^n(S^n, x^n | h^n) = \begin{cases} n/N & \text{if } S^n = 1 \text{ and } x^n = 1, \\ 0 & \text{otherwise.} \end{cases}$$

With this information, we can now set up the optimality equations

$$V^n(s^n) = \max_{x^n \in \mathcal{X}} \mathbb{E}\{C^n(s^n, x^n | h^n) + V^{n+1}(S^{n+1}) | s^n\}.$$

12.2.2 Solution

The solution to the problem is quite elegant, but the technique is unique to this
particular problem. Readers interested in the elegant answer but not the particular
proof (which illustrates dynamic programming but otherwise does not generalize to
other problem classes) can skip to the end of the section.

Let $V^n(s)$ be the probability of choosing the best candidate out of the entire
population, given that we are in state s after interviewing the nth candidate. Recall

that implicit in the definition of our value function is that we are behaving optimally from time period t onward. The terminal reward is

$$
\begin{aligned}
V^N(1) &= 1, \\
V^N(0) &= 0, \\
V^N(\Delta) &= 0.
\end{aligned}
$$

Let

$$
\begin{aligned}
C^{stop,n} &= \left(C^n(1, \text{stop}) + V^{n+1}(\Delta) \right), \\
C^{continue,n} &= \left(C^n(1, \text{continue}) + \sum_{s' \in \{0,1\}} p(s'|s) V^{n+1}(s') \right).
\end{aligned}
$$

The optimality recursion for the problem is given by

$$
V^n(1) = \max \left\{ C^{stop,n}, C^{continue,n} \right\}.
$$

Noting that

$$
\begin{aligned}
C^n(1, \text{continue}) &= 0, \\
C^n(1, \text{stop}) &= \frac{n}{N}, \\
V^{n+1}(\Delta) &= 0, \\
p(s'|s) &= \begin{cases} 1/(n+1) & s' = 1, \\ n/(n+1) & s' = 0, \end{cases}
\end{aligned}
$$

we get

$$
V^n(1) = \max \left\{ \frac{n}{N}, \frac{1}{n+1} V^{n+1}(1) + \frac{n}{n+1} V^{n+1}(0) \right\}. \tag{12.3}
$$

Similarly, it is easy to show that

$$
\begin{aligned}
V^n(0) &= \max \left\{ 0, \frac{1}{n+1} V^{n+1}(1) + \frac{n}{n+1} V^{n+1}(0) \right\} \\
&= \frac{1}{n+1} V^{n+1}(1) + \frac{n}{n+1} V^{n+1}(0). \tag{12.4}
\end{aligned}
$$

Comparing (12.4) and (12.3), we can rewrite (12.3) as

$$
V^n(1) = \max \left\{ \frac{n}{N}, V^n(0) \right\}. \tag{12.5}
$$

From this we obtain the inequality

$$
V^n(1) \geq V^n(0), \tag{12.6}
$$

which seems pretty intuitive (you are better off if the last candidate you interviewed was the best you have seen so far).

At this point, we are going to suggest a policy that seems to be optimal. We are going to interview the first $\bar{n}$ candidates, without hiring any of them. Then, we will stop and hire the first candidate who is the best we have seen so far. The decision rule can be written as

$$x^n(1) = \begin{cases} 0 \text{ (continue)}, & n \le \bar{n}, \\ 1 \text{ (quit)}, & n > \bar{n}. \end{cases}$$

To prove this, we are going to start by showing that if $V^m(1) > m/N$ for some m (or alternatively if $V^m(1) = m/N = V^m(0)$), then $V^{m'}(1) > m'/N$ for $m' < m$. If $V^m(1) > m/N$, then it means that the optimal decision is to continue. We are going to show that if it was optimal to continue at set m, then it was optimal to continue for all steps $m' < m$.

Assume that $V^m(1) > m/N$. This means, from equation (12.5), that it was better to continue, which means that $V^m(1) = V^m(0)$ (or there might be a tie, implying that $V^m(1) = m/N = V^m(0)$). This allows us to write

$$
\begin{aligned}
V^{m-1}(0) &= \frac{1}{m}V^m(1) + \frac{m-1}{m}V^m(0) \\
&= V^m(1) & (12.7) \\
&\ge \frac{m}{N}. & (12.8)
\end{aligned}
$$

Equation (12.7) is true because $V^m(1) = V^m(0)$, and equation (12.8) is true because $V^m(1) \ge m/N$. Stepping back in time, we get

$$
\begin{aligned}
V^{m-1}(1) &= \max\left\{\frac{m-1}{N}, V^{m-1}(0)\right\} \\
&\ge \frac{m}{N} & (12.9) \\
&> \frac{m-1}{N}. & (12.10)
\end{aligned}
$$

Equation (12.9) is true because $V^{m-1}(0) \ge m/N$. We can keep repeating this for $m-1, m-2, \ldots$, so it is optimal to continue for $m' < m$.

Now we have to show that if $N > 2$, then $\bar{n} \ge 1$. If this is not the case, then for all n, $V^n(1) = n/N$ (because we would never continue). This means that, from equation (12.4),

$$
\begin{aligned}
V^n(0) &= \left(\frac{1}{n+1}\right)\left(\frac{n+1}{N}\right) + \left(\frac{n}{n+1}\right)V^{n+1}(0) \\
&= \frac{1}{N} + \left(\frac{n}{n+1}\right)V^{n+1}(0). & (12.11)
\end{aligned}
$$

Using $V^N(0) = 0$, we can solve (12.11) by backward induction:

$$
\begin{aligned}
V^N(0) &= 0, \\
V^{N-1}(0) &= \frac{1}{N} + \frac{N-1}{N-1+1} V^N(0) \\
&= \frac{1}{N}, \\
V^{N-2}(0) &= \frac{1}{N} + \frac{N-2}{N-2+1} \left(\frac{1}{N} \right) \\
&= \frac{N-2}{N} \left(\frac{1}{N-2} + \frac{1}{N-1} \right).
\end{aligned}
$$

In general, we get

$$
V^m(0) = \frac{m}{N} \left[\frac{1}{m} + \frac{1}{m+1} + \cdots + \frac{1}{N-1} \right].
$$

We can easily see that $V^1(0) > \frac{1}{N}$; since we were always quitting, we had found that $V^1(1) = \frac{1}{N}$. Finally, equation (12.6) tells us that $V^1(1) \geq V^1(0)$, which means we have a contradiction.

This structure tells us that, for $m \leq \bar{n}$, we have

$$
V^m(0) = V^m(1),
$$

and for $m > \bar{n}$ we obtain

$$
\begin{aligned}
V^m(1) &= \frac{m}{N}, \\
V^m(0) &= \frac{m}{N} \left[\frac{1}{m} + \frac{1}{m+1} + \cdots + \frac{1}{N-1} \right].
\end{aligned}
$$

It is optimal to continue as long as $V^m(0) > m/N$, so we want to find the largest value for m such that

$$
\frac{m}{N} \left[\frac{1}{m} + \frac{1}{m+1} + \cdots + \frac{1}{N-1} \right] > \frac{m}{N},
$$

or

$$
\left[\frac{1}{m} + \frac{1}{m+1} + \cdots + \frac{1}{N-1} \right] > 1.
$$

If $N = 5$, then we can solve by enumeration:

$$
\begin{aligned}
\bar{n} = 1 \quad & \tfrac{1}{1} + \tfrac{1}{2} + \tfrac{1}{3} + \tfrac{1}{4} && = 2.08 \\
\bar{n} = 2 \quad & \tfrac{1}{2} + \tfrac{1}{3} + \tfrac{1}{4} && = 1.08 \\
\bar{n} = 3 \quad & \tfrac{1}{3} + \tfrac{1}{4} && = 0.58
\end{aligned}
$$

So for $N = 5$, we would use $\bar{n} = 2$. This means interview (and discard) two candidates, and then take the first candidate that is the best to date.

For large N, we can find a neat approximation. We would like to find m such that

$$1 \approx \frac{1}{m} + \frac{1}{m+1} + \cdots + \frac{1}{N-1}$$
$$\approx \int_M^N \frac{1}{x}\, dx$$
$$= \log N - \log m$$
$$= \log\left(\frac{N}{m}\right).$$

Solving for m means finding $\log(N/m) = 1$ or $N/m = e$ or $m/N = e^{-1} = 0.368$. So, for large N, we want to interview 37 percent of the candidates, and then choose the first candidate that is the best to date.

The secretary problem is a classic, partly because it illustrates an interesting information collection problem, and partly because it yields such an elegant solution. In real applications, we can translate the result to a rough rule that says "look at a third of the candidates, and then choose the first candidate that is better than all the others."

12.3 BIBLIOGRAPHIC NOTES

Section 12.1 - The sequential probability ratio test is due to Wald & Wolfowitz (1948).

Section 12.2 - The secretary problem was first introduced in Cayley (1875). Our presentation is based on Puterman (1994). Vanderbei (1980) provides an elegant generalization of the secretary problem to one of finding the best subset. See also Bruss (1984) for extensions.

PROBLEMS

12.1 Download a spreadsheet illustrating the sequential probability ratio test from

http://optimallearning.princeton.edu/exercises/SPRT.xls

In the initial spreadsheet, the standard deviation of a measurement has been set to 5.

a) Translate the cells in row 12 to mathematics. Identify the equation in the book corresponding to each cell starting in column D.

b) Change the probability of missing from .02 to .10. How does this change the hypotheses $H0$ and $H1$? Now change the probability of a false alarm from

.02 to .10? How does this change the hypotheses? When you are done this question, restore both probabilities back to .02.

c) The sequential probability ratio test stops with a conclusion that either hypothesis $H0$ is true or $H1$ is true when the red line crosses one of the blue lines. If it does not cross either line within the measurement budget, then the test is inconclusive. Hit the F9 key 20 times, and count the number of times the red line crosses one of the blue lines.

d) Now change the standard deviation (cell B9) to 3. Again perform 20 simulations and count how many times the red line crosses one of the blue lines. You should see that with a lower standard deviation, SPRT is more effective at declaring that one of the two hypotheses is true. Why does a smaller standard deviation make it easier to come to a conclusion?

12.2   Assume you think you can look at up to 20 bids for the house you are selling. Completely unknown to you, the bids can be modeled as being drawn from a uniform distribution between $380,000 and $425,000. Use the policy that you are going to reject the first seven bids (which is 35 percent of 20), and then accept the first bid that is better than these seven. If none are better, you have to accept the very last bid. Start the process by generating all 20 bids in advance, so that when you are done, you can compare the bid you accepted against the best that you might have accepted with perfect foresight.

a) Repeat this policy 100 times and report on: i) how many times you accepted the very best bid out of the 20, and ii) the bid you accepted as a percentage of the very best bid.

b) Repeat (a), but now reject only the first four bids before you are ready to accept the bid.

c) Finally, repeat (a), but now you decide to reject the first 10 bids before you are ready to accept a bid.

d) Compare the results of the policies you simulated in (a), (b) and (c). Do you see evidence that one policy outperforms the others?

12.3   You have to choose the best out of up to 30 bids, where the ith bid, R_i, follows an exponential distribution given by

$$f_R(y) = .02e^{-.02y}.$$

You can generate random observations from an exponential distribution using

$$R = -50 \log U,$$

where U is a random variable that is uniformly distributed between 0 and 1.

a) Use the policy where you look at the first 11 bids, and then pick the best bid which outperforms all previous bids. Repeat this policy 100 times and report on: i) how many times you accepted the very best bid out of the 20, and ii) the bid you accepted as a percentage of the very best bid.

b) Repeat (a), but now reject only the first six bids before you are ready to accept the bid.

c) Finally, repeat (a), but now you decide to reject the first 16 bids before you are ready to accept a bid.

d) Compare the results of the policies you simulated in (a), (b), and (c). Do you see evidence that one policy outperforms the others?

CHAPTER 13

ACTIVE LEARNING IN STATISTICS

We have focused our attention on problems where we are learning in order to make a better decision in some sort of optimization problem, which gives us an economic value of information. It is often the case, however, that we are just trying to fit a statistical model which might be used for a variety of decision problems. We may not know how the model may be used; we are simply interested in carefully choosing what to measure to get the best model.

Most applications of statistics involve problems where we may have to fit a model from a fixed set of observations. This is often referred to as batch statistics. We may also have a process where observations arrive over time, but where we do not have any control over what we observe. This is known as *passive learning*.

There are many situations where we can control the inputs of a process. For example, we may be able to set the price of a product and then observe sales. We may set user preferences on Netflix, thus affecting the movies that are displayed to us; we then choose a movie to rent, which in turn affects what Netflix observes. After we choose the inputs (these might be referred to as independent variables or covariates) x^n, we then observe the response y^{n+1} which is then used to fit the parameters of a model. The machine learning community refers to this process as *active learning*.

Optimal Learning. By Warren B. Powell and Ilya O. Ryzhov
Copyright © 2012 John Wiley & Sons, Inc.

In statistics, active learning refers to the ability to control the choice of independent variables. In Chapter 1, we made a distinction between the broad umbrella of active learning, where we make choices with the intent to learn, and the subset of policies we call optimal learning, where our choice is guided by a well-defined objective function. We retain this distinction in this chapter, but cover a variety of heuristic and optimal policies for deciding what observations to make when fitting a function. We divide these methods into the following classes:

1) Deterministic policies - These determine all the points to observe in advance, without the benefit of learning the results of any measurements. We consider a special case where these policies are optimal.

2) Heuristic sequential policies - These are active learning policies where the choice of what to observe uses a rule that is not based on any particular performance metric.

3) Variance minimizing policies - These are sequential policies designed to minimize the variance of an estimator.

We specifically exclude the consideration of the economic value of a better statistical model in the form of better decisions, which has been the focus of most of this volume. For the purpose of this chapter, we focus on general metrics for the quality of a statistical model without regard to the decisions that it is used to support.

13.1   DETERMINISTIC POLICIES

There is an extensive literature in statistics that goes under the name "design of experiments." The problem is to choose a set of independent variables $x^1, \ldots, x^n$, where x^m is an F-dimensional vector of features which generates an observation $\hat{y}^m$. Let x^n be the vector of independent variables (features), given by

$$x^n = \begin{pmatrix} x_1^n \\ x_1^n \\ \vdots \\ x_F^n \end{pmatrix},$$

where $|\mathcal{F}|$ is a set of features, with $F = |\mathcal{F}|$. Our goal is to fit a linear model

$$y = \sum_{f \in \mathcal{F}} \theta_f x_f + \varepsilon,$$

where ε is an error term.

We want to choose θ to minimize the total errors squared given by

$$\min_\theta F(\theta) = \sum_{m=1}^n \left(y^m - \sum_{f \in \mathcal{F}} \theta_f x_f^m \right)^2. \tag{13.1}$$

To find the optimal solution, we begin by defining the matrix X^n as

$$X^n = \begin{pmatrix} x_1^1 & x_2^1 & & x_F^1 \\ x_1^2 & x_2^2 & & x_F^2 \\ \vdots & \vdots & \cdots & \vdots \\ x_1^n & x_2^n & & x_F^n \end{pmatrix}.$$

The vector of observations $\hat{y}^1, \ldots, \hat{y}^n$ is represented using

$$Y^n = \begin{pmatrix} \hat{y}^1 \\ \hat{y}^2 \\ \vdots \\ \hat{y}^n \end{pmatrix}.$$

As in Chapter 8, the vector θ^n that solves (13.1) is given by

$$\theta^n \;=\; [(X^n)^T X^n]^{-1} (X^n)^T Y^n. \tag{13.2}$$

Equation (13.2) gives us an easy way to find the variance of our estimate θ^n. Let v be an n-dimensional random vector, let A be a $F \times n$ deterministic matrix, and let $u = Av$. Let $Cov(v)$ be the covariance matrix of v. Then we can use the identity

$$Cov(u) = A Cov(w) A^T,$$

where $Cov(w)$ is the covariance matrix of w. Recall that for matrices A and B, $AB^T = (BA^T)^T$, and that $[(X^n)^T X^n]^{-1}$ is a symmetric matrix. We can use these observations to find the covariance matrix of θ^n if we let $A = [(X^n)^T X^n]^{-1}(X^n)^T$, giving us

$$\begin{aligned} Cov(\theta^n) &= [(X^n)^T X^n]^{-1}(X^n)^T Cov(Y^n) \left([(X^n)^T X^n]^{-1}(X^n)^T \right)^T \\ &= [(X^n)^T X^n]^{-1}(X^n)^T Cov(Y^n)(X^n)[(X^n)^T X^n]^{-1}. \end{aligned}$$

Since the elements of Y^n are independent, $Cov(Y^n) = \sigma_\epsilon^2 I$ where I is the identity matrix and σ_ϵ^2 is the variance of our measurement error . This allows us to write

$$\begin{aligned} Cov(\theta^n) &= [(X^n)^T X^n]^{-1}(X^n)^T X^n [(X^n)^T X^n]^{-1}\sigma_\epsilon^2 \\ &= [(X^n)^T X^n]^{-1}\sigma_\epsilon^2. \end{aligned}$$

The observation error σ_ϵ^2 is fixed; we cannot change it through our choice of x^n. However, we have direct control over the covariance of θ^n by our choice of X^n. Furthermore, we quickly see that the matrix $Cov(\theta^n)$ is a deterministic function of X^n, which means we do not gain anything by observing $\hat{y}^n$. It is for this reason that we can determine the precision of θ^n without making any observations. This is the theoretical foundation of deterministic policies.

So this leaves the question, what do we want to minimize? θ^n is a vector, and $Cov(\theta^n)$ is a matrix. While we would like to minimize all the elements of $Cov(\theta^n)$,

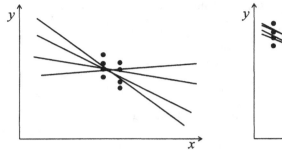 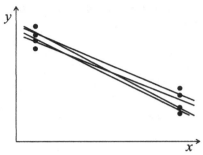

Figure 13.1 Learning a line with closely spaced measurements (a) and measurements with more separation (b).

we have to choose a single metric. This issue has created a variety of metrics that have come to be known as alphabet-optimality, for reasons that will become clear shortly. These metrics include the following:

A-optimality - Minimize the average (or trace, which is the sum) of the diagonal elements of $[(X^n)^T X^n]^{-1}$. This is the same as minimizing the average of the variances of each element of θ^n.

C-optimality - Given a weighting vector c, minimize $c^T[(X^n)^T X^n]c$.

D-optimality - Minimize $|[(X^n)^T X^n]^{-1}|$, or, equivalently, maximize the determinant of $[(X^n)^T X^n]$.

E-optimality - Maximize the minimum eigenvalue of $[(X^n)^T X^n]$.

G-optimality - Maximize the largest element in the diagonal of $X^n[(X^n)^T X^n](X^n)^T$, which has the effect of minimizing the largest variance of θ^n.

I-optimality - Minimize the average prediction variance over a particular region.

T-optimality - Maximize the trace of $[(X^n)^T X^n]$.

V-optimality - Minimize the average prediction variance over a specific set of points.

All of these methods aim at making the matrix $[(X^n)^T X^n]^{-1}$ smaller in some way, or, equivalently, making the matrix $[(X^n)^T X^n]$ bigger.

We can illustrate the central intuition behind these strategies using the simple example of fitting a line. Figure 13.1(a), (copied here for convenient reading from Figure 8.6), shows the lines that we might estimate if we make a small number of observations that are close to each other. Figure 13.1(b) illustrates the estimates that we might obtain if the measurements are spaced farther apart. It is not surprising that the more widely spaced observations provide a better estimate.

We can quantify this intuitive behavior by computing the matrix $[(X^n)^T X^n]$. Assume we make two observations each at two different locations. An observation

x^n might be $(1, 5)$, where the 1 corresponds to the constant term, and the 5 is the value we are measuring. The closely spaced points might be

$$X^n = \begin{pmatrix} 1 & 5 \\ 1 & 5 \\ 1 & 6 \\ 1 & 6 \end{pmatrix}. \tag{13.3}$$

The matrix $[(X^n)^T X^n]$ is given by

$$[(X^n)^T X^n] = \begin{bmatrix} 4 & 22 \\ 22 & 122 \end{bmatrix}.$$

Now assume we measure at more extreme points (but where the average is still the same), given by

$$X^n = \begin{pmatrix} 1 & 1 \\ 1 & 1 \\ 1 & 10 \\ 1 & 10 \end{pmatrix}.$$

The matrix $[(X^n)^T X^n]$ is given by

$$[(X^n)^T X^n] = \begin{bmatrix} 4 & 22 \\ 22 & 202 \end{bmatrix}.$$

The trace of the matrix $[(X^n)^T X^n]$ for the closely spaced points is $4 + 122 = 126$, while the trace for the points that are more spread out is 206. We would say that the second matrix has more "information," and the result is an estimate of θ with lower variance.

It is important that the data be scaled so that the average value of the measured values x remains the same (as we did above). The best way to do this is to simply average the measurements, and compute a corrected value $\bar{X}^n$ for each dimension, using

$$\bar{X}_i^n = X_i^n - \frac{1}{n} \sum_{m=1}^{n} X_i^m.$$

We then compute the matrix using $[(\bar{X}^n)^T \bar{X}^n]$. If we do this to the measurements in equation (13.3), we would obtain

$$\bar{X}^n = \begin{pmatrix} 0 & -0.5 \\ 0 & -0.5 \\ 0 & 0.5 \\ 0 & 0.5 \end{pmatrix}. \tag{13.4}$$

Regardless of which type of optimality we use, we now have a metric that we can use to help decide which measurements to make. Given a set of potential measurements, we generally want to choose measurements that are distributed around a center, but as far from the center as possible.

It is important to realize that we can choose the best set of potential measurements before making any observations. This property is a unique byproduct of the property that all statistics relating to the reliability of our estimates of θ are purely a function of the measurements and not the observations. We have not enjoyed this property in our previous applications.

13.2 SEQUENTIAL POLICIES FOR CLASSIFICATION

An important class of problems in machine learning is known as classification problems. In this problem class, we are given a set of features x^m for the mth object (this might be a document, an email or a website), and we are asked to place this document into one of a set of discrete classifications, such as {dangerous, threatening, suspicious, safe}. Often, collecting information about the classification of a document for training purposes is expensive. For example, we may need to ask a security expert to assess the threat level of an email or website. However, given the sheer volume of these sources, we cannot ask a trained expert to provide this information on a large number of documents.

This section describes a series of primarily heuristic search policies for efficiently collecting information for classification problems.

13.2.1 Uncertainty Sampling

Let $\hat{y}$ be a discrete quantity that indicates the classification of a document with attributes described by x, and let $P_\theta(\hat{y}|x)$ be the probability of the particular classification $\hat{y}$ if we observe x, given a parameter vector θ. Our goal is to observe the document where our prediction has the highest level of uncertainty. If the classification is binary (for example, dangerous or not), then we want to sample the documents where $P_\theta(\hat{y}|x)$ is as close as possible to 0.5.

If there are more than two outcomes, an alternative strategy is to choose to query the document whose prediction offers the lowest level of confidence, which we compute using

$$x^{LC} = \arg\max_x \left(1 - P_\theta(\hat{y}|x)\right),$$

where

$$\hat{y} = \arg\max_y P_\theta(y|x)$$

is the most likely classification. The idea here is that if the most likely classification in a set is small, then this represents a document (more precisely, a class of documents) where we have a lot of uncertainty, and would benefit from more information. We are trying to mitigate the effect of a worst-case scenario by learning about the document for which we are most likely to make an error in prediction.

This strategy focuses on the most probable classification, and as a result ignores the probabilities of other classifications. For example, we may be much more certain about one classification, implying that more testing is unlikely to change the classification. An alternative strategy is to focus on documents whose most likely classification is close to the second most likely classification. For fixed θ, let

$$\hat{y}_1 = \arg\max_y P_\theta(y|x)$$

and

$$\hat{y}_2 = \arg\max_{y \neq \hat{y}_1} P_\theta(y|x)$$

be the most likely and second most likely classification for a document. The marginal sampling policy chooses the document x where the two highest classification probabilities are the closest. We state this policy using

$$x^M = \arg\min_x \left(P_\theta(\hat{y}_1|x) - P_\theta(\hat{y}_2|x) \right).$$

The previous two policies look at the level of uncertainty indicated by the most certain document or the difference between the two most certain documents. An idea that takes this thinking a step further uses entropy as a measure of uncertainty. The entropy maximization policy is computed using

$$x^H = \arg\max_x - \sum_i P_\theta(y_i|x) \log P_\theta(y_i|x),$$

where y_i represents a single classification. Entropy looks at all the potential classifications, and represents an information-theoretic measure of the information required to encode the classification probability distribution.

Empirical research with these policies has produced mixed results. Clearly there will be differences in behavior as the number of potential classifications changes. Not surprisingly, performance depends on the true utility function. The margin and level of confidence policies work better when the goal is to get the classification right, while the entropy maximization policy works if the objective is to minimize the log of the loss from an incorrect classification ("log-loss").

13.2.2 Query by Committee

Imagine that we have several competing models for estimating the classification of a document. We would refer to this family of models as a "committee," where $c \in C$ is a particular model in the set C. Let $P_{\theta^{(c)}}(y_i|x)$ be the probability that model c (parameterized by $\theta^{(c)}$) returns classification y_i given the attributes x of a document.

There are several ways to perform active learning in this setting. One is to let each model vote for a classification. We might record a vote using the indicator function

$$I_c(y_i|x) = \begin{cases} 1 & \text{if } y_i = \arg\max_i P_{\theta^{(c)}}(y_i|x), \\ 0 & \text{otherwise.} \end{cases}$$

The indicator function $I_c(y_i|x)$ simply captures if model c thinks that y_i is the most likely classification. We can then count the number of votes using

$$V(y_i) = \sum_{c \in C} I_c(y_i|x).$$

Alternatively, we could compute "soft votes" using

$$V(y_i) = \sum_{c \in C} P_{\theta(c)}(y_i|x).$$

We can then choose to sample the document with the highest *vote entropy*, giving us the policy

$$x^{VE} = \arg\max_x - \sum_i \frac{V(y_i)}{C} \log \frac{V(y_i)}{C}.$$

Another measure is the Kullback–Leibler divergence metric, which is a way of measuring the differences between two distributions. We first compute an average probability of classification across the competing models using

$$P_C(y_i|x) = \frac{1}{|C|} \sum_{c \in C} P_{\theta(c)}(y_i|x).$$

The KL divergence is then computed using

$$D(P_{\theta(c)} \| P_C) = \sum_i P_{\theta(c)}(y_i|x) \log \frac{P_{\theta(c)}(y_i|x)}{P_C(y_i|x)}.$$

The Kullback–Leibler divergence is a type of distance metric since it is measuring the degree of similarity between two probability distributions: the distribution (across potential classifications) obtained when we average all the probabilities across the competing models, versus the distribution produced by each model. The quantity $D(P_{\theta(c)} \| P_C)$ is the KL divergence for a particular model c. Our policy for deciding which document to evaluate is obtained by maximizing the average KL divergence across all the models, given by

$$x^{KL} = \arg\max_x \frac{1}{C} \sum_{c \in C} D(P_{\theta(c)} \| P_C).$$

The idea with this policy is to choose to evaluate the document with the greatest disagreement among the different models. If the competing models largely agree, then it is unlikely that more information will change this consensus. Additional information is likely to have the greatest impact when competing models disagree.

13.2.3 Expected Error Reduction

An interesting policy uses an estimate of the degree to which information reduces the likelihood of being incorrect. We are going to assume that we have a set of unlabeled documents $\mathcal{U}$, which means that we have not solicited a classification from a domain expert. Assume we are considering the possibility of collecting information on the document x. We do not know its classification, but our current estimate of the probability that it will be classified as y_i is $P_\theta(y_i|x)$. If we choose document x and observe the classification y_i, we can use this information to update our classification probability, which we represent using $P_{\theta+,(x,y_i)}(y|x)$. This is analogous to a posterior belief.

Now, we are going to use this updated probability model on each document $u \in \mathcal{U}$, with feature vector x^u. We let

$$\hat{y}^u = \arg\max_y P_\theta(y|x^u)$$

be the most likely classification for a particular document x^u in our set $\mathcal{U}$. If $P_{\theta+,(x,y_i)}(\hat{y}^u|x^u)$ is the probability of this most likely classification, then $1 - P_{\theta+,(x,y_i)}(\hat{y}^u|x^u)$ represents the probability that the document u does not belong to the class that we think is the most likely. This is analogous to our posterior belief about how likely we are to make a mistake about document u after observing a result for x. We then choose x to minimize this probability of error in expectation over all possible classifications y_i of document x. The policy, then, is given by

$$x^{ER} = \arg\min_x \sum_i P_\theta(y_i|x) \left(\sum_{u \in \mathcal{U}} (1 - P_{\theta+,(x,y_i)}(\hat{y}^u|x^u)) \right).$$

13.3 A VARIANCE-MINIMIZING POLICY

We are going to describe an information collecting policy with the goal of reducing variance, but this time we are going to address a richer set of models and issues than we encountered in Section 13.1 when the goal was to minimize variance measures of the regression vector θ.

We start by assuming that there exists a model $y(x) = f(x) + \epsilon$ where $f(x)$ is the true model and ϵ is a source of measurement noise over which we have no control. Our goal is to collect a training dataset $\mathcal{D}^n = \{(x^0, y^1), (x^1, y^2), \ldots, (x^{n-1}, y^n)\}$. We group the choice $(x^0, \ldots, x^{n-1})$ of what to measure with the corresponding observations $(y^1, \ldots, y^n)$ into a single training dataset. We note that the standard notation in statistics is to let x^1 be the measurement that produces y^1, but as the development below illustrates, it is cleaner to let the superscript capture the information content; when we choose x^m, y^{m+1} is a random variable.

We assume we are using a sequential policy to determine $x^0, x^1, x^2, \ldots$ (starting with x^0) which may depend on the outcomes $y^1, y^2, \ldots$. This means that $\mathcal{D}^n$ is a random set, constructed sequentially. We use this information to fit an approximation

$\bar{y}^n = \bar{f}^n(x|\mathcal{D}^n)$. Below, we are going to write the prediction as $\bar{y}^n(x|\mathcal{D}^n)$ to express its dependence on the query point x, and the data $\mathcal{D}^n$.

The goal is to design a policy that minimizes the variance in our predictions $\bar{y}^n(x|\mathcal{D}^n)$, but this depends on the points x where we *might* query the function. We did not have to deal with this issue in Section 13.1, but now we are going to assume that we are given a distribution $P(x)$ that gives the probability that we will want an estimate of the function at x. In practice, $P(x)$ may be chosen to be uniform over some region, or we may specify a normal distribution with some mean μ_x and a spread σ_x. Alternatively, we may have a sequence of observations $\tilde{x}^1, \ldots, \tilde{x}^k$ from history which can serve as a probability distribution. Either way, $P(x)$ serves as a weighting function that tells us the region in which we are interested. However, it is important to recognize that the probability that x is in the random set $\mathcal{D}^n$, which is influenced by our learning policy, may be completely different than $P(x)$.

If we make an observation at x, we are going to observe

$$y(x) = f(x) + \epsilon,$$

where $f(x)$ is our true (but unknown) function, and ϵ is the inherent noise in the observation. Our approximation is going to give us the estimate $\bar{y}^n(x|\mathcal{D}^n)$. A reasonable goal is to design a policy for choosing a measurement x to minimize the prediction error $(\bar{y}^n(x|\mathcal{D}^n) - y(x))^2$ for a single realization of $y(x)$ (which is random because of ϵ), and a single estimate $\bar{y}^n(x|\mathcal{D}^n)$ from a dataset $\mathcal{D}^n$ (which is random because $\mathcal{D}^n$ is random). To formalize this idea, we need to take expectations. We let $\mathbb{E}_T(\cdot)$ be the total expectation over the observation noise imbedded in $y(x)$ and the observation of $\mathcal{D}^n$. These are independent, so we can write them as

$$\mathbb{E}_T(\bar{y}^n(x|\mathcal{D}^n) - y(x))^2 = \mathbb{E}_y \mathbb{E}_\mathcal{D}(y(x) - \bar{y}^n(x|\mathcal{D}^n))^2, \tag{13.5}$$

where $\mathbb{E}_\mathcal{D}$ is the expectation over all the possible outcomes of $\mathcal{D}^n$, and $\mathbb{E}_y$ is the expectation over all possible realizations of $y(x)$. We can break down the expected total variation (for a given x) in (13.5) using

$$\begin{aligned}
\mathbb{E}_T(\bar{y}^n(x|\mathcal{D}^n) - y(x))^2 &= \mathbb{E}[(y(x) - \mathbb{E}[y(x)])^2] \\
&\quad + \big(\mathbb{E}_\mathcal{D}[\bar{y}(x|\mathcal{D}^n)] - \mathbb{E}[y(x)]\big)^2 \\
&\quad + \mathbb{E}_\mathcal{D}\big[(\bar{y}(x|\mathcal{D}^n) - \mathbb{E}_\mathcal{D}[\bar{y}(x|\mathcal{D}^n)])^2\big]. \tag{13.6}
\end{aligned}$$

The first term reflects the pure noise due to ϵ, which will not be affected by the measurement policy. The second term captures the bias in the model, which is purely a function of the structural form of the underlying model, as well as the choice of x. Again, this is not affected by the choice of $\mathcal{D}^n$. The right-hand side of equation (13.6) is also known as a *bias-variance decomposition*.

The third term is the variance due to the estimation of the model from the training data. This is where we capture the variation in the estimated model (after n observations) due to the different variations in $\mathcal{D}^n$ that might be produced by following a specific measurement policy. The variations in $\mathcal{D}^n$ arise because of differences in realizations in the observations $y^1, \ldots, y^n$ that lead to different decisions $x^2, \ldots, x^n$. This is the term that we wish to minimize by choosing a good learning policy.

For compactness, we are going to let

$$\sigma_y^{2,n}(x) = \mathbb{E}_\mathcal{D}\big(\bar{y}^n(x|\mathcal{D}^n) - \mathbb{E}_\mathcal{D}[\bar{y}^n(x|\mathcal{D}^n)]\big)^2.$$

Now assume that we are considering the possibility of adding (x^n, y^{n+1}) to create an expanded dataset

$$\mathcal{D}^{n+1} = \mathcal{D}^n \cup (x^n, y^{n+1}).$$

The choice x^n is a deterministic quantity (given $\mathcal{D}^n$) that we are thinking of measuring, while y^{n+1} is the random observation that we have not yet observed when we choose x^n. Let $\mathbb{E}_\mathcal{D}^n$ be the conditional expectation given $\mathcal{D}^n$. The choice of observation x^n is a deterministic function of $\mathcal{D}^n$. The new information in $\mathcal{D}^{n+1}$ is y^{n+1}. If we choose x^n, let $\tilde{\sigma}_y^{2,n}(x|x^n)$ be the conditional variance (given $\mathcal{D}^n$) of our estimate if we choose x^n, which is given by

$$\tilde{\sigma}_y^{2,n}(x|x^n) = \mathbb{E}_\mathcal{D}^n\big(\bar{y}^{n+1}(x|\mathcal{D}^{n+1}) - \mathbb{E}_\mathcal{D}^n[\bar{y}^{n+1}(x|\mathcal{D}^{n+1})]\big)^2. \qquad (13.7)$$

The variance $\tilde{\sigma}_y^{2,n}(x|x^n)$ is defined in exactly the same way as $\tilde{\sigma}$ in (2.8). However, in Chapter 2, we also proved that this quantity coincided with the change in variance between two measurements, under a Bayesian learning model. This equivalence does *not* hold in the frequentist model we are considering here. In (13.7), $\tilde{\sigma}^{2,n}$ represents only the conditional variance of our estimate given a choice of measurement, and minimizing this quantity is equivalent to minimizing the future variance of our prediction (similar to the posterior variance considered in the Bayesian models). If we were to adopt a Bayesian approach here, $\mathbb{E}[y(x)]$ would itself be random (since we do not know the true mean of the observation), and we would need to consider all the terms in (13.6) together in order to minimize the posterior variance.

In the frequentist model, it is enough to minimize the last term in (13.6). We want to choose x^n so as to minimize (13.7). However, we do not know that we are going to want to observe the function at x, so we need to take the expectation over potential observations using our marginal distribution $P(x)$, giving us

$$\bar{\sigma}_y^{2,n}(x^n) = \int_x \tilde{\sigma}_y^{2,n}(x|x^n) P(x)\, dx. \qquad (13.8)$$

The variance $\bar{\sigma}_y^{2,n}(x^n)$ is the expected variance given that we make the measurement x^n. Our policy will be to choose the value x^n that produces the greatest reduction in variance. We state this policy using

$$x^n = \arg\min_{x'} \bar{\sigma}_y^{2,n}(x'). \qquad (13.9)$$

The value of this policy is that it is fairly easy to compute. We illustrate the calculations for an approximation architecture that uses mixtures of Gaussians.

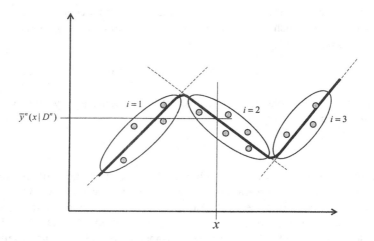

Figure 13.2   Illustration of fitting data using mixtures of Gaussians, based on Cohn et al. (1994).

13.4 MIXTURES OF GAUSSIANS

In Section 13.1, we considered the case of estimating a single regression function. Here, we turn to a more general approximation architecture where we assume that observations come from one of a set of populations that we index $i = 1, \ldots, N$. We assume that each population produces a different behavior that can be reasonably approximated by a line, as illustrated in Figure 13.2.

13.4.1 Estimating Parameters

We need to estimate the mean and variance of the measurements (independent variables) x, given by $\mu_{x,i}$ and $\sigma^2_{x,i}$, the mean and variance of the observations y, given by $\mu_{y,i}$ and $\sigma^2_{y,i}$, and the covariance between x and y, $\sigma_{xy,i}$. Temporarily assume that for each population i, we know the expected value of the observation x, $\mu_{x,i}$, and its variance $\sigma^2_{x,i}$. We are going to begin by presenting calculations for the remaining moments, after which we show how to compute $\bar{\sigma}^2(x^n)$.

We can represent the joint distribution of x and y for a population i. First define

$$z = \left[\begin{array}{c} x \\ y \end{array} \right], \quad \mu_i = \left[\begin{array}{c} \mu_{x,i} \\ \mu_{y,i} \end{array} \right], \quad \Sigma_i = \left[\begin{array}{cc} \sigma^2_{x,i} & \sigma_{xy,i} \\ \sigma_{xy,i} & \sigma^2_{y,i} \end{array} \right].$$

We can now write the joint distribution of x and y for group i using

$$P(x,y|i) = \frac{1}{2\pi\sqrt{|\Sigma_i|}} \exp\left[-\frac{1}{2}(z - \mu_i)^T \Sigma_i^{-1}(z - \mu_i) \right].$$

We will not know the true means and variances, but we can estimate them from an initial sample.

The conditional variance of y given x is given by

$$\sigma^2_{y|x,i} = \sigma^2_{y,i} - \frac{\sigma_{xy,i}}{\sigma^2_{x,i}}.$$

The conditional expectation $\bar{y}^n_i(x|\mathcal{D}^n)$ and variance $\sigma^2_{\hat{y},i}$ given x are

$$\bar{y}^n_i(x|\mathcal{D}^n) = \mu_{y,i}(x) + \frac{\sigma_{xy,i}}{\sigma^2_{x,i}}(x - \mu_{x,i}),$$

$$\sigma^2_{y,i} = \frac{\sigma^2_{y|x,i}}{n_i}\left(1 + \frac{(x - \mu_{x,i})^2}{\sigma^2_{x,i}}\right).$$

The scalar n_i can be viewed as the weight that group i should be given, calculated using

$$n_i = \sum_{j=1}^{m} \frac{P(x_j, y_j|i)}{\sum_{k=1}^{N} P(x_j, y_j|k)}.$$

We next compute the probability that population i contributes to the observation corresponding to x using

$$h_i(x) = \frac{P(x|i)}{\sum_{j=1}^{N} P(x|j)},$$

where

$$P(x|i) = \frac{1}{\sqrt{2\pi\sigma^2_{x,i}}} \exp\left[-\frac{(x - \mu_{x,i})^2}{2\sigma^2_{x,i}}\right]. \qquad (13.10)$$

Given the observation x, the expectation across all the populations of $\bar{y}^n_i(x|\mathcal{D}^n)$ and its variance are given by

$$\bar{y}^n(x|\mathcal{D}^n) = \sum_{i=1}^{N} h_i(x)\bar{y}^n_i(x|\mathcal{D}^n),$$

$$\sigma^2_y(x|\mathcal{D}^n) = \sum_{i=1}^{N} \frac{h_i^2\sigma^2_{y|x,i}}{n_i}\left(1 + \frac{(x - \mu_{x,i})^2}{\sigma^2_{x,i}}\right).$$

We now have the foundation we need to estimate the expected value of an additional measurement.

13.4.2 Active Learning

We have to calculate the variance given a measurement x^n, where we continue to assume that $P(x)$ is known. We have assumed that $\mu_{x,i}$ and $\sigma^2_{x,i}$ are known for each population i, but we cannot derive these statistics directly from the marginal distribution $P(x)$. For example, we may approximate $P(x)$ by assuming that it is uniformly distributed over some region, or we may represent it from a sample obtained from history.

We could try to estimate $\mu_{x,i}$ and $\sigma^2_{x,i}$ from training data, but these observations are chosen according to our learning policy, and may not reflect the true distribution that reflects the likelihood of actually needing to calculate the function at some x. If we use the training sample, we are effectively using the joint distribution $P(x^n, y|i)$ instead of $P(x, y|i)$. We can correct for the difference in the sampling distribution (at least in principle) using

$$ P(x, y|i) = P(x^n, y|i) \frac{P(x|i)}{P(x^n|i)}. $$

We compute the conditional distribution $P(x|i)$ by using equation (13.10) with the mean and variance $\mu_{x,i}$ and $\sigma^{2,n}_{x,i}$ computing using data sampled from $P(x)$ or a reference sample. The distribution $P(x^n, y|i)$ is also computed using equation (13.10) with mean and variance calculated from a training dataset.

We now have to compute $\bar{\sigma}^{2,n}(x^n)$ which we use in equation (13.9) to determine the next point to measure. We first calculate $\tilde{\sigma}^{2,n}_y(x|x^n)$ for a particular query point x. The distribution of y^{n+1} given x^n can be calculated as a mixture of normal distributions, given by

$$
\begin{aligned}
P(y^{n+1}|x^n) &= \sum_{i=1}^{N} h_i(x^n) P(y^{n+1}|x^n, i) \\
&= \sum_{i=1}^{N} h_i(x^n) \mathcal{N}\big(\bar{y}^n_i(x^n|\mathcal{D}^n), \sigma^{2,n}_{y|x,i}(x^n)\big),
\end{aligned}
$$

where $\mathcal{N}(\mu, \sigma^2)$ represents the normal distribution, where $\bar{y}^n_i(x^n|\mathcal{D}^n)$ is the mean of y^{n+1} based on data for group i, and $\sigma^2_{y|x,i}(x^n)$ is the variance. The conditional variance of our prediction of x given the choice of x^n can be found using

$$ \tilde{\sigma}^{2,n}_y(x|x^n) = \sum_{i=1}^{n} \frac{h_i^2(x)\tilde{\sigma}^{2,n}_{y|x,i}(x|x^{n+1})}{n_i + h_i(x|x^{n+1})} \left(1 + \frac{(x - \mu_{x,i})}{\sigma^2_{x,i}}\right), $$

where

$$\tilde{\sigma}_{y,i}^{2,n}(x|x^{n+1}) = \frac{n_i \sigma_{y,i}^{2,n}}{n_i + h_i(x|x^{n+1})}$$
$$+ \frac{n_i + h_i(x|x^{n+1})\left(\sigma_{y|x,i}^2 + (\bar{y}_i^{n+1}(x|x^{n+1}) - \mu_{y,i})^2\right)}{(n_i + h_i(x|x^{n+1}))^2},$$

$$\tilde{\sigma}_{y|x,i}^{2,n}(x|x^{n+1}) = \tilde{\sigma}_{y,i}^{2,n}(x|x^{n+1}) - \frac{\tilde{\sigma}_{xy,i}^{2,n}}{\sigma_{x,i}^2},$$

$$\tilde{\sigma}_{xy,i}(x|x^{n+1}) = \frac{n_i \sigma_{xy,i}}{n_i + h_i(x|x^{n+1})}$$
$$+ \frac{n_i h_i(x|x^{n+1})(x^{n+1} - \mu_{x,i})(\bar{y}_i^{n+1}(x|x^{n+1}) - \mu_{y,i})}{(n_i + h_i(x|x^{n+1}))^2},$$

$$\tilde{\sigma}_{xy,i}^2(x|x^{n+1}) = (\tilde{\sigma}_{xy,i})^2 + \frac{n_i^2(h_i(x|x^{n+1}))^2 \sigma_{y|x^{n+1},i}^2(x^{n+1} - \mu_{x,i})^2}{(n_i + h_i)^4}.$$

If we are estimating $\mu_{x,i}$ and $\sigma_{x,i}^2(x|x^{n+1})$ from data, we can use the following equations

$$\mu_{x,i}^{n+1}(x^{n+1}) = \frac{n_i \mu_{x,i}^n + h_i(x|x^{n+1})x^{n+1}}{n_i + h_i(x|x^{n+1})},$$

$$\sigma_{x,i}^{2,n+1}(x|x^{n+1}) = \frac{n_i \sigma_{x,i}^{2,n}}{n_i + h_i(x|x^{n+1})} + \frac{n_i h_i(x|x^{n+1})(x^{n+1} - \mu_{x,i}^n)^2}{(n_i + h_i(x|x^{n+1}))^2}.$$

With $\tilde{\sigma}_y^{2,n}(x|x^n)$ in hand, we use equation (13.8) to integrate over the different query points x to obtain $\bar{\sigma}_y^{2,n}(x^n)$.

13.5 BIBLIOGRAPHIC NOTES

Section 13.1 - This section covers classic material from experimental design as it is known within the statistics community (see DeGroot 1970, Wetherill & Glazebrook 1986, Montgomery 2008).

Section 13.2 - This section is based on Settles (2009).

Sections 13.3 - This section is from Geman et al. (1992), with material from Cohn et al. (1994), Cohn et al. (1996) and Settles (2009). See Hastie et al. (2005) for a nice discussion of bias-variance decomposition.

Section 13.4 - This section is based on Settles (2009).

CHAPTER 14

SIMULATION OPTIMIZATION

At the highest level of abstraction, the ranking and selection problem can be written as

$$\max_{x \in \mathcal{X}} f(x),$$

where $\mathcal{X} = \{1, ..., M\}$ and $f(x) = \mu_x$. In most applications, we cannot compute $f(x)$ exactly, but instead we are forced to obtain noisy estimates of the function through simulation. Let W represent a set of random variables that can be used to compute an estimate of the function, which we represent as $\tilde{f}(x, W)$, where

$$f(x) = \mathbb{E}\tilde{f}(x, W).$$

For ranking and selection, we might have $\tilde{f}(x, W) = W_x$. The point, however, is that $f(x)$ represents the mean of a random quantity that also depends on x. In ranking and selection problems, we cannot compute $f(x)$ directly because we simply do not know μ_x. In other applications, $\tilde{f}(x, W)$ may represent the results of a discrete event simulation, where it is computationally intractable to compute its expectation.

This general problem is extensively studied by the *simulation optimization* community. Our presentation throughout this book has considered applications where the observations W can be simulated or obtained through a field experiment (e.g.,

Optimal Learning. By Warren B. Powell and Ilya O. Ryzhov
Copyright © 2012 John Wiley & Sons, Inc.

lab testing of a medical treatment). The simulation optimization community focuses on problems where observations come from discrete event simulations such as those used to model manufacturing systems. Typically, we have a simulation model of a complex system, such as a factory floor, a supply chain, or a transportation network. The "output" of the simulator is some sort of aggregate performance measure, such as throughput for a manufacturing plant. The simulator also has many inputs, which can be stochastic. For example, our supply chain can be subject to randomly arriving demand for a product at different locations. We generally assume that the interplay between inputs is so complex that it is hopeless to attempt to analytically compute expected performance. Instead, we simulate the behavior of the system across many sample realizations for the inputs, and use the resulting output to estimate the mean.

If the simulation model is large enough (e.g., modeling operations for multiple manufacturing plants), even a single simulation run will require significant computing resources. We may need to spend several hours just to get a single sample realization of the mean performance. We give a brief list of applications where this issue may arise:

- Simulation calibration – Suppose that we have a computer simulation of business operations for a trucking company or an airline. Before we can use the model to make decisions and recommendations (e.g., which trucks to assign to which demand, or how to price airline tickets), we first have to be sure that the simulation model is accurate to begin with. That is, the output of our model needs to match the historical performance of the real-world system. We can achieve this by tuning a relatively small set of parameters, such as a vector of bonuses and penalties to encourage the simulated decision-maker to behave in a certain way.

- Manufacturing throughput – A production line in a manufacturing plant consists of a series of machines. A new part is completed when it has passed through all the stages of production in order. To prevent the machines from becoming blocked or idle, we may place an intermediate buffer between two consecutive machines. If a job passes through the ith stage, only to find that machine $i + 1$ is busy, the job will wait its turn in the buffer, while machine i is free to work on the next job. The problem is to find the optimal size for the buffers. More capacity means improved production rates, but also increased space and storage costs. We can evaluate the performance of a particular buffer size via a simulation model of the entire factory floor.

- Cost management in health care – Patients with different priority levels arrive randomly at an emergency room. Higher priority levels must be treated quickly. The emergency room is staffed by a number of doctors, nurses, and other personnel. The number of beds is also limited. Suppose that we wish to improve service (e.g. change the scheduling policy to reduce bed block), or to reduce operational costs while minimizing the impact on service. We can evaluate the effectiveness of a particular scheduling policy or staff size in a simulator.

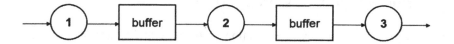

Figure 14.1 Diagram of a manufacturing process where three machines are separated by two buffers. We can use simulation optimization to find the right buffer capacity.

The contextual applications which have motivated the simulation optimization community have resulted in the development of algorithms which share some common features:

- Problems are most often motivated by applications where x is a design and where we have to choose from among a discrete set $\{1, \ldots, M\}$ where M is not too large (say, less than 100).

- The belief model is typically frequentist, reflecting the lack of any prior belief about the performance of any particular design.

- The process of running a discrete event simulation model involves a certain startup cost (loading the software, reading the data and initializing parameters) before the simulation starts. This complicates the process of collecting a discrete measurement of a design x, as we have tended to assume throughout this volume. For this reason, it is easier to decide on a run length in advance, and then simply execute this run.

- The simulation optimization community has introduced a number of alternative objective functions (originally discussed in Chapter 7) as alternatives to expected opportunity cost.

It is difficult to summarize the entire simulation optimization literature in a single chapter. Our goal is to give an overview of those methods which are closest in spirit to the theme of this book – the idea of adaptive learning and changing our solution strategy over time as we improve some form of belief about the simulation output. Unfortunately, the distinction between optimal learning and stochastic optimization can become blurred at times. In particular, we discuss the methodology of optimal computing budget allocation (OCBA), which can be used in a Bayesian setting like the LL method discussed in Chapter 5. We do not, for example, touch on the vast literature on stochastic approximation or sample average approximation (SAA). In our view, these approaches are closer to traditional conceptions of stochastic optimization with fixed (though unknown) distributions, rather than the idea of shifting distributions of belief that is our focus here. However, these are very important topics and we invite interested readers to examine such references as Spall (2003), Ruszczynski & Shapiro (2003), or Ben-Tal et al. (2009) for different perspectives.

14.1 INDIFFERENCE ZONE SELECTION

Let us return to the ranking and selection problem, with M alternatives. An alternative might be a design for a manufacturing system, a policy for dispatching ambulances, or a set of parameters for tuning a simulation. Our goal is still to pick the alternative x with the highest value μ_x, and we still make N sequential measurement decisions. However, we evaluate the quality of our learning policy through the *probability of correct selection* (PCS). Denote by $\mu_{(k)}$ the kth biggest value, such that $\mu_{(1)} \geq \mu_{(2)} \geq \dots \geq \mu_{(M)}$. A policy is considered to be "good" if

$$P\left(\arg\max_{x'} \mu_x = \arg\max_{x'} \theta_{x'}^N \mid \mu_{(1)} \geq \mu_{(2)} + \delta\right) \geq 1 - \alpha. \tag{14.1}$$

The *indifference-zone* parameter δ represents the smallest difference in values that is significant. In other words, if the true best value is greater than the second best by at least δ, we want to correctly choose the best alternative with some high probability $1 - \alpha$. Otherwise, if the top two alternatives are similar in value (the difference is less than δ), we are indifferent between them.

This formulation originates from the frequentist view of statistics. In this view, the values μ_x are fixed, though unknown, and we look for a policy that would perform well on any fixed configuration of values. In the Bayesian view, the values μ_x are random variables, and so the event $\{\mu_{(1)} \geq \mu_{(2)} + \delta\}$ has a probability that may be difficult to calculate. In our discussion of indifference-zone methods, we shall revert to the frequentist notation from Section 2.1, where $\bar{\theta}_x^n$ denotes our estimate of μ_x, while $\hat{\sigma}^{2,n}$ is our estimate of the variance of the observations.

14.1.1 Batch Procedures

The simplest batch procedure dates back to Bechhofer (1954). This method assumes that the observations are normal and independent, with means μ_x and a common, known variance σ_W^2. We proceed by computing a number h satisfying

$$P(Z_i \geq h) = 1 - \alpha, \quad i = 1, \dots, M - 1, \tag{14.2}$$

where the vector $Z_1, \dots, Z_{M-1}$ has a multivariate normal distribution with means equal to zero, variances equal to 1, and $Cov(Z_i, Z_j) = 1/2$ for any Z_i and Z_j. We then calculate

$$k^{IZ} = \frac{2h^2 \sigma_W^2}{\delta^2}, \tag{14.3}$$

and round this quantity to the nearest integer. We then sample every alternative exactly x^{IZ} times, and implement the alternative with the highest sample mean.

This procedure is based on a worst-case analysis. We suppose that $\mu_{(2)} = \dots = \mu_{(M)} = \mu(1) - \delta$. We view this as the "least favorable" scenario among all possible alternative values that still satisfy our modeling assumption $\mu_{(1)} \geq \mu_{(2)} + \delta$. Intuitively, this particular scenario is unfavorable because it makes it as difficult as possible (under the modeling assumption) to distinguish between the true best alternative and any of the others. Without loss of generality, we can assume that

$\mu_{(2)} = \ldots = \mu_{(M)} = 0$ and $\mu_{(1)} = \delta$. Because the measurement noise is common for all alternatives, we can also suppose that $\sigma_W^2 = \frac{1}{2}$. The quantity Z_i in (14.2) then represents the deviation of the difference in the sample means from the expected difference,

$$Z_i = \left(\bar{\theta}_{(i)} - \bar{\theta}_{(1)}\right) - (0 - \delta), \qquad (14.4)$$

where (1) through (M) denote the indices of the alternatives arranged in order of decreasing μ_x. We can see that the variances and correlations between Z_i and Z_j are 1 and $1/2$.

When computing the sampling allocation, we rescale by $2\sigma_W^2$. The sampling allocation is chosen so that the sample variance $\hat{\sigma}_x^2$ is small enough to ensure

$$P\left(\bar{\theta}_{(1)} > \bar{\theta}_{(2)}, \ldots, \bar{\theta}_{(1)} > \bar{\theta}_{(M)}\right) = 1 - \alpha,$$

where (1) through (M) denote the indices of the alternatives arranged in order of decreasing μ_x.

Table 14.1 provides approximate values of h (rounded to the nearest 0.05) for some different problem sizes M and values of α. In MATLAB, we can approximate h by using the function mvncdf to evaluate the multivariate normal cdf for $(Z_1, \ldots, Z_{M-1})$ repeatedly. Larger values of α lead to smaller h, since we need to increase the size of the region $[h, \infty)^{M-1}$ over which we are calculating the cdf in order to ensure a greater level of coverage. The same is true when we increase the problem size. Unfortunately, numerical approximations of the multivariate normal cdf are much less accurate than the univariate version, and it is very time-consuming to find h in many dimensions. Later, we will discuss how h can be approximated without resorting to a brute-force search.

When the variances of the observations are different (but still known), that is, the variance of W_x is λ_x, computing the analog of (14.2) is quite complicated (and requires us to solve an integral equation), because the variances of the multivariate normal vector $(Z_1, \ldots, Z_{M-1})$ are now different. However, we can compute sampling allocations to make the sample variances equal, thereby allowing ourselves to use the original procedure. We let $s = \sum_x \lambda_x$ and compute k^{IZ} assuming that $\sigma_W^2 = s$. We then let

$$k_x = \frac{\lambda_x}{s} k^{IZ}$$

Table 14.1 Calculations of h for Bechhofer's procedure in some small problems.

$M - 1$	$\alpha = 0.1$	$\alpha = 0.05$	$\alpha = 0.01$	$\alpha = 0.001$
5	-1.95	-2.25	-2.85	-3.55
10	-2.15	-2.45	-3.05	-3.70
15	-2.30	-2.60	-3.15	-3.85
20	-2.35	-2.65	-3.25	-3.85
25	-2.45	-2.75	-3.30	-3.95

be the sampling allocation for k_x.

When the variances of the observations are *unknown*, indifference-zone methods are typically divided into two stages. The first stage allocates some number k^0 simulations to each alternative, from which we obtain the first-stage sample means $\bar{\theta}_x^1$ and variances $\hat{\sigma}_x^{2,1}$. The second-stage sampling allocation follows a Bechhofer-like argument. We ensure the indifference-zone condition (14.1) by considering the more general condition that $P\left(\left|(\bar{\theta}_x^2 - \bar{\theta}_y^2) - (\mu_x - \mu_y)\right| < \delta\right) = 1 - \alpha$ for all x and y. The deviation from (14.4) is now defined as

$$Z_{xy} = \frac{(\bar{\theta}_x^2 - \bar{\theta}_y^2) - (\mu_x - \mu_y)}{\sqrt{\frac{\lambda_x}{k_x^1} + \frac{\lambda_y}{k_x^1}}},$$

and we solve an integral equation to find h_y such that

$$P\left(Z_{xy} \geq \frac{h_y}{\sqrt{\frac{\lambda_x}{\hat{\sigma}^{2,1}} + \frac{\lambda_y}{\hat{\sigma}^{2,1}}}},\ x \neq y\right) = 1 - \alpha.$$

The second-stage sampling allocation k_y^1 is then set to $k_y^1 = \frac{2h_y^2 \hat{\sigma}_y^{2,1}}{\delta^2}$.

In all of these cases, the fundamental reasoning is essentially the same: We want to ensure the indifference-zone condition for a worst-case set of true values, thereby ensuring it for all possible truths. This is a different philosophy from the one underlying our more familiar Bayesian procedures, where we maximize an expected value of a performance measure rather than a worst-case value.

14.1.2 Sequential Procedures

In simulation optimization, the word "sequential" can be used to mean that we are simulating one sample per alternative at a time. We are not, however, restricted to simulating only one alternative. The most basic sequential procedure, due to Paulson (1964), simulates from multiple alternatives in each time step, and gradually eliminates alternatives that do not look promising. We make the same assumptions as in Bechhofer's procedure, namely that observations are independent and normal with common variance σ_W^2.

We first choose a number $0 < q < \delta$ and define

$$a = \log\left(\frac{M-1}{\alpha}\right) \frac{\sigma_W^2}{\delta - q}.$$

The set $\mathcal{P}^0 = \{1, ..., M\}$ keeps track of the promising alternatives (initially all alternatives appear promising). At time n, we collect one observation W_x^{n+1} for every $x \in \mathcal{P}^n$. We then update

$$\mathcal{P}^{n+1} = \left\{x \in \mathcal{P}^n : \bar{\theta}_x^{n+1} \geq \left(\max_{x' \in \mathcal{P}^n} \bar{\theta}_{x'}^{n+1}\right) - \max\left(0, \frac{a}{n-q}\right)\right\}.$$

In other words, an alternative is promising only if its sample mean continues to be "close" to the current biggest sample mean. The procedure ends when a single alternative remains in the promising set. We are familiar with this notion: recall that the expected improvement factor in Chapter 5 depended on the difference between our beliefs about a given alternative and the current best.

The same idea can be extended to the setting where the variances of the observations are unequal and unknown. Once again, we employ a two-stage algorithm in which the first stage provides us with estimates of the variances, and the second stage is used to thin out the set of promising alternatives. The resulting method is a hybrid of Bechhofer's and Paulson's procedures: we sequentially update the promising set as in Paulson's procedure, but use the batch sampling allocation as a limit on our simulation budget.

In the first stage, we simulate k^0 samples from each alternative, obtaining independent observations W_{xj}^1 for $j = 1, ..., k^0$ and $x = 1, ..., M$ with the relevant sample means $\bar{\theta}_x^1$. We then calculate

$$\hat{\sigma}_{xy}^{2,1} = \frac{1}{k^0 - 1} \sum_{j=1}^{k_0} \left[(W_{xj}^1 - W_{yj}^1) - (\bar{\theta}_x^1 - \bar{\theta}_y^1) \right]^2,$$

the sample variances of the pairwise differences between alternatives. We then let

$$k_x^{batch} = \max_{y \neq x} \frac{h^2 \hat{\sigma}_{xy}^{2,1}}{\delta^2}$$

be the maximum allowable number of samples for alternative x. This calculation is reminiscent of (14.3), where h is the solution to an integral equation ensuring that the sample variance will be small enough to allow us to differentiate between alternatives. We can approximate

$$h \approx 2\eta(k^0 - 1),$$

where

$$\eta = \frac{1}{2} \left[\left(\frac{2\alpha}{M-1} \right)^{-\frac{2}{k^0-1}} - 1 \right].$$

We now create the set $\mathcal{P}^1 = \{1, ..., M\}$ and turn to the sequential part of the procedure. At time n, we collect one observation W_x^{n+1} for every $x \in \mathcal{P}^n$ and update

$$\mathcal{P}^{n+1} = \left\{ x \in \mathcal{P}^n : \bar{\theta}_x^{n+1} \geq \max_{x' \in \mathcal{P}^n} \bar{\theta}_{x'}^{n+1} - \max\left(0, \frac{\delta}{2n} \left(\frac{h^2 \hat{\sigma}_{x,x'}^{2,1}}{\delta^2} - n \right) \right) \right\}.$$

The procedure ends either when only one alternative remains in the promising set, or $n > \max_x k_x^{batch}$. In the latter case, we implement the alternative with the highest sample mean among the promising set.

14.1.3 The 0–1 Procedure: Connection to Linear Loss

There is also a Bayesian batch procedure for indifference-zone selection that closely follows the linear loss method introduced in Section 5.6. Let us pause to remind ourselves of the Bayesian notation. The $LL(N)$ method is designed for ranking and selection with independent but unknown means as well as unknown sampling variances. As in Sections 2.3.5 and 5.6, we place a joint prior on the unknown value μ_x of alternative x, and the unknown precision β_x^W of our observations for this alternative. We say that (μ_x, β_x^W) follows a normal-gamma distribution with parameters $\theta_x^0, \tau_x^0, a_x^0, b_x^0$, meaning that $\beta_x^W \sim Gamma(a_x^0, b_x^0)$ and the conditional distribution of μ_x given $\beta_x^W = r$ is $\mathcal{N}\left(\theta_x^0, \tau_x^0 r\right)$. The equations (2.37)–(2.40) are used to update our beliefs.

The $LL(N)$ procedure was designed to choose a sampling allocation $k = (k_1, ..., k_M)$, with $k_1 + ... + k_M = N$, to maximize the objective (5.29) with $F^k\left(\mu, \beta^W, W\right) = \max_x \theta_x^N$. For indifference-zone selection, we simply replace this objective by

$$F^k\left(\mu, \beta^W, W\right) = 1_{\{\arg\max_x \mu_x = \arg\max_x \theta_x^N\}},$$

which is known as the 0–1 objective function. The expected value of this function is precisely the probability of selecting the true best alternative. Instead of specifying an explicit indifference-zone parameter, we simply seek to maximize the probability of correct selection.

Section 5.6 describes how (5.29) can be approximated by an expression that has a closed-form optimal solution, using the fact that the predictive distribution of θ_x^N is the t-distribution. Analogous approximations for the 0–1 objective function lead us to the sampling allocation

$$k_x = \frac{N + \sum_{x'=1}^{M} \tau_{x'}^0}{\sum_{x'=1}^{M} \sqrt{\frac{b_{x'}^0, \kappa_{x'}^0}{b_x^0 \kappa_x^0}}} - \tau_x^0, \tag{14.5}$$

which is identical to (5.32, but with

$$\kappa_x^0 = \begin{cases} (\lambda_{x,x^*}^0)^{\frac{3}{2}} (\theta_{x^*}^0 - \theta_x^0)\, \phi_{2a_x^0}\left(\sqrt{\lambda_{x,x^*}^0}\, (\theta_{x^*}^0 - \theta_x^0)\right), & x \neq x^*, \\ \sum_{x' \neq x^*} \kappa_{x'}^0 & x = x^*, \end{cases}$$

and λ_{x,x^*}^0 exactly as in (5.33).

The right procedure to use depends on how much we know about the problem. In the case of known, common variances, we may prefer to use the simple computations of Bechhofer's procedure. If the variances are unknown, and it is important for us to ensure a certain level of performance in the worst case, the two-stage sequential procedure may be preferable. If the variances are unknown, and we simply wish to do as well as possible, the 0–1 procedure is computationally straightforward.

14.2 OPTIMAL COMPUTING BUDGET ALLOCATION

The value of the indifference zone strategy is that it focuses on achieving a specific level of solution quality, being constrained by a specific budget. However, it is often the case that we are trying to do the best we can within a specific computing budget (as in linear loss or the 0–1 method). This setting has inspired an entire methodology under the name *optimal computing budget allocation*, or OCBA.

Just like the procedures in Section 14.1, OCBA ensures the indifference-zone criterion by requiring the more restrictive condition $P\left(\left|\left(\bar{\theta}_x^2 - \bar{\theta}_y^2\right) - (\mu_x - \mu_y)\right| < \delta\right) = 1 - \alpha$ for all alternatives $x \neq y$ at the end of the second stage. Unlike Bechhofer's procedure, we do not assume that the true values follow the worst-case configuration, but rather work with the distribution of $\bar{\theta}_x^2$ directly. Unlike Paulson's procedure, we do not gradually reduce the choice set. Rather, the OCBA methodology is closer to the 0–1 procedure, assuming instead that the simulation budget is fixed, but that any alternative can be simulated.

14.2.1 Indifference-Zone Version

Figure 14.2 illustrates a typical version of an OCBA algorithm. The algorithm proceeds by taking an initial sample $N_x^0 = n_0$ of each alternative x, which means we use $B^0 = Mn_0$ measurements from our budget B. Letting M be the number of alternatives, we divide the remaining budget of measurements $B - B^0$ into equal increments of size Δ, so that we do $N = (B - Mn_0)\Delta$ iterations.

After n iterations, assume that we have measured alternative x N_x^n times, and let W_x^m be the mth observation of x, for $m = 1, \ldots, N_x^n$. The updated estimate of the value of each alternative x is given by

$$\bar{\theta}_x^n = \frac{1}{N_x^n} \sum_{m=1}^{N_x^n} W_x^m.$$

Let $x^n = \arg\max \bar{\theta}_x^n$ be the current best option.

After using Mn_0 observations from our budget, at each iteration we increase our allowed budget by $B^n = B^{n-1} + \Delta$ until we reach $B^N = B$. After each increment, the allocation N_x^n, $x = 1, ..., M$ is recomputed using

$$\frac{N_x^{n+1}}{N_{x'}^{n+1}} = \frac{\hat{\sigma}_x^{2,n}/(\bar{\theta}_{x^n}^n - \bar{\theta}_x^n)^2}{\hat{\sigma}_{x'}^{2,n}/(\bar{\theta}_{x^n}^n - \bar{\theta}_{x'}^n)^2}, \qquad x \neq x' \neq x^n, \tag{14.6}$$

$$N_{x^n}^{n+1} = \hat{\sigma}_{x^n}^n \sqrt{\sum_{i=1, i\neq x^n}^{M} \left(\frac{N_x^{n+1}}{\hat{\sigma}_i^n}\right)^2}. \tag{14.7}$$

We use equations (14.6)–(14.7) to produce an allocation N_x^n such that $\sum_x N_x^n = B^n$. Note that after increasing the budget, it is not guaranteed that $N_x^n \geq N_x^{n-1}$ for some x. If this is the case, we would not measure these alternatives at all in the next iteration. We can solve these equations by writing each N_x^n in terms of some fixed alternative

Step 0. Initialization:

> **Step 0a.** Given a computing budget B, let n^0 be the initial sample size for each of the M alternatives. Divide the remaining budget $T - Mn_0$ into increments so that $N = (T - Mn_0)/\delta$ is an integer.
>
> **Step 0b.** Obtain samples W_x^m, $m = 1, \ldots, n_0$ samples of each $x = 1, \ldots, M$.
>
> **Step 0c.** Initialize $N_x^1 = n_0$ for all $x = 1, \ldots, M$.
>
> **Step 0d.** Initialize $n = 1$.

Step 1. Compute

$$\bar{\theta}_x^n = \frac{1}{N_x^n} \sum_{m=1}^{N_x^n} W_x^m.$$

Compute the sample variances for each pair using

$$\hat{\sigma}_x^{2,n} = \frac{1}{N_x^n - 1} \sum_{m=1}^{N_x^n} \left(W_x^m - \bar{\theta}_x^n\right)^2.$$

Step 2. Let $x^n = \arg\max_{x=1,\ldots,M} \bar{\theta}_x^n$.

Step 3. Increase the computing budget by Δ and calculate the new allocation $N_1^{n+1}, \ldots, N_M^{n+1}$ so that

$$\frac{N_x^{n+1}}{N_{x'}^{n+1}} = \frac{\hat{\sigma}_x^{2,n}/(\bar{\theta}_{x^n}^n - \bar{\theta}_{x'}^n)^2}{\hat{\sigma}_{x'}^{2,n}/(\bar{\theta}_{x^n}^n - \bar{\theta}_{x'}^n)^2}, \qquad x \neq x' \neq x^n,$$

$$N_{x^n}^{n+1} = \hat{\sigma}_{x^n}^n \sqrt{\sum_{i=1, i \neq x^n}^{M} \left(\frac{N_x^{n+1}}{\hat{\sigma}_i^n}\right)^2}.$$

Step 4. Perform $\max\left(N_x^{n+1} - N_x^n, 0\right)$ additional simulations for each alternative x.

Step 5. Set $n = n + 1$. If $\sum_{x=1,\ldots,M} N_x^n < B$, go to step 1.

Step 6. Return $x^n \arg\max_{x=1,\ldots,M} \bar{\theta}_x^n$.

Figure 14.2   Optimal computing budget allocation procedure.

(other than x^n), such as N_1^n (assuming $x^n \neq 1$). After writing N_x^n as a function of N_1^n for all x, we then determine N_1^n so that $\sum N_x^n \approx B^n$ (within rounding).

In a sense, OCBA can be viewed as a frequentist version of the 0–1 procedure, using different approximations for the probability of correct selection. The one difference is that 0-1 allocates the entire simulation budget in one go, whereas OCBA divides it into small batches (see Steps 4 and 5). Because the probability of correct selection can only be computed approximately, we simulate one small batch at a time to see if the approximate PCS changes in the way we expect.

14.2.2 Linear Loss Version

There is also a Bayesian version of OCBA for the linear-loss objective function (5.29), a very close relative of the $LL(N)$ policy discussed in Section 5.6. For simplicity, let us consider the normal-normal learning model with known sampling variance. That is, each observation $W_x \sim \mathcal{N}\left(\mu_x, \beta^W\right)$, and $\mu_x \sim \mathcal{N}\left(\theta_x^0, \beta_x^0\right)$, and we learn according to (2.5) and (2.6).

Let $x^* = \arg\max_x \theta_x^0$ be the alternative that seems to be the best. Now define the quantities

$$
\begin{aligned}
\delta_x^0 &= \theta_{x^*}^0 - \theta_x^0, \\
\lambda_{x,x'}^0 &= \sqrt{\frac{1}{\beta_x^0} + \frac{1}{\frac{B}{m} + \beta_{x'}^0}}, \\
\ell_x^0 &= \sqrt{\frac{1}{\beta_x^0} + \frac{1}{\beta_{x^*}^0}}.
\end{aligned}
$$

Here, δ_x^0 represents an estimate of the opportunity cost resulting from choosing alternative x instead of x^*. The other two quantities are standard errors of a pairwise comparison of two alternatives, but ℓ_x^0 represents a prior comparison, based only on the time-0 beliefs, and $\lambda_{x,x'}^0$ is a posterior comparison assuming that one of the alternatives has been simulated B/m times.

We then define

$$
D_x = \begin{cases}
\lambda_{x^*,x}^0 f\left(\frac{\delta_x^0}{\lambda_{x^*,x}^0}\right) - \ell_x^0 f\left(\frac{\delta_x^0}{\ell_x^0}\right), & x \neq x^*, \\
\sum_{x' \neq x^*} \lambda_{x',x^*}^0 f\left(\frac{\delta_{x'}^0}{\lambda_{x',x^*}^0}\right) - \ell_{x'}^0 f\left(\frac{\delta_{x'}^0}{\ell_{x'}^0}\right), & x = x^*,
\end{cases}
$$

analogously to η_x^0 in $LL(N)$ or κ_x^0 in 0-1. The function f is the familiar $f(z) = z\Phi(z) + \phi(z)$. Finally, for some number $m \leq M$, we allocate $\frac{B}{m}$ samples for the m alternatives with the lowest values of D_x. We do not simulate any other alternatives.

This version of OCBA should be viewed as an easily computable alternative to $LL(N)$. Essentially, we are computing an expected improvement factor with the approximation that only one alternative is simulated B/m times. While the derivation of the policy assumes that the measurement noise is known, the approximations used for the expected value of the batch are such that the measurement noise never appears in the calculations. The policy can thus be easily applied to problems with unknown measurement noise.

14.2.3 When Does It Work?

The simulation optimization literature contains numerous empirical studies comparing the performance of different procedures. A landmark study by Branke et al. (2007) made several general conclusions:

- Linear loss and OCBA procedures tended to perform efficiently in a variety of settings: the "slippage configuration," when there is one best alternative, and the others all have the same true value; the "monotone decreasing means" configuration, when the true values are evenly spaced; and randomly generated configurations.

- Indifference-zone methods are sensitive to the choice of the indifference-zone parameter δ. They can also "over-deliver," that is, produce a better PCS than what we actually require. Indifference-zone methods also perform better when the number of alternatives is relatively small.

- The performance of Bayesian methods can be improved by using an adaptive stopping rule, rather than cutting off simulation after a fixed time N. For example, we may recalculate the estimated PCS at each time step (accounting for the new observations we have collected), and stop only when this quantity is high enough.

14.3 MODEL-BASED SIMULATED ANNEALING

A recent approach that has attracted attention within the simulation optimization community is known variously as annealing adaptive search (AAS) or model-based annealing random search (MARS). It is essentially a novel take on a familiar heuristic, the Boltzmann exploration policy from Section 17.2. Recall that, under this policy, we would randomly choose to sample alternative x at time n with probability

$$p_x^n = \frac{e^{\rho \bar{\theta}_x^n}}{\sum_{x'} e^{\left(\rho \bar{\theta}_{x'}^n\right)}}.$$

where ρ is a tunable parameter. In simulated annealing, the probability that x is the best alternative is estimated as

$$p_x^{ann,n} = \frac{e^{\rho^n \mu_x}}{\sum_{x'} e^{\left(\rho^n \mu_{x'}\right)}}. \tag{14.8}$$

The definition in (14.8) is based on the frequentist philosophy. If we were to adopt a Bayesian model, the right-hand side of (14.8) would be a random variable with a very complicated distribution. In the frequentist philosophy, we view (14.8) as simply an unknown number that we would like to approximate.

Notice that the tunable parameter is now allowed to depend on n. In simulated annealing, the reciprocal $1/\rho^n$ is referred to as the "temperature" at time n. This is an analogy to materials science. Annealing is a process in metallurgy whereby a metal is alternately heated and cooled to improve the internal energy of its atomic configuration. In simulation, increasing the temperature (decreasing ρ^n) will encourage random exploration, while decreasing the temperature promotes exploitation. In simulated annealing, we choose ρ^n to eventually converge to a large value ρ^* such that our policy will eventually prefer to sample from the true best alternative almost all of the time.

Of course, the main issue is that we cannot sample from the Boltzmann distribution in (14.8). The probabilities depend on the true values μ_x, and if we knew those, there would be no need for random sampling to begin with. Therefore, MARS focuses on approximating (14.8) with what is known as a *surrogate* distribution.

The procedure works as follows. We wish to create surrogate probabilities that are as close to (14.8) as possible. For time n, we define

$$q^n = \arg\min_{q:\sum_{j=1}^{M} q_j=1} \mathbb{E}_{p^{ann,n}}\left[\log \frac{p_X^{ann,n}}{q_X}\right] \tag{14.9}$$

$$= \arg\max_{q:\sum_{j=1}^{M} q_j=1} \sum_{x=1}^{M} e^{(\rho^n \mu_x)} \log q_x. \tag{14.10}$$

In (14.9), X represents a discrete random variable whose probability mass function is given by $P(X = x) = p_x^{ann,n}$ for $x = 1, ..., M$. The expectation $\mathbb{E}_{p^{ann,n}}$ is taken over this discrete distribution. The quantity inside the expectation is known as the *Kullback–Leibler divergence* between $p^{ann,n}$ and q, a measure of the difference between these distributions. (Recall that we have already encountered the KL divergence in Section 13.2.) We minimize the KL divergence, so we are choosing q^n to be as close as possible to the desired Boltzmann probabilities $p^{ann,n}$.

Since μ_x in (14.10) continues to be unknown, we replace the theoretical KL divergence by its sample average and compute an approximation. We suppose that, at time n, we have access to a current set of surrogate probabilities $\tilde{q}^{n-1}$. Then, we randomly sample K alternatives $\tilde{x}_1^n, ..., \tilde{x}_K^n$ from the probabilities $\tilde{q}^{n-1}$, simulate an observation $\tilde{W}_k^n$ for alternative $\tilde{x}_k^n$, and let

$$\tilde{q}^n = \arg\max_{q:\sum_{j=1}^{M} q_j=1} \frac{1}{K} \sum_{k=1}^{K} e^{\rho^n \tilde{W}_k^n} \tilde{q}_{x_k^n}^{n-1} \log q_{x_k^n}. \tag{14.11}$$

We can make our simulations more efficient and increase the accuracy of our approximation by averaging them together with the results of past simulations. It turns out that we can rewrite (14.11) as

$$\tilde{q}^n = \arg\max_{q:\sum_{j=1}^{M} q_j=1} \sum_{x=1}^{M} S_x^n \log q_x, \tag{14.12}$$

where

$$S_x^n = \frac{n-1}{n} S_x^{n-1} + \frac{1}{n}\left[\frac{1}{K} \sum_{k=1}^{K} \exp^{(\rho^n \tilde{W}_k^n)} \tilde{q}_{x_k^n}^{n-1}\right] \tag{14.13}$$

is a recursive sample average of the simulated KL divergences over all time steps up to n. Note that we can define S^0 any way we want (or not define it at all), as no weight is placed on it in the calculation of S^1.

The payoff of this technique is that the optimization problem in (14.12) has a very clean solution

$$\tilde{q}_x^n = \frac{S_x^n}{\sum_{x'=1}^{M} S_{x'}^n}. \tag{14.14}$$

All we really need to do is to keep track of the recursive sample average in (14.13). From there, the surrogate probabilities $\tilde{q}^n$ follow automatically with (14.14).

Just as in Paulson's procedure, we are allowed to simulate the values of K different alternatives at each time step. In fact, if our simulation budget permits, we could simulate multiple replications for each alternative, and replace $\tilde{W}_k^n$ by a sample average. We still need to set the temperatures ρ^n; practitioners of simulated annealing recommend $\rho^n = \sum_{j=1}^n \frac{1}{j}$, which increases to infinity over time.

14.4   OTHER AREAS OF SIMULATION OPTIMIZATION

Our presentation has only scratched the surface of simulation optimization. The field encompasses a wide variety of problems and solutions. For example, a very popular and general technique is stochastic approximation, often applied to problems where the set of possible alternatives or decisions is continuous and possibly multi-dimensional, and we do not know enough about the problem to make some of the strong modeling assumptions used by the other techniques in this chapter (such as known variance). Stochastic approximation follows a recursion

$$x^{n+1} = x^n + \alpha_n \hat{\nabla} f(x),$$

where $\hat{\nabla} f(x)$ is an estimate of the gradient of f, and α_n is a stepsize. In this manner, we gradually improve our guess of the optimal alternative or decision. The literature on stochastic approximation focuses on developing good stepsizes and gradient esti-mation methods. We do not discuss this field here because it takes us away from our domain of adaptively learning a distribution; however, interested readers are referred to Spall (2003). We do, however, discuss optimal learning in the setting of continuous decision spaces in Chapter 16 of this book.

Another popular school of thought studies what is known as *evolutionary* or *genetic algorithms*. As their name suggests, these methods work by analogy to biology. The properties of a solution or alternative are encoded in a *genetic representation*. Different representations are then compared via a *fitness function*. We then select a subset of the decision space where the solutions appear to be fit, and generate new solutions with similar genetic representation (hence the use of the term "evolution"). In short, we select solutions that appear to have high potential, then find similar solutions, focusing more attention on parts of the decision space that appear to be promising.

The key to the algorithm is that the genetic representation takes the form of a fixed-length string of numbers. The fitness value for a string is the simulation output for the corresponding alternative. Optimizing the function may be difficult, but we can generate a few candidate solutions, then focus on the ones with the highest fitness. To generate a new solution, we can take numbers from two different strings, and randomly change (or "mutate") some of them to encourage exploration. One way of doing this is to splice two strings together, that is, combine the first half of one string and the second half of another string to create a new candidate solution.

The method allows us to handle potentially immense problems. Consider a manufacturing application where we have to route a new part through a production process. At each stage of the process, we can choose one of several machines to handle the part, and we can also choose a layout for the whole system. The genetic representation of a single decision is a string containing the chosen machine for each stage, as well as the characteristics of the chosen routing policy. If we have six possible alternatives for each of 13 components of the decision (six machines per stage, six routing decisions, six factory layouts), the number of possible decisions is 6^{13}. The algorithm never has to enumerate these alternatives.

At the same time, the method can be slow, and it is difficult to guarantee good performance. The algorithm may converge to an arbitrary solution, not necessarily even locally optimal. There is no explicit belief model on the fitness of previously untested solutions; rather, we simply hope that combining two solutions that seem to be good, in a heuristic way, will produce an even better solution. Thus, one might view the genetic algorithm as falling outside the scope of optimal learning, but it is an important technique that can make progress on otherwise intractable problems.

14.5 BIBLIOGRAPHIC NOTES

Section 14.1 - A good introduction to simulation optimization is available in Hong & Nelson (2009), which briefly surveys the most important classes of algorithms in this field. Applications are routinely surveyed in the proceedings of the Winter Simulation Conference; see Fu et al. (2005) for a recent example. Better et al. (2008) discusses some applications in risk management, including the health care example used in the beginning of this chapter. Bechhofer's procedure originates from Bechhofer (1954), and is generalized to the case of unequal variances by Rinott (1978). The two-stage batch procedures for the case of unknown variance are set forth in Matejcik & Nelson (1995) and Nelson & Matejcik (1995). The sequential procedures come from Paulson (1964) and Kim & Nelson (2001). See also Kim & Nelson (2006) for further extensions. The monograph by Gosavi (2003) offers a nice introduction to some of these methods. The 0-1 procedure is due to Chick & Inoue (2001).

Section 14.2 - There is a rich literature on OCBA methods; for an introduction, see the monograph by Chen & Lee (2010). The basic algorithm in Figure 14.2 is described, for instance, by Chen et al. (2008). Variations and improvements on the OCBA concept were developed by Chen et al. (2000) and Chen et al. (2006). The Bayesian version of OCBA for linear loss is due to Chick et al. (2007), with a theoretical analysis in Frazier & Powell (2011). There are also extensions of OCBA for handling correlated observations; see Fu et al. (2004) for an example. Branke et al. (2007) undertook an extensive empirical comparison of OCBA to other policies, including linear loss and indifference-zone methods.

Section 14.3 - Annealing adaptive search was introduced by Romeijn & Smith (1994). The model-based adaptive variant that we discuss here is due to Hu &

Hu (2010), Hu & Hu (2011), and Hu & Wang (2011). We chose to focus on this version because of the connection to Boltzmann exploration. However, there are also other versions of simulated annealing; for example, see Alrefaei & Andradóttir (1999). Other adaptive random search methods include the model reference adaptive search approach of Hu et al. (2007), also described in Chang et al. (2007).

Section 14.4 - Stochastic approximation first appeared with the seminal paper of Robbins & Monro (1951). A good introduction to stochastic approximation and search is the book by Spall (2003). Kushner & Yin (2003) provides an in-depth theoretical treatment. See also Pasupathy & Kim (2011) for an overview from the perspective of stochastic root-finding. A powerful tool for practical simulation optimization, based on random search methods, is the COMPASS algorithm of Hong & Nelson (2006). With regard to genetic algorithms, a general overview can be found in Davis & Mitchell (1991). See also Goldberg & Deb (1991) for an overview of selection methods, i.e. how to choose the most fit solutions for reproduction. Applications of genetic algorithms to simulation optimization include Azadivar & Tompkins (1999), Azadivar & Wang (2000), Hedlund & Mollaghasemi (2001), and Can et al. (2008). An overview of applications in operations management is given in Chaudhry & Luo (2005). The manufacturing example of a genetic algorithm is from Azadivar & Tompkins (1999).

CHAPTER 15

LEARNING IN MATHEMATICAL PROGRAMMING

There are applications where we have to collect information to be used in a larger optimization problem such as a shortest path problem or linear program. Ranking and selection can be viewed as the solution to a simple linear program

$$v^n = \max_x \sum_{i=1}^{M} \mu_i x_i,$$

subject to

$$\sum_{i=1}^{M} x_i = 1, \quad i \in \mathcal{I},$$
$$x_i \geq 0.$$

The solution to this linear program requires sorting (μ_i) for all i and choosing the index i^* with the largest value of μ_i, which gives us a solution where $x_{i^*} = 1$, and $x_i = 0$, $i \neq i^*$. Our learning challenge is to choose an element j to measure, producing a value $v^{n+1}(j)$ computed using the vector μ_i. The goal is to choose the index j that maximizes the expected value of v^{n+1}. Note for this discussion that we

Optimal Learning. By Warren B. Powell and Ilya O. Ryzhov
Copyright © 2012 John Wiley & Sons, Inc.

are using indices i and j for our choices, and we are switching (for this chapter) to x as the vector of implementation decisions.

We can take this perspective one step further. Assume now that we have a more general linear program which we write as

$$\min_x \sum_{i \in \mathcal{I}} c_i^T x_i,$$

subject to

$$Ax = b,$$
$$x_i \geq 0, \quad i \in \mathcal{I}.$$

Here, x is a vector with potentially hundreds or even tens of thousands of dimensions. More significantly, the A matrix might be general, although there are specific problem structures that are of interest to us.

Linear programming has numerous applications, including a broad class of problems on networks, where the matrix A is used to represent flow conservation equations. The objective can be to minimize the travel distance across the graph (shortest-path problems) or to efficiently move resources between nodes (such as the "transportation problem" or general network flow problems). Many other optimization problems can also be expressed as linear programs. A classic example is the problem of formulating a production plan to maximize profit. A company produces M products, and wishes to produce x_i units of product $i = 1, ..., M$ to maximize the total profit $c^T x$ with c_i being the profit from selling one unit of product i. The products are made using J different resources, with b_j being the total amount of resource j available. The matrix A denotes the production constraints, with $A_{j,i}$ being the amount of resource j needed to create one unit of product i.

In the basic LP model, we assume that the parameters A, b, and c are known. In reality, we are unlikely to know them exactly, just as we do not know μ in ranking and selection. Suppose that our company has developed a new product, and now needs to decide how much of it should be produced. We may have an estimate of c_i based on some preliminary data. Perhaps our sales figures for our old line of MP3 players may give us some idea of the profitability of the new line. At the same time, we are still quite uncertain about the true value of c_i. But now, suppose that we have the ability to collect additional information about c_i before we commit to a production plan x. Perhaps we have a chance to run a test market where we can get a sense of how well product i might perform. The results might change our estimate of c_i, thus affecting our final production plan. Given our limited time and money, which products should be included in the test market? Or, in other words, which information will help us make the best possible decision?

Optimal learning has a role to play here, but it is not a simple matter of picking up and applying the formulas and algorithms from Chapters 4 and 5. In mathematical programming applications, our *measurement decision* (say, measuring the coefficient c_i) is distinct from our *implementation decision*, which we now represent as the vector x. We may learn about a single product or a single region in the network, but our overall goal is to solve an optimization problem. The coefficients $i = 1, ..., M$ in our LP are a bit like the "alternatives" from Chapter 4, in that we have a choice of which coefficients to learn about, but we are not simply interested in finding the largest coefficient. The part should teach us about the whole; we should learn about those coefficients that contribute the most to our ability to optimize. Running a test market for product i should lead us to a better production plan.

Although we cannot recycle the knowledge gradient formulas from Chapter 5 in a straightforward way, we can still apply the concept of the knowledge gradient. In this chapter, we show how the value of information can be expressed in problems with uncertain objective functions. After describing several applications, we reinterpret the knowledge gradient concept, first for simple shortest path problems, and then for more general linear programs.

15.1 APPLICATIONS

In this section, we introduce three models where the implementation decision requires solving a mathematical program. We begin with a model for piloting a hot air balloon, where the challenge is moving the balloon up and down to find the fastest wind. Given a current set of estimates about the wind, the balloon can solve a linear program to find the best path to the destination. Then, we describe a common problem in the investment community where it is necessary to visit different companies to learn about their potential for future profits. The challenge is determining which companies to visit (we do not have the time to visit all of them), after which we have to take estimates of expected returns, variances and covariances to determine the best portfolio. We close with a discussion of graph problems.

15.1.1 Piloting a Hot Air Balloon

Consider the problem faced by a balloonist trying to get from one point to another in the presence of changing winds that can be different at different altitudes. Her ability to make progress depends on finding wind that moves in the right direction. She observes the speed and direction of the wind at her current location, but she may wonder if the conditions are better at a different altitude. Observing the wind at one altitude provides some information about the wind at other altitudes, as well as the wind at later points in time. But it takes time and (if she has to raise her altitude) energy to make these observations.

The Model Optimizing the trajectory of a hot air balloon is a fairly difficult control problem, even when we do not include learning as a dimension. For our purposes, we are going to simplify the problem by assuming that our balloon can only move in two dimensions: forward and up or down. Our interest is in finding a path that balances distance against changing (and uncertain) wind speeds so that we get to the destination as quickly as possible.

Our view of the problem is depicted in Figure 15.1, where our balloon is currently at horizontal location x and vertical location y and is faced with one of three decisions:

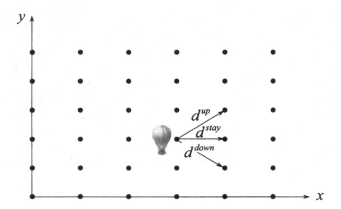

Figure 15.1   Configuration of a two-dimensional hot air balloon trajectory.

d^{up} takes the balloon on an upward trajectory, d^{stay} holds the balloon at its current altitude while d^{down} takes the balloon on a downward trajectory. Although we would like to model energy expenditure, we are only going to capture the time required to complete each decision. We assume that the balloon moves 30 percent slower moving up, and 20 percent faster moving down.

The physical state $R^n = (x^n, y^n)$ captures the location of the balloon. Our state of knowledge about the wind is given by $K^n = (\theta^n_{xy}, \sigma^{2,n}_{xy})_{x,y}$ where θ^n_{xy} is what we think the wind speed is at time n, location x, y, and $\sigma^{2,n}_{xy}$ is the variance in our distribution of belief. When the balloon is at location (x, y) at time n, the true wind speed is μ^n_{xy} which we measure with noise according to

$$W^n_{xy} = \mu^n_{xy} + \varepsilon^n.$$

We assume measurements are independent across space and time, and are normally distributed with mean 0 and variance λ^ϵ. The true wind speed also evolves over time according to

$$\theta^{n+1}_{xy} = \mu^n_{xy} + \hat{\mu}^{n+1}_{xy},$$

where $\hat{\mu}^{n+1}_{xy}$ describes the actual change in the wind from time period to time period. The random changes in the wind, $\hat{\mu}^n_{xy}$, are correlated across both space and time, since if the wind is high at time n, location (x, y), then it is likely to be high at nearby locations, and at the same location at later points in time. The covariance between wind observations at two different altitudes at the same point in time is given by

$$Cov(W^n_{xy}, W^n_{xy'}) = \sigma^2_W e^{-\beta_y |y-y'|} + \lambda_{xy}.$$

Similarly, we assume the covariance between a location x and a location $x' > x$ at the same altitude is given by

$$Cov(W^n_{xy}, W^n_{xy'}) = \sigma^2_W e^{-\beta_x |x-x'|} + \lambda_{xy}.$$

We combine these two to give us a general covariance structure

$$Cov(W_{xy}^n, W_{xy'}^n) = \sigma_W^2 e^{-\beta_y|y-y'|-\beta_x|x-x'|} + \lambda_{xy}.$$

Using this function, we can construct a covariance matrix Σ^n with elements $\Sigma_{xy,x'y'}^n$. We use this structure to initialize our covariance matrix Σ_0. Next let e_{xy} be a vector of 0's with a 1 corresponding to the location (x, y). The general formulas for updating the vector of estimates of the mean and covariance matrix for our distribution of belief about the wind is given by

$$
\begin{aligned}
\theta^{n+1} &= \Sigma^{n+1}\left((\Sigma^n)^{-1}\theta^n + (\lambda_{xy})^{-1}W^{n+1}e_{xy}\right), \\
\Sigma^{B,n+1} &= \left((\Sigma^n)^{-1} + (\lambda_{xy})^{-1}e_{xy}(e_{xy})'\right)^{-1}.
\end{aligned}
$$

This gives us the updated estimate of the mean and variance, where the variance takes into account the measurement error. But this ignores the changing velocity due to $\hat{\mu}^n$. When we take this into account, our updated covariance matrix is given by

$$\Sigma^{n+1} = \Sigma^{B,n+1} + \Sigma^{\mu}.$$

So, $\Sigma^{B,n+1}$ will shrink relative to Σ^n, but it will grow due to the effect of Σ^μ. We note that this is not a very realistic model for wind. For example, if we did not do any measurements (and possibly even if we do), the covariance matrix will grow without bound. However, this model will illustrate an interesting learning problem without becoming buried in the details of the physics.

Planning a Trip The next step is to plan a trip given what we know now. We use our current estimates of speeds at each altitude. Without additional information, our best estimate of the speed in the future is the same as it is now, for each altitude, which means that $\theta_{xy}^n = \theta_{x'y}^{n'}$ for $n' \geq n$ and $x' \geq x$. Using the assumption that speeds are 30 percent slower going up and 20 percent faster going down, we can construct a speed graph over the entire distance, similar to what is depicted in Figure 15.2. If we convert speeds to travel times, we now have a simple shortest path problem to find the fastest path from the current location to the end (presumably at an altitude of 0).

This problem can be formulated as a linear program by defining the decision variable

$$
u_{xy,x'y'} = \begin{cases} 1 & \text{if we decide to make the move from } (x, y) \text{ to } (x', y'), \\ 0 & \text{otherwise.} \end{cases}
$$

This entire vector is chosen at time n and can be thought of as the current planned trajectory given what we know now. From a particular point (x, y), we assume that there are only three points that can be reached, by using the decisions up, down or stay. Let $c_{xy,x'y'}(\mu^n)$ be the time required to move from (x, y) to (x', y'). Our linear program would be given by

$$F^n(\mu^n) = \min_u \sum_{x,y} \sum_{x',y'} c_{xy,x'y'}(\mu^n) u_{xy,x'y'}, \tag{15.1}$$

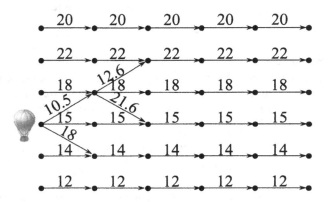

Figure 15.2   Speeds for each decision at each (x, y) location given the current state of knowledge.

subject to

$$\sum_{x'y'} u_{xy,x'y'} = R_{0y}, \tag{15.2}$$

$$\sum_{x'y'} u_{xy,x'y'} - \sum_{xy} u_{xy,x'y'} = 0, \tag{15.3}$$

$$u_{xy,x'y'} \geq 0. \tag{15.4}$$

Here, $R_{0y} = 1$ if the balloon starts the process at an altitude y. Equation (15.3) requires that if the balloon arrives to (x, y), then it must also leave. Equation (15.4) requires that the flows be nonnegative. We would also like the decision u_t to be 0 or 1, but we will get this as a natural result of the structure of the problem.

We are assuming that we are solving the problem deterministically, using our current estimates of the speeds. We could put probability distributions around the speeds and solve a stochastic optimization problem, but this would not contribute to our understanding of how to handle the learning aspects of the problem.

Learning While Flying   Imagine that the wind at your altitude has dropped significantly. You are, of course, wondering if the wind is better at a different altitude. If we move to (x', y'), we are going to observe $W_{x'y'}^{n+1}$. Because of the covariance structure, this observation will allow us to update our estimate of the entire vector θ^n giving us $\theta^{n+1}(W_{x'y'}^{n+1})$. This would allow us to solve the problem $F^{n+1}(\theta^{n+1}(W_{x'y'}^{n+1}))$ using our new state of knowledge. We have three possible decisions (up, down, or stay). If we want to use pure exploitation, we would solve the problem $F^n(\theta^n)$, which would tell us which action to take. But this ignores the potential value of learning. As an alternative, we could choose the best value of $(x'y')$ (corresponding to one of the three decisions) to solve

$$\max_{(x'y')} \mathbb{E}^n \left\{ F^{n+1}(\theta^{n+1}(W_{x'y'}^{n+1})) - F^n(\theta^n) \right\}. \tag{15.5}$$

We have written this function in the form of the knowledge gradient, recognizing that $F^n(\theta^n)$ is a constant (given what we know at time n) which does not affect the solution.

When we first introduced the knowledge gradient, we presented it in the context of ranking and selection where we had to choose from among a small, discrete set of choices. Now, we have to solve a linear program for each possible measurement, which means we can no longer compute the expectation exactly. But we can take advantage of the fact that we have a small number of potential measurement decisions (up, down or stay), which means we effectively face a ranking and selection problem. We cannot compute the expectation exactly, but we can use Monte Carlo methods. Let $W_{x'y'}^{n+1}(\omega)$ be a sample observation of the wind if we were to choose to go to location $(x'y')$. This is not a true observation - it is a simulated observation for the purpose of approximating the expectation.

Given this observation, we can update the mean and covariance matrix and solve the linear program. Note that this observation does not just change the speed out of node $(x'y')$ (take another look at Figure 15.2); it changes estimates of speeds over the entire network, partly because of the presence of correlations, and partly because we solve our linear program by assuming that as we step forward in time, we do not collect additional information (which means that speeds at the same altitude for larger values of x' are the same). Let $F^{n+1}(x'y'|\omega)$ be the resulting "sample observation" of $\mathbb{E}^n F^{n+1}(\theta^{n+1}(W_{x'y'}^{n+1}))$. We could then repeat this, say, 100 times, for each possible decision and take an average. We could then make the decision that has the highest estimated value.

Sound familiar? This is precisely the ranking and selection problem. Solving the linear program 100 times and taking an average is a bit clumsy (famously known as the "brute force" solution). A more elegant approach would be to use (drum roll please) the knowledge gradient algorithm to choose the best decision. This would allow us to avoid making 100 measurements of a choice that does not look promising. We can build a prior by initially evaluating each choice assuming that we do not learn anything (the wind does not change from the prior), so that we start with a solution built around the pure exploitation policy.

Optimal Learning Versus Stochastic Optimization It is useful to contrast our learning algorithm versus what we might have done if we were solving this as a stochastic optimization problem. In a stochastic optimization model, we would acknowledge that we do not know what wind we will face when we arrive to a location $(x'y')$ (above, we solved the problem deterministically). For example, a deterministic solution might take us closer to the ground if the wind seems to be fastest there. However, a stochastic solution might recognize that the wind near the ground might be much lower than expected, and if we are close to the ground, the only place to go is up (at a high cost). A stochastic solution to this problem is relatively difficult, and probably requires the techniques of approximate dynamic programming.

The issue of whether to use a deterministic or stochastic algorithmic strategy did not arise with the simpler problems such as ranking and selection, because the optimal solution was to pick the index $j*$ with the highest expected value of θ_j^N. Our stochastic shortest path problem, however, is different because it involves finding a path over a sequence of steps with uncertain costs. Even if we fix our distribution of belief, it is still a stochastic optimization problem.

The essential difference between a stochastic optimization algorithm and a learning algorithm is that with a learning algorithm, we accept that observations of the wind (the external measurement) come from an exogenous distribution that may be different from our true distribution of belief. As a result, we use these observations to update our distribution of belief. With a classical stochastic optimization formulation, we acknowledge that we do not know the actual wind, but we do know the probability distribution, which means that observations do not change our distribution of belief about the distribution.

15.1.2   Optimizing a Portfolio

Imagine that you are looking at investing in a group of companies. The companies are organized by industry segment (retailing, financial services, transportation), so we will refer to the jth company within the ith industry segment. We start by assuming that θ_{ij}^0 is the expected return (over some period of time) for company j in segment i. We also let Σ^0 be the matrix that captures the covariances in our beliefs about μ_{ij}, where the diagonal elements $\Sigma_{ij,ij}$ represent the variance.

In general, we assume that the covariance between two companies has the structure

$$\Sigma_{ij,i'j'} = \sigma_0^2 + \sigma_i^2 1_{\{i=i'\}}.$$

Thus, σ_0^2 captures the common covariance (e.g. the extent to which companies respond to the general economy), and σ_i^2 captures the common covariance for companies within industry segment i.

Before we invest in a company, we have the ability to visit the company and learn more about the management team, marketing plans and facilities. We might assume that our observation W of each company is independent with variance λ_{ij}. Let $x_{ij} = 1$ if we visit company (i,j). If we choose to visit company (i,j), we can update the mean and variance using our standard formulas, giving us an updated $\theta_{ij}^1(x)$. If $x_{ij} = 0$, then $\theta_{ij}^1(x) = \theta_{ij}^0$, and we let $\Sigma^1(x)$ be the updated covariance matrix.

Given the vector x of visits (presumably chosen to a constraint on how many visits we can make), we then have to find an optimal portfolio. Let y_{ij} be the amount that we are going to invest in company (ij). We do this by solving the standard quadratic programming problem

$$F(x) = \mathbb{E}F(x, W) = \min_y \left(\theta^1(x)y + \theta y^T \Sigma^1(x)y \right), \tag{15.6}$$

subject to

$$\sum_{ij} y_{ij} = R^I, \tag{15.7}$$

$$y_{ij} \geq 0. \tag{15.8}$$

Here, x is our measurement decision, W captures what we observe if we choose to make the measurement, and y is our implementation decision. R^I is the amount of money we have to invest. The measurement problem is given by

$$\min_x \mathbb{E} F(x, W), \tag{15.9}$$

subject to

$$\sum_{ij} c_{ij} x_{ij} = R^M, \tag{15.10}$$

$$x_{ij} \geq 0 \ \text{ and integer.} \tag{15.11}$$

Here, R^M is our measurement budget. The measurement problem is hard to solve because it requires integer solutions, but primarily because of the problems computing the expectation in (15.9). We suggest simply finding the marginal value of making each visit, and then assigning visits in order of decreasing marginal value until the budget is exhausted. This is a type of batch ranking and selection problem, where instead of finding the best choice, we find the best set of choices.

Even this simple heuristic is hard to implement. Imagine that there are 100 companies that we are considering, and we have the budget to make 20 visits. What are the best 20 companies to visit? We are not able to compute the expectation $\mathbb{E} F(x, W)$ in (15.6) either. Instead, we can choose a company to visit, then randomly sample what we might learn $(W_{ij}(\omega))$. We can perform 100 samples of each company and take an average to obtain an estimate of the value of visiting each company, but this may be computationally demanding (and 100 samples may not be enough). Or, we can use the knowledge gradient algorithm to allocate our computational budget across the 100 companies.

15.1.3 Network Problems

Optimization on a graph is a well-known class of mathematical programming problems, with applications in transportation, project management, telecommunication and other areas. A graph is described by a set $\mathcal{V}$ of nodes (or "vertices") and a set $\mathcal{E} \subseteq \mathcal{V} \times \mathcal{V}$ of edges, where each edge connects two nodes. Figure 15.3 provides a visual illustration. It is easiest to think of a graph as a map of physical locations: For example, nodes could represent cities and edges could stand for major highways connecting those cities. An edge $(i, j) \in \mathcal{E}$ between nodes i and j carries a cost (or "length") c_{ij}. The cost can be expressed in terms of money or time. For example, we might focus on the economic cost of transporting a shipment of goods from one city to another, or on the time needed for the shipment to reach its destination.

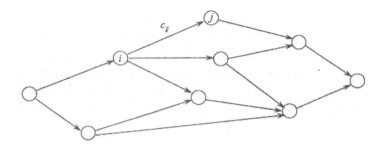

Figure 15.3   A generic network, with nodes i and j connected by an edge of length c_{ij}.

Implementation Decisions in Networks The shipment problem is a well-studied application of *minimum-cost flows*, a model for optimizing the route of resources across a network. A number b_i represents either the supply or demand for a commodity at node i. If $b_i > 0$, then i is a supply node (perhaps we have a production facility in city i, or we purchase from a manufacturer located in this city). If $b_i < 0$, then i is a demand node (perhaps retailers in this city are selling our products). If $b_i = 0$, the node is a transit node and our product can be transported through it en route to a demand node.

The cost of transporting a single unit of product from node i to node j is given by c_{ij}. The objective function can be written as

$$\min_x \sum_{(i,j)\in\mathcal{E}} c_{ij}x_{ij},$$

where x_{ij} represents the quantity of product that we wish to ship from i to j. Obviously we require $x_{ij} \ge 0$ for all i, j.

Each node is subject to a flow conservation constraint. Product can only enter the network through a supply node and leave through a demand node. Thus, the total product leaving node j has to be equal to the amount that entered node j, plus the amount that was supplied (or, minus the amount that was used to satisfy demand). We can write this constraint as

$$b_j + \sum_{i:(i,j)\in\mathcal{E}} x_{ij} = \sum_{i:(j,i)\in\mathcal{E}} x_{ji}, \quad j \in \mathcal{V}.$$

The implementation decision in this problem is the flow schedule, the values of x_{ij} that we choose for all edges (i,j). The objective function and constraints are both linear in x_{ij}, making this a classic application of linear programming. There are many variations of the problem. For example, we might add capacity constraints $x_{ij} \le d_{ij}$ if there is a cap on the amount that we can ship along a particular route (perhaps only a limited number of trucks is available). We may also add more flexibility to our model by giving ourselves the option to use only part of the available supply, in a situation where we have a choice of suppliers. In this case, for $b_j > 0$, we can add new decision variables $0 \leq y_j \leq b_j$ denoting the amount of supply we used, and replacing b_j with y_j in the corresponding flow constraints. We would also incorporate additional purchase costs into our objective function.

In some applications, we may not always know the costs c_{ij} exactly. In a supply chain, costs may depend on numerous factors. They are affected by purchase and production costs, but also by more vague factors like the reliability of a supplier. Perhaps a particular supplier has a high chance of experiencing shortages, or a particular transit route is more likely to encounter delays. Eventually, the shortages and delays will be passed down to us, increasing our own costs. The distribution of the costs, or even their expected value, may be difficult to quantify, and has to be learned by doing business with a particular supplier or shipping along a particular transit route. We will need to model our beliefs about the costs and think about which suppliers we would like to experiment with in order to arrive at the best shipment schedule.

A special case of minimum-cost flows is the well-known shortest path problem. In this setting, there is a single supplier s with $b_s = 1$, and a single demand node t with $b_t = -1$. All other nodes are transit nodes, and the cost c_{ij} represents the "length" of edge (i, j). Our goal is to find a route from s to t with the shortest possible length. Like all the other network problems discussed here, the shortest-path problem can be formulated and solved as a linear program. There are also many optimization algorithms that are specifically tailored to shortest-path problems. We give one well-known example, the Bellman–Ford algorithm. Given a cost vector c and specified source and destination nodes s, t, the algorithm calculates the distance $V(i; c)$ from any node i to t in the following way:

1) Let $V(s; c) = 0$. For all other nodes $i \neq s$, let $V(i; c) = \infty$.

2) For each edge $(i, j) \in \mathcal{E}$, if $V(i; c) + c_{ij} < V(j; c)$, let $V(j; c) = V(i; c) + c_{ij}$.

3) Repeat step 2) a total of $|\mathcal{V}| - 1$ times.

4) If $V(i; c) + c_{ij} < V(j; c)$ for any $(i, j) \in \mathcal{E}$, the graph contains a negative-cost cycle and the shortest path length is $-\infty$ (corresponding to an unbounded LP). Otherwise, $V(t; c)$ gives the length of the shortest path from s to t.

Optimal learning becomes an issue when the lengths c_{ij} are unknown. Length is often interpreted as travel time. For example, a GPS navigation system finds the quickest way to get from a specified origin to a specified destination. However, in reality, travel times are highly variable. GPS relies on certain estimates of travel time, but they may be outdated or inaccurate when a particular query is received. However, in the limited time available to make a decision, we may have an opportunity to collect a very small amount of information to help us choose a travel route. Some drivers located in different regions of the traffic network may be signed up to participate in a program where their smartphone can send information on traffic congestion (using the ability of the phone to tell how fast it is moving) to the GPS. We can choose to query a small number of smartphones, receive real-time observations of traffic congestion, and use them to update our estimates and improve our implementation decision (the route we choose).

Other Network Representations The graph does not have to represent a physical structure. Graphs are used in project management to represent precedence relations between different tasks or assignments. Consider the problem of managing a software project that consists of a series of tasks. Some tasks must be completed in a specific order, whereas others can be tackled in parallel. Figure 15.4 displays precedence relationships for the following set of tasks:

A	Build preliminary demo	F	Compatibility testing/QA
B	Alpha testing, user feedback	G	Train support staff
C	Preliminary market study	H	Final marketing campaign
D	Finalize design	I	Begin production
E	Plan production/distribution	J	Ship to retailers and launch

Each node in the graph represents a task, whereas an edge between tasks i and j indicates that i must be completed before j. In addition, each task has a duration d_i. Given T tasks, we can schedule the tasks in the following way. Let x_i be the starting time of task i and solve

$$\min_x x_T,$$

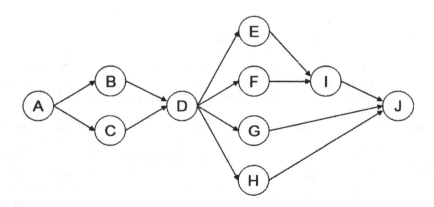

Figure 15.4   Graph of precedence relations for tasks in software development.

subject to

$$t_j \geq t_i + d_i \quad \text{for all } (i, j) \in \mathcal{E},$$
$$t_i \geq 0 \quad \text{for } i = 1, ..., T.$$

We minimize the starting time of the last job. Another problem that often arises in project scheduling is known as the critical path problem. This is essentially the same as finding the longest path (the path with the largest total duration) from the first to the last job. The critical path is of interest to managers because it shows which jobs can be delayed without impacting key deadlines.

The duration of a particular task is rarely known exactly. Preliminary estimates are often inaccurate, leading to unplanned delays. Before beginning the project, we may wish to improve our estimates by going into our archives and analyzing data from earlier projects that involved similar tasks. The data may be cumbersome to access (many such archives have yet to be digitized), so our time is limited. It then becomes important to use our learning budget effectively by singling out the tasks that seem most important, and focusing on creating good estimates of their duration.

15.1.4 Discussion

We have just seen several problems where the decision problem (we sometimes call this the implementation problem) is a linear or nonlinear program. Finding the expected value of a measurement can be computationally difficult. But we have seen that we can apply our optimal learning ideas within a learning algorithm to help us find a good measurement policy.

Both of these problems could be formulated in the general framework of finding a measurement decision x to minimize $\mathbb{E}F(x, W)$. This is a very classical optimization problem with a rich algorithmic history. There are many algorithms which depend on the structure of the problem. An excellent reference for this problem class is Spall (2003) and the references cited there (the literature is extremely large).

15.2 LEARNING ON GRAPHS

We show how optimal learning can be incorporated into network optimization using, the shortest-path problem from Section 15.1.3 as an example. To make our analysis cleaner, we assume that the graph has no cycles, as in Figure 15.3.

Optimal learning begins to play a role when we no longer know the cost vector c. Let us apply our standard modeling assumptions from Chapter 5. Suppose that, for each edge $(i, j) \in \mathcal{E}$, we have $c_{ij} \sim \mathcal{N}\left(\theta_{ij}^0, \beta_{ij}^0\right)$, where β^0 denotes precision. We put a Gaussian prior on the length of each edge, and assume that the edges are independent. This is a strong assumption: For example, if the graph represents a traffic network, as in some of the examples discussed in Section 15.1.3, one would expect the travel times on neighbouring streets to be heavily correlated. We will allow correlations in Section 15.4. For now, we focus on the insights that can be obtained if we make the independence assumption.

As in ranking and selection, the knowledge state in this problem can be written as $S = (\theta, \beta)$, a vector of means and variances. A set of beliefs about the edges induces a belief about the best path. Define $V^n = V(s; \theta^n)$ to be the estimated length of the shortest path, given our time-n beliefs about the edge lengths. In other words, we simply run the Bellman–Ford algorithm using the estimates θ^n as the costs, and use the quantity V^n as our time-n estimate of the shortest path. Bellman–Ford can also provide us with the path that achieves this estimated length. We refer to this path as p^n.

Now suppose that, before we commit to a path through the network, we have N opportunities to collect information about individual edges. If we choose to learn about (i^n, j^n) at time n, we collect an observation $W^{n+1} \sim \mathcal{N}\left(c_{ij}, \beta_{ij}^W\right)$ and our beliefs evolve according to the familiar updating equations

$$\theta_{ij}^{n+1} = \begin{cases} \frac{\beta_{ij}^n \theta_{ij}^n + \beta_{ij}^W W_{ij}^{n+1}}{\beta_{ij}^n + \beta_{ij}^W} & \text{if } (i, j) = (i^n, j^n), \\ \theta_{ij}^n & \text{otherwise}, \end{cases} \tag{15.12}$$

$$\beta_x^{n+1} = \begin{cases} \beta_{ij}^n + \beta_{ij}^W & \text{if } (i, j) = (i^n, j^n), \\ \beta_{ij}^n & \text{otherwise}. \end{cases} \tag{15.13}$$

Just as in ranking and selection, we change our beliefs about one edge at a time. However, a small change in our beliefs about a single *edge* can have a profound effect on our beliefs about the shortest *path*. Figure 15.5(a) shows the graph from Figure 15.3, with the addition of prior beliefs about the edge lengths, represented by bell

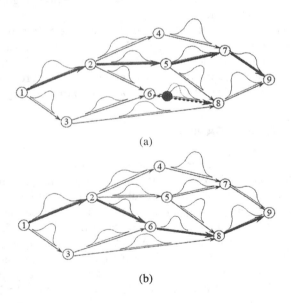

(a)

(b)

Figure 15.5   (a) The effect of a measurement of a single edge (dotted line) on our beliefs about the shortest path (solid line). (b) The updated shortest path given a particular measurement.

curves. The solid line in Figure 15.5(a) represents the path from node 1 to node 9 that we currently believe is the shortest. When we measure the edge $(6, 8)$, the outcome of our observation is smaller than expected, leading us to shift that particular bell curve to the left. Because this edge now seems shorter than before, we prefer to route our path through it, leading to the new solution in Figure 15.5(b). We only changed our beliefs about one edge, but the shortest path now looks very different from before.

Our objective is to choose a policy π for allocating the N measurements in a way that minimizes

$$\min_{\pi} \mathbb{E}F^{\pi}(c, W), \qquad (15.14)$$

where $F^{\pi}(c, W) = V^N$, the final time-N estimate of the length of the shortest path. Equation (15.14) reflects the key difference between learning on a graph and ranking and selection: we measure individual edges ("alternatives") just as before, but now we are no longer interested in finding the single alternative with the best value. We seek to measure alternatives that help us improve our solution to the shortest-path problem.

To see how this can be done, let us examine the impact of a single measurement on our solution, that is, the difference between V^n and V^{n+1} brought about by measuring (i, j). For a fixed edge (i, j) define two paths as follows. The path p_{ij}^n is defined to be the shortest path *containing* (i, j), based on the time-n beliefs, whereas $\bar{p}_{ij}^n$ is the shortest path *not containing* (i, j), again according to the time-n beliefs. Both paths are easy to find. For p_{ij}^n, we can run Bellman–Ford to find the shortest path from s to i, and again for the shortest path from i to t. We then use (i, j) to connect these two paths. To find $\bar{p}_{ij}^n$, we merely run Bellman–Ford once on a modified graph with the edge (i, j) removed. Figure 15.6 gives a visual illustration.

Keep in mind that these two paths also depend on our beliefs at time n. When we run Bellman–Ford to find p_{ij}^n and $\bar{p}_{ij}^n$, we use θ^n for the edge costs. The algorithm will also provide us with V_{ij}^n and $\bar{V}_{ij}^n$, the estimated lengths of both paths. A length of a path is simply the sum of the estimates θ_{ij}^n on every edge (i, j) in the path.

The key insight of this section is the following. If we measure (i, j) at time-n, the path p^{n+1} that will be optimal under the time-$(n+1)$ beliefs (that is, the path that we will get by running Bellman–Ford with the costs c^{n+1} at time $n+1$) will necessarily be either p_{ij}^n or $\bar{p}_{ij}^n$. To understand this result, it helps to think in the following way.

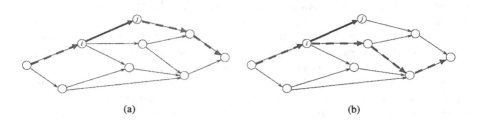

(a)   (b)

Figure 15.6 Illustration of (a) the best path containing (i, j) and (b) the best path not containing (i, j).

Suppose that (i, j) is already part of the current optimal path p^n. Then, if we measure it, our beliefs about it will either improve (in which case p^n will become even better than before) or get worse to the point where it is no longer advisable to route a path through (i, j). In the latter case, the best choice among the remaining paths is $\bar{p}_{ij}^n$. Similarly, if (i, j) is not part of p^n, our beliefs about it will either get worse (in which case we will keep the current optimal solution), or get better to the point where we will choose to route our path through (i, j). In the latter case, p_{ij}^n will be the best such route.

This argument allows us to calculate

$$\nu_{ij}^{KG,n} = \mathbb{E}\left[V^n - V^{n+1} \mid S^n, (i^n, j^n) = (i, j)\right], \qquad (15.15)$$

the expected improvement contributed to our estimated length of the shortest path by a single measurement of (i, j). We reprise the knowledge gradient notation of Chapter 5 because $\nu_{ij}^{KG,n}$ is the exact analog of (5.1), the marginal value of a single measurement. We merely replace $\max_{x'} \theta_{x'}^{n+1} - \max_{x'} \theta_{x'}^n$, the improvement in the estimated value of the best alternative, with $V^n - V^{n+1}$, the improvement in the estimated value of our current implementation decision, which is the solution to a shortest-path problem. Our measurement decision can then be written as

$$(i^n, j^n) = \arg\max_{(i,j)} \nu_{ij}^{KG,n},$$

the edge with the highest value of information.

In fact, (15.15) has a closed-form solution

$$\nu_{ij}^{KG,n} = \tilde{\sigma}_{ij}^n f\left(-\frac{|V_{ij}^n - \bar{V}_{ij}^n|}{\tilde{\sigma}_{ij}^n}\right), \qquad (15.16)$$

which is remarkably similar to (5.10), the KG formula for ranking and selection. As before, the quantity

$$\tilde{\sigma}_{ij}^n = \sqrt{\frac{1}{\beta_{ij}^n} - \frac{1}{\beta_{ij}^n + \beta_{ij}^W}}$$

represents the reduction in our uncertainty about (i, j) brought about by a single measurement of this edge. The function $f(z) = z\Phi(z) + \phi(z)$ is exactly the same as before. The main difference is in the computation of the normalized influence. Where (5.8) uses the quantity $|\theta_x^n - \max_{x' \neq x} \theta_{x'}^n|$, (15.16) uses $|V_{ij}^n - \bar{V}_{ij}^n|$. This can be thought of as a generalization of (5.8). In ranking and selection, the implementation decision was the same as the measurement decision. We measured individual alternatives in order to find a good alternative, and the influence was determined by the distance between the alternative we chose and the best of the others. In this setting, the influence depends on the distance between the best implementation decision containing the chosen edge (alternative), and the best implementation decision not containing that edge.

15.3 ALTERNATIVE EDGE SELECTION POLICIES

The value of the KG concept, aside from giving us an algorithm for making decisions, is that it allows us to think about new problems. It is not immediately clear how to carry over ideas like Gittins indices, interval estimation, or UCB methods to a problem where the measurement decision is distinct from the implementation decision. On the other hand, the concept of the marginal value of information can be easily extended to such problems. In (15.15), we defined KG as the expected improvement made in our estimate of the value of the optimal implementation decision. We will see in Section 15.4 that this same idea can be used together with more general optimization models.

Still, there are always alternatives. For example, it is possible to shoehorn the graph problem into the framework of ranking and selection, using the concepts we developed in Chapter 9 for subset selection. A shortest-path problem can be viewed as a type of subset selection, because a path is just a subset of the edge set $\mathcal{E}$. We can convert the problem of learning on a graph into a ranking and selection problem where each alternative is a path. The only issue is that a graph is typically described by a set of nodes and edges. The set of all possible paths is typically very large and difficult to enumerate. Fortunately, we can use Monte Carlo simulation to create a small choice set of paths, as in Section 9.3.

Given the current knowledge state (θ^n, β^n), fix an integer K. For every $(i,j) \in \mathcal{E}$, generate samples $\bar{c}_{ij}^n(\omega_k)$, $k = 1, ..., K$ from the prior distribution $\mathcal{N}\left(\theta_{ij}^n, \beta_{ij}^n\right)$. Now, for every $k = 1, ..., K$, run the Bellman–Ford algorithm using the sampled costs $\bar{c}^n(\omega_k)$. The resulting path $\bar{p}_k^n$ is the optimal solution for the kth sample.

Our beliefs about the edges now induce a set of beliefs $\left(\theta^{paths,n}, \Sigma^{paths,n}\right)$ on the K paths we generated, according to the equations

$$\theta_k^{paths,n} = \sum_{(i,j)\in\bar{p}_k^n} \theta_{ij}^n, \quad k = 1, ..., K,$$

$$\Sigma_{k,l}^{paths,n} = \sum_{(i,j)\in\bar{p}_k^n\cap\bar{p}_l^n} \frac{1}{\beta_{ij}^n}, \quad k, k' = 1, ..., K.$$

We can now apply any standard ranking and selection algorithm (including KG with correlated beliefs!) to this prior. The result will be some path $\bar{p}^*$. We can then use a simple heuristic to pick a single edge to measure from this path. One sensible choice is

$$(i^n, j^n) = \arg \min_{(i,j)\in\bar{p}^*} \beta_{ij}^n,$$

the edge with the lowest precision on the path of interest. This approach allows us to avoid having to re-derive a KG formula every time we work on a new problem class. Instead, we can simply use the old framework of ranking and selection a little more flexibly.

15.4   LEARNING COSTS FOR LINEAR PROGRAMS*

We now extend the learning problem described in Section 15.2 to the broader setting of a general LP. This section is intended for readers with an interest in more advanced topics. Some familiarity with LP geometry and duality theory will be very helpful for understanding this material; Chvátal (1983) or Vanderbei (2008) are useful LP references.

We return to the optimization problem

$$V(c) = \max_x \; c^T x, \tag{15.17}$$

subject to

$$Ax = b,$$
$$x \geq 0,$$

with which we started this chapter. Recall that, by the strong duality property of linear programming,

$$V(c) = \min_{y,s} \; b^T y, \tag{15.18}$$

subject to

$$A^T y - s = c,$$
$$s \geq 0.$$

That is, the dual LP has the same optimal value. We can let $x(c)$, $y(c)$ and $s(c)$ represent the optimal primal and dual solutions, that is, the choices of x, y, s in (15.17) and (15.18) that achieve the value $V(c)$.

Now, we assume that c is unknown (the other parameters A and b will still be known), and begin with a multivariate Gaussian prior $c \sim \mathcal{N}\left(\theta^0, \Sigma^0\right)$. As in Section 15.2, we have a budget of N measurements of individual objective coefficients which we can use to improve our solution. Measuring the jth coefficient will give us an observation from the distribution $\mathcal{N}(c_j, \lambda_j)$ where λ_j denotes a known variance. Measuring j^n at time n will cause our beliefs to change according to the equations

$$\theta^{n+1} = \theta^n + \frac{W_{j^n}^{n+1} - \theta_{j^n}^n}{\lambda_{j^n} + \Sigma_{j^n j^n}^n} \Sigma^n e_{j^n},$$

$$\Sigma^{n+1} = \Sigma^n - \frac{\Sigma^n e_{j^n} e_{j^n}^T \Sigma^n}{\lambda_{j^n} + \Sigma_{j^n j^n}^n},$$

exactly as in (2.22) and (2.23). We can see that the graph problem in Section 15.2 is just a special case of this more general problem. The graph structure can be encoded in the matrix A, and we no longer require independent beliefs. Our knowledge state $S^n = (\theta^n, \Sigma^n)$ consists of the parameters of the multivariate Gaussian distribution.

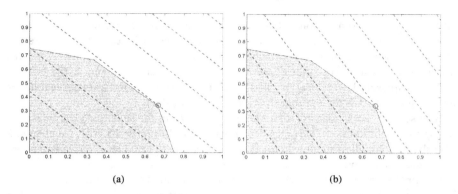

(a)   (b)

Figure 15.7 The optimal solution $V(\theta^n + \tilde{\sigma}(j) z)$ remains unchanged over a range of values of z.

At time n, our beliefs about the optimal value of the LP are given by $V(\theta^n)$. We simply solve the deterministic LP (using the simplex method or any standard software package such as Excel) using the vector θ^n for the objective coefficients. Our objective is to choose a measurement policy π to maximize

$$\max_{\pi} \mathbb{E} F^{\pi}(c, W)$$

with $F^{\pi}(c, W) = V(c^N)$. The KG factor in this problem can be defined as

$$\nu_j^{KG,n} = \mathbb{E}\left[V(\theta^{n+1}) - V(\theta^n) \mid S^n, j^n = j\right]. \qquad (15.19)$$

Next, recall from (5.16) that the conditional distribution of θ^{n+1}, given that we measure j at time n, can be written as

$$\theta^{n+1} = \theta^n + \tilde{\sigma}(j) Z^{n+1},$$

where $Z^{n+1} \sim \mathcal{N}(0, 1)$ and $\tilde{\sigma}(j) = \frac{\Sigma^n e_j}{\sqrt{\lambda_j + \Sigma_{jj}^n}}$ is a vector of variance reductions. Thus, the challenge of computing the expectation in (15.19) reduces to solving

$$\mathbb{E}\left[V(\theta^{n+1}) \mid S^n, j^n = j\right] = \mathbb{E} V(\theta^n + \tilde{\sigma}(j) Z^{n+1}), \qquad (15.20)$$

which is the expected value of a linear program with a stochastic objective function. In general, this problem is computationally intractable. However, in this specific case, the randomness in the objective function depends on a single scalar random variable. The expression $V(\theta^n + \tilde{\sigma}(j) Z^{n+1})$ can be interpreted as the optimal value of an LP whose objective function θ^n is perturbed by a single stochastic parameter Z^{n+1}. If Z^{n+1} were to be replaced by a fixed quantity z, this would be a standard problem in LP sensitivity analysis. In our case, we have to analyze the sensitivity of the LP over the entire distribution of Z^{n+1}.

Let us recall two facts from linear programming. First, the optimal solution always occurs at a corner point (or "extreme point") of the feasible region. Figure 15.7 shows an example in two dimensions where the shaded region represents the set of feasible solutions x. For some fixed value z, the dashed lines in Figure 15.7(a) represent the level curves of the objective function, or points where $(\theta^n + \tilde{\sigma}(j)z)^T x = a$ for different values of a. The value of a that makes the level curve tangent to the feasible region is precisely $V(\theta^n + \tilde{\sigma}(j)z)$, and the corner point where the level curve touches the feasible region is the optimal solution $x(c^n + \tilde{\sigma}(j)z)$.

The second fact is that, for small enough changes in the sensitivity parameter z, the optimal solution $x(c^n + \tilde{\sigma}(j)z)$ does not change. In Figure 15.7(b), we change z slightly, rotating all of the level curves. However, the optimal level curve is tangent to the feasible region at the same point as before. The range of values of z for which the optimal solution does not change is sometimes called the *invariant support set*. We can rewrite $\mathbb{E}V(\theta^n + \tilde{\sigma}(j) Z^{n+1})$ as a sum of integrals over the different possible invariant support sets. Suppose that $z_1, ..., z_I$ are a finite set of points, arranged in increasing order, such that $x(\theta^n + \tilde{\sigma}(j) z)$ is constant for all $z_i < z < z_{i+1}$. Then, for every i, there is a single corner point x_i satisfying $x_i = x(\theta^n + \tilde{\sigma}(j) Z^{n+1})$ for $z_i < Z^{n+1} < z_{i+1}$. Consequently, taking the expectation over the distribution of Z^{n+1}, we obtain

$$
\begin{aligned}
\mathbb{E}V(\theta^n + \tilde{\sigma}(j)Z^{n+1}) &= \sum_i \int_{z_i}^{z_{i+1}} (\theta^n + \tilde{\sigma}(j)z)^T x_i \phi(z) dz \\
&= \sum_i a_i(\Phi(z_{i+1}) - \Phi(z_i)) + b_i(\phi(z_i) - \phi(z_{i+1})),
\end{aligned}
$$

where $a_i = (\theta^n)^T x_i$ and $b_i = \tilde{\sigma}(j)^T x_i$. It turns out that this expression is equivalent to

$$
\mathbb{E}V(\theta^n + \tilde{\sigma}(j)Z^{n+1}) = \left(\max_i a_i\right) + \sum_i (b_{i+1} - b_i) f(-|z_i|),
$$

where f is once again the familiar function $f(z) = z\Phi(z) + \phi(z)$. We now make the observation that

$$
\max_i a_i = \max_i (\theta^n)^T x_i = V(\theta^n),
$$

because $\max_i a_i$ is the largest time-n objective value at all of the extreme points that could ever be optimal for any perturbation z, including $z = 0$. Therefore, $x(\theta^n)$ is one of the points x_i and its value is precisely $V(\theta^n)$. Therefore,

$$
\mathbb{E}V(\theta^n + \tilde{\sigma}(j)Z^{n+1}) = V(\theta^n) + \sum_i (b_{i+1} - b_i) f(-|z_i|). \tag{15.21}
$$

Combining (15.19), (15.20), and (15.21) gives us

$$
\nu_j^{KG,n} = \sum_i (b_{i+1} - b_i) f(-|z_i|), \tag{15.22}
$$

which looks identical to the computation of KG in ranking and selection with correlated beliefs (Section 5.3), with the important difference that the breakpoints z_i now represent the values of Z^{n+1} that change our implementation decision, which in this case is a corner point of the LP.

In the remainder of this section, we explain how the breakpoints z_i and the corresponding corner points x_i can be computed. After this is done, computing the KG factor is a simple matter of applying (15.22). We can find the breakpoints by starting with $x(\theta^n)$, which is the optimal solution when $z = 0$, finding the breakpoints that are nearest to $z = 0$, and expanding outward until all the breakpoints have been found.

Our first order of business is to determine whether $z = 0$ is itself a breakpoint. We can do this by calculating the smallest and largest values of z for which $x(\theta^n)$ is an optimal solution of the perturbed LP. These quantities, which we denote as z_- and z_+, are the optimal values of the LPs

$$z_- = \min_{y,s,z} \; z, \tag{15.23}$$

subject to

$$\begin{aligned}
A^T y - s - z\tilde{\sigma}(j) &= \theta^n, \\
x(\theta^n)^T s &= 0, \\
s &\geq 0,
\end{aligned}$$

and

$$z_+ = \max_{y,s,z} \; z, \tag{15.24}$$

subject to

$$\begin{aligned}
A^T y - s - z\tilde{\sigma}(j) &= \theta^n, \\
x(\theta^n)^T s &= 0, \\
s &\geq 0.
\end{aligned}$$

The feasible region, which is the same in both (15.23) and (15.24), merely ensures that $x\,(\theta^n)$ continues to be an optimal solution of the perturbed LP. Recall that there are three conditions for x to be an optimal solution to the primal LP (15.17) and y, s to be an optimal solution to the dual LP (15.18). First, x has to be primal-feasible. Second, y and s have to be dual-feasible. Third, we must have $x^T s = 0$, known as the complementary slackness property. In our case, $x\,(\theta^n)$ is automatically feasible for the perturbed LP, because it is optimal for the unperturbed LP

$$V(\theta^n) = \max_x \; (\theta^n)^T x, \tag{15.25}$$

subject to

$$\begin{aligned}
Ax &= b, \\
x &\geq 0,
\end{aligned}$$

which has the same feasible region as the perturbed LP. The constraints $A^T y - s - z\tilde{\sigma}(j) = \theta^n$ and $x(\theta^n)^T s = 0$ in (15.23) and (15.24) simply ensure that we are able to find a solution to the dual of the perturbed LP that maintains complementary slackness with $x\,(\theta^n)$. We are looking for the smallest and largest values of the perturbation z for which this is still possible. We now analyze four possible outcomes of this procedure.

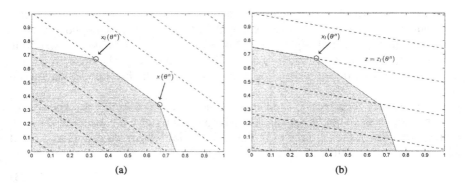

Figure 15.8 When $z_- = 0$, the optimal level curve is tangent to a face of the feasible region. The endpoints of this face are $x(\theta^n)$ and $x_l(\theta^n)$.

Case 1: $z_- < 0 < z_+$ In this case, $z = 0$ is not a breakpoint. This is the case illustrated in Figure 15.7(a). The level curve of θ^n is tangent to a single corner point. However, both z_- and z_+ are breakpoints.

Case 2: $z_- = 0, z_+ > 0$ When zero is a breakpoint, the optimal level curve is tangent to a face of the feasible region, rather than to a single corner point, and the solution $x(\theta^n)$ is located at one corner of the face. Figure 15.8(a) provides a visual illustration. If we apply a positive perturbation $0 < z < z_+$, the level curves will rotate clockwise, but $x(\theta^n)$ will still be the optimal solution. However, any negative perturbation will rotate the level curves counterclockwise and change the optimal solution.

To find the next breakpoint $z_l(\theta^n) < 0$, we first need to locate the other corner $x_l(\theta^n)$ of the face. This solution will still be optimal for the unperturbed LP, but it will also be optimal for the perturbed LP with $z = z_l(\theta^n)$.

We can do this by solving two LPs. First, $x_l(\theta^n)$ is the optimal solution to the problem

$$V_l(\theta^n) = \quad \min_x \quad \tilde{\sigma}(j)^T x, \tag{15.26}$$

subject to

$$\begin{aligned} Ax &= b, \\ s(\theta^n)^T x &= 0, \\ x &\geq 0. \end{aligned}$$

Any feasible solution to this problem continues to be optimal for the unperturbed LP in (15.25). In Figure 15.8(a), such a solution would be located somewhere on the face of the feasible region. Of all such points, the x with the lowest $\tilde{\sigma}(j)^T x$ corresponds to the opposite corner of the face. The optimal value of the LP

$$z_l(\theta^n) = \quad \min_{y,s,z} \quad z, \tag{15.27}$$

subject to

$$A^T y - s - z\tilde{\sigma}(j) = \theta^n,$$
$$x_l(\theta^n)^T s = 0,$$
$$s \geq 0,$$

corresponds to the next breakpoint, the most negative perturbation for which the perturbed LP still has $x_l(\theta^n)$ as an optimal solution. The relationship between $x(\theta^n)$, $x_l(\theta^n)$, and $z_l(\theta^n)$ can be seen in Figure 15.8(b).

Case 3: $z_- < 0, z_+ = 0$ Zero is still a breakpoint, but $x(\theta^n)$ is now at the opposite corner of the face. We can now rotate the level curves counterclockwise without changing the optimum, but not clockwise. The next corner point $x_u(\theta^n)$ is the optimal solution to the problem

$$V_u(\theta^n) = \max_x \quad \tilde{\sigma}(j)^T x, \tag{15.28}$$

subject to

$$Ax = b,$$
$$s(\theta^n)^T x = 0,$$
$$x \geq 0.$$

The next breakpoint $z_u(\theta^n) > 0$ is the optimal value of the LP

$$z_u(\theta^n) = \max_{y,s,z} \quad z,$$

subject to

$$A^T y - s - z\tilde{\sigma}(j) = \theta^n,$$
$$x_u(\theta^n)^T s = 0,$$
$$s \geq 0.$$

Case 4: $z_- = z_+ = 0$ Zero is a breakpoint, so the optimal level curve for the vector θ^n is still tangent to a face of the feasible region. However, $x(\theta^n)$ is no longer a corner point. Rather, it is somewhere in the middle of the face. We can sometimes discover optimal solutions like this if we use an interior-point algorithm to solve (15.25) instead of the simplex method. Fortunately, this case is easy to deal with: We simply solve (15.26)–(15.29) all together to find both corner points of the face at the same time.

Putting It All Together We can repeat this analysis to find sequences of positive and negative breakpoints. For example, if $z_- = 0$ and $z_+ = 0$, as in Case 2, we can find $z_l(\theta^n)$ and then repeat the analysis of Case 2 again, this time with $\theta^n + \tilde{\sigma}(j) z_l(\theta^n)$ as the "unperturbed" objective function, in place of θ^n. This will give us the next negative breakpoint after $z_l(\theta^n)$, and so on, until (15.27) becomes unbounded, which means that the same solution will continue to be optimal regardless of how much we perturb the objective function. The procedure to find a vector $\bar{z}$ of breakpoints and a vector $\bar{x}$ of corresponding corner points can be summarized as follows.

1) Solve (15.25) to obtain $x(\theta^n)$, $y(\theta^n)$ and $s(\theta^n)$.

2) Solve (15.23) and (15.24) to obtain z_-, z_+.

    2a) If $z_- < 0 < z_+$, let $\bar{z} = (z_-, z_+)$ and $\bar{x} = (x(\theta^n))$.
    2b) Otherwise, let $\bar{z} = (0)$ and let $\bar{x}$ be an empty vector.

3) While $\min \bar{z} > -\infty$, do the following:

    3a) Let $\theta' = \theta^n + \tilde{\sigma}(j)\min \bar{z}$.
    3b) Find $x_l(\theta')$ and $z_l(\theta')$ using (15.26) and (15.27).
    3c) Update $\bar{z} \leftarrow (\min \bar{z} + z_l(\theta'), \bar{z})$ and $\bar{x} \leftarrow (x_l(\theta'), \bar{x})$.

4) While $\max \bar{z} < \infty$, do the following:

    4a) Let $\theta' = \theta^n + \tilde{\sigma}(j)\max \bar{z}$.
    4b) Find $x_u(\theta')$ and $z_u(\theta')$ using (15.28) and (15.29).
    4c) Update $\bar{z} \leftarrow (\bar{z}, \max \bar{z} + z_u(\theta'))$ and $\bar{x} \leftarrow (\bar{x}, x_u(\theta'))$.

Learning the coefficients of a mathematical programming model is a relatively new setting, which we present as a way of hinting at the breadth of potential applications of optimal learning.

15.5 BIBLIOGRAPHIC NOTES

Section 15.1 - For other important mathematical programming applications (with a focus on the network and LP models considered in this chapter), see Vanderbei (2008).

Sections 15.2 - 15.3 - These sections are based on Ryzhov & Powell (2011b). Prior to this paper, several authors in the computer science community addressed the "bandit shortest path" problem, which would be the online version of the shortest path problem that we consider. For example, Abernethy et al. (2008) describes an algorithm that requires enumerating all the paths, ignoring the structure that arises from overlapping paths. Takimoto & Warmuth (2003) describes the use of a hedging algorithm for the online shortest path problems. Stochastic shortest path problems have been studied even earlier (Kulkarni 1986, Snyder & Steele 1995, Peer & Sharma 2007), but usually with the assumption of a known distribution for the rewards.

Section 15.4 - This material is due to Ryzhov & Powell (2011a). The procedure for finding breakpoints comes from Ghaffari-Hadigheh & Terlaky (2006).

CHAPTER 16

OPTIMIZING OVER CONTINUOUS MEASUREMENTS

There are many applications where we have to choose the best value of a continuous, multidimensional vector x. One example arises in the electricity sector, where utilities have to deal with the high variability of electricity spot prices. One strategy is to use a battery to store energy when prices are low, and release energy when prices are high. The idea is illustrated in Figure 16.2, where we store energy when the price goes below x^{store}, and we release energy when the price goes above $x^{withdraw}$. We now face the problem of deciding how to choose $x = (x^{store}, x^{withdraw})$. Let $\mu(x)$ be the expected profits per day that we obtain from using a policy fixed by x, where we assume we can only obtain noisy estimates of $\mu(x)$. Our challenge, as with elsewhere in this volume, is finding the best value of x as quickly as possible.

While we do not know the true function $\mu(x)$, we imagine it might look like the surface shown in Figure 16.2. We are going to assume it is smooth, but not necessarily concave (or convex, if we were solving a minimization problem). If x had only one or two dimensions, we could discretize it and use the techniques presented in the earlier chapters, although even two dimensions starts to become problematic if, for example, we would like to discretize each dimension into 100 intervals. If we have three or more dimensions, discretization quickly becomes cumbersome.

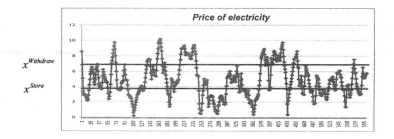

Figure 16.1 A battery charging problem, where we storage energy when the price goes below x^{store}, and withdraw it when the price goes above $x^{withdraw}$.

There are a number of problems which require tuning continuous but relatively low-dimensional parameter vectors. Some examples include:

- Finding the best parameters that govern a business simulator - We might have a model that simulates the use of ambulances, or a manufacturing system that depends on the speed of different machines. A model of freight transportation may require assumptions on the time that drivers have to rest or the maximum time that a customer may be served early or late.

- Tuning the concentrations of different chemicals - We may be trying to maximize the yield of a chemical process that depends on the concentrations of different additives (we may also have to vary the temperature of different steps of the process).

- Design of devices - The design of an aerosol spray can requires tuning the diameter of the feed-in tube, the spacing between the feed-in tube and the

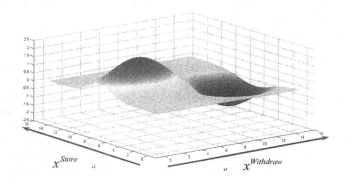

Figure 16.2 Illustration of the surface generated by varying x^{store} and $x^{withdraw}$.

aerosol spray tube, the diameter of the aerosol spray tube, and the pressure in the can.

In some cases, these parameters have to be tuned using field experiments or physical lab experiments, while in other cases they can be done using computer simulations. Computer simulations may run in a few minutes, but in some instances they can require hours or even days to complete a single run. Laboratory and field experiments can be even more time-consuming, and are generally much more expensive. For this reason, we want to choose our measurements carefully so that we learn as much as possible from each function evaluation.

We are going to continue to use our notation that $\mu(x)$ is the true value of the function at x. The only difference is that x is now a continuous vector. We are going to make a series of measurements $x^0, x^1, \ldots, x^n$, from which we make noisy observations of our function of the form

$$\hat{y}^{n+1} = \mu(x^n) + \varepsilon^{n+1}.$$

We assume that ε is a random variable that is normally distributed with mean 0 and variance λ. We wish to design a sequential search policy π that guides the selection of x^i, giving us an estimate of the value of our function which we represent as

$$F^\pi = \mu(x^N),$$

where $x^N = \arg\max_{x \in \mathcal{X}} \mu^N(x)$, and $\mu^N(x)$ is our estimate of $\mu(x)$ after N measurements obtained by following policy π. Our challenge is to find the policy π^* that solves

$$F^* = \max_\pi F^\pi.$$

This chapter addresses the problem of tuning these multidimensional parameter vectors for problems where we do not have access to derivatives. As always, we assume that observing $\mu(x)$ is time consuming and/or expensive, and noisy. Also, while we focus on multidimensional parameter vectors, our experimental work as of this writing is for vectors where the number of dimensions is less than 10.

16.1 THE BELIEF MODEL

In previous chapters, we have considered different ways of representing our belief in the function. We started with a lookup table, where we assumed that x was discrete, and there was an estimate θ_x^n for each x (see, for example, Chapters 4, 5, and 6). We have also considered models where we used linear regression to approximate $\mu(x)$ (as in Chapter 8). In this chapter, we need a different belief model since we have almost no idea about the structure of our function, but we do know it is continuous. For this reason, we are going to use an approach known as *Gaussian process regression* which is a class of nonparametric models which creates an approximation based on all prior observations.

We begin with a prior $\theta^0(x)$, which is now in the form of a continuous function. We are going to use our policy to generate a sequence of measurements $x^0, \ldots, x^{n-1}$, and we are going to use these points to approximate our function. Thus, after these n measurements $\hat{y}^1, \ldots, \hat{y}^n$, we will have a belief $\theta^n(x)$ at each of these points. We also need to capture the covariance in our beliefs about the function at each of these points. We define the $n \times n$ matrix Σ^n, where $\Sigma^n_{ij} = Cov^n(\mu(x^i), \mu(x^j))$ is the covariance in our belief about μ at x^i and x^j after n iterations. Note that each time we make a new measurement, θ^n grows by one element, while the matrix Σ^n adds a row and a column.

We are going to take advantage of the continuous structure of the problem and assume that we can write the covariance in our belief between any two points using the function

$$Cov^0(\mu(x^i), \mu(x^j)) = \beta \exp(- \sum_{m=1}^{p} \alpha_i (x^i_m - x^j_m)^2), \quad \alpha > 0, \beta > 0. \quad (16.1)$$

It is common to refer to the p-dimensional vector α as the activity of μ, while the scalar parameter β captures the uncertainty of our belief about μ. The activity parameter α controls the degree of smoothness, where the function becomes smoother as α becomes smaller. For larger values of α, the covariance between two points shrinks with α, which means that we learn less about $\mu(x^i)$ and $\mu(x^j)$ as x^i and x^j become farther apart.

16.1.1 Updating Equations

We have to learn how to update our beliefs about $\mu(x^i)$, $i = 1, \ldots, n$ as we make more observations. We are going to use the property that the vector θ^n follows a multivariate normal distribution, where $\mathbb{E}^n \mu(x^i) = \theta^n(x^i)$ and the covariance matrix is given by Σ^n. Throughout this chapter $\mathbb{E}^n$ refers to the conditional information given the history of observations $\hat{y}^1, \ldots, \hat{y}^n$.

We start by calculating a vector $\tilde{y}^n$ which we call *measurement residuals* using

$$\tilde{y}^n = \begin{bmatrix} \hat{y}^1 \\ \vdots \\ \hat{y}^n \end{bmatrix} - \begin{bmatrix} \theta^0(x^0) \\ \vdots \\ \theta^0(x^{n-1}) \end{bmatrix}. \quad (16.2)$$

This is the difference between our observations $\hat{y}^i$, $i = 1, \ldots, n$, and our original belief $\theta^0(x^i)$, $i = 1, \ldots, n$. We also define the *residual covariance* S^n which is updated using

$$S^n = \Sigma^0 + \text{diag}([\lambda(x^0), ..., \lambda(x^{n-1})]). \quad (16.3)$$

Expanded into matrices, this is the same as

$$
\begin{bmatrix}
S_{11}^n & \cdots & S_{1i}^n & \cdots & S_{1n}^n \\
\vdots & \vdots & \vdots & \vdots & \vdots \\
S_{i1}^n & \cdots & S_{ii}^n & \cdots & S_{in}^n \\
\vdots & \vdots & \vdots & \vdots & \vdots \\
S_{n1}^n & \cdots & S_{ni}^n & \cdots & S_{nn}^n
\end{bmatrix}
=
\begin{bmatrix}
\Sigma_{11}^0 & \cdots & \Sigma_{1i}^0 & \cdots & \Sigma_{1n}^0 \\
\vdots & \vdots & \vdots & \vdots & \vdots \\
\Sigma_{i1}^0 & \cdots & \Sigma_{ii}^0 & \cdots & \Sigma_{in}^0 \\
\vdots & \vdots & \vdots & \vdots & \vdots \\
\Sigma_{n1}^0 & \cdots & \Sigma_{ni}^0 & \cdots & \Sigma_{nn}^0
\end{bmatrix}
$$

$$
+
\begin{bmatrix}
\lambda(x^0) & 0 & \cdots & \cdots & 0 \\
0 & \ddots & 0 & \cdots & \vdots \\
\vdots & 0 & \lambda(x^i) & 0 & \vdots \\
\vdots & \vdots & 0 & \ddots & 0 \\
0 & \cdots & \cdots & 0 & \lambda(x^{n-1})
\end{bmatrix}.
$$

Finally, we are going to compute the *gain matrix*, denoted K^n using

$$
K^n = \Sigma^0 [S^n]^{-1}. \tag{16.4}
$$

As we see shortly below, the gain matrix is used to weigh new information, which plays the same role as a stepsize.

The updating equations for the mean vector θ^n and the covariance matrix S^n are now given by

$$
\begin{bmatrix}
\theta^n(x^0) \\
\vdots \\
\theta^n(x^{n-1})
\end{bmatrix}
=
\begin{bmatrix}
\theta^0(x^0) \\
\vdots \\
\theta^0(x^{n-1})
\end{bmatrix}
+ K^n \tilde{y}^n, \tag{16.5}
$$

$$
\Sigma^n = (I_n - K^n)\Sigma^0. \tag{16.6}
$$

where I_n is the $n \times n$ identity matrix.

Equations (16.5) and (16.6) update our beliefs around the points $x^0, \ldots, x^n$ that we have already measured, but we also need a belief for an arbitrary point x, since we need to search over all possible points to determine the next point that we *might* wish to evaluate. We begin by defining

$$
\bar{\Sigma}^0 = \Sigma^0([x^0, ..., x^{n-1}, x]),
$$
$$
\bar{\Sigma}^n = \Sigma^n([x^0, ..., x^{n-1}, x]).
$$

Let $\vec{0}$ be a column vector of zeroes. Our new gain matrix is given by

$$
\bar{K}^n = \bar{\Sigma}^0 \begin{bmatrix} I_n \\ - \\ \vec{0}^T \end{bmatrix} [S^n]^{-1}. \tag{16.7}
$$

We can now find θ^0 and $\bar{\Sigma}^0$ for an arbitrary $(n+1)$st point x using

$$
\begin{bmatrix} \theta^n(x^0) \\ \vdots \\ \theta^n(x^{n-1}) \\ \theta^n(x) \end{bmatrix} = \begin{bmatrix} \theta^0(x^0) \\ \vdots \\ \theta^0(x^{n-1}) \\ \theta^0(x) \end{bmatrix} + \bar{K}^n \tilde{y}^n, \tag{16.8}
$$

$$
\bar{\Sigma}^n = (I_{n+1} - \bar{K}^n \begin{bmatrix} I_n & | & \vec{0} \end{bmatrix}) \bar{\Sigma}^0. \tag{16.9}
$$

If we want the distribution of $\mu(x)$ given our n observations at some arbitrary decision x, we can use (16.8) and (16.9) to obtain

$$
\theta^n(x) = \theta^0(x) + \begin{bmatrix} \Sigma^0(x^0, x) & , \ldots, & \Sigma^0(x^{n-1}, x) \end{bmatrix} [S^n]^{-1} \tilde{y}^n, \tag{16.10}
$$

$$
\Sigma^n(x, x) = \Sigma^0(x, x) - \begin{bmatrix} \Sigma^0(x^0, x) & , \ldots, & \Sigma^0(x^{n-1}, x) \end{bmatrix} [S^n]^{-1} \begin{bmatrix} \Sigma^0(x^0, x) \\ \vdots \\ \Sigma^0(x^{n-1}, x) \end{bmatrix}. \tag{16.11}
$$

Equation (16.10) is known as *Gaussian process regression* (GPR) in some communities, and regression kriging in others.

We have to choose the point $x^n = x$ to measure before we have observed the outcome $\hat{y}^{n+1}$. The updated regression function, given the observations $\hat{y}^1, \ldots, \hat{y}^n$, is normally distributed with distribution

$$
\begin{bmatrix} \theta^{n+1}(x^0) \\ \vdots \\ \theta^{n+1}(x^{n-1}) \\ \theta^{n+1}(x^n) \end{bmatrix} = \begin{bmatrix} \theta^n(x^0) \\ \vdots \\ \theta^n(x^{n-1}) \\ \theta^n(x^n) \end{bmatrix} + \tilde{\sigma}(\bar{\Sigma}^n, x^n) Z^{n+1}, \tag{16.12}
$$

where $Z^{n+1} = \left(\hat{y}^{n+1} - \theta^n(x^n) \right) / \sqrt{\lambda(x^n) + \Sigma^n(x^n, x^n)}$, with

$$
\tilde{\sigma}(\Sigma, x) = \frac{\Sigma e_x}{\sqrt{\lambda(x) + e_x^T \Sigma e_x}}. \tag{16.13}
$$

Here e_x is a column vector of zeroes with a 1 at the row corresponding to decision x. It can be shown that Z^{n+1} is a standard normal random variate (mean 0, variance 1) because $\text{Var}(\hat{y}^{n+1} - \theta^n(x^n) | \mathcal{F}^n) = \lambda(x^n) + \Sigma^n(x^n, x^n)$.

16.1.2 Parameter Estimation

Our model is characterized by the p-dimensional vector α, the scalar β and the noise λ, as well as our prior $\theta^0(x)$. There may be problems where we feel we can assume these are known, but in practice we are going to have to estimate them from data. As is so often the case with statistical estimation, there is more than one way to solve this problem, but a popular method that we have found works quite well is maximum likelihood estimation (MLE).

MLE starts by creating a *likelihood function* which is the product of the density (using the normal distribution) of all the prior observations given the unknown parameters. We take the log of this product (giving us the log-likelihood), and then find the values of the parameters to maximize the resulting sum.

We form the likelihood function using

$$L_{\hat{y}}\left(\alpha, \beta, \lambda(x^0), ..., \lambda(x^{n-1}), \theta^0(x^0), ..., \theta^0(x^{n-1})\right)$$

$$= (2\pi)^{-n/2} |S^n|^{-1/2} \exp\left(-\frac{1}{2} \begin{bmatrix} \hat{y}^1 - \theta^0(x^0) \\ \vdots \\ \hat{y}^n - \theta^0(x^{n-1}) \end{bmatrix}^T (S^n)^{-1} \begin{bmatrix} \hat{y}^1 - \theta^0(x^0) \\ \vdots \\ \hat{y}^n - \theta^0(x^{n-1}) \end{bmatrix}\right).$$

Now, if we assume that the variance of the observation noise, $\lambda(\cdot)$, is a constant λ and $\theta^0(\cdot)$ is a constant θ^0, we can write the likelihood function as

$$L_{\hat{y}}(\alpha, \beta, \lambda, \theta^0) = (2\pi)^{-n/2}|\Sigma^0 + \lambda I_n|^{-1/2}\exp\left(-\frac{1}{2}(\hat{y} - \theta^0 1)^T(\Sigma^0 + \lambda I_n)^{-1}(\hat{y} - \theta^0 1)\right),$$

where 1 is a $n \times 1$ column vector of ones and $\hat{y} = \begin{bmatrix} \hat{y}^1 & \cdots & \hat{y}^n \end{bmatrix}^T$. Note that in this case we are estimating $p + 3$ parameters using n observations. We can write the log-likelihood function as

$$\begin{aligned} \ell_{\hat{y}}(\alpha, \beta, \lambda, \theta^0) &= -\frac{n}{2}\log(2\pi) - \frac{1}{2}\log(|\Sigma^0 + \lambda I_n|) \\ &\quad - \frac{1}{2}(\hat{y} - \theta^0 1)^T(\Sigma^0 + \lambda I_n)^{-1}(\hat{y} - \theta^0 1). \end{aligned} \quad (16.14)$$

We can approximately maximize the likelihood over the parameters by using the function `patternsearch` in MATLAB started at multiple points chosen by a Latin hypercube sampling (LHS) design using the command `lhsdesign`. Also, in the above log-likelihood we can easily solve for θ^0 in terms of α, β, and λ, giving us the estimate

$$\hat{\theta^0} = \frac{\hat{y}^T(\Sigma^0 + \lambda I_n)^{-1}1}{1^T(\Sigma^0 + \lambda I_n)^{-1}1}.$$

Finally, to prevent numerical issues, if $|\Sigma^0 + \lambda I_n|$ is very small in (16.14), a useful equivalent expression to $\log(|\Sigma^0 + \lambda I_n|)$ is `trace(logm(`$\Sigma^0 + \lambda I_n$`))` where `logm` is the matrix logarithm.

We typically will have to perform an initial sample of measurement points x to obtain a starting estimate of α, β and λ. If d is the dimensionality of our parameter vector (the sum of the dimensions of α, β and λ), a common rule of thumb is to perform a Latin hypercube design with $2d$ plus 2 points. In MATLAB, we can use `lhsdesign` to choose the points $x^1, \ldots, x^{2d+2}$. However, as the number of dimensions grows, the value of a LHS design diminishes, and it is better to simply choose the starting measurement points at random.

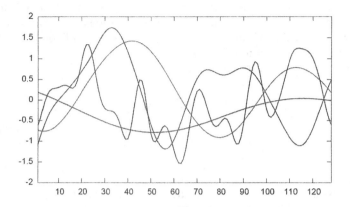

Figure 16.3   Illustration of sample realizations of one-dimensional Gaussian surfaces.

16.2   SEQUENTIAL KRIGING OPTIMIZATION

Sequential kriging uses a form of meta-modeling that assumes the surface is represented by a linear model, a bias model and a noise term, which we can write using

$$\mu(x) = \sum_{f \in \mathcal{F}} \theta_f \phi_f(x) + Z(x) + \varepsilon.$$

Here, $\mathcal{F}$ is a set of features, $\phi_f(x)$ is a basis function (also known as an independent variable) corresponding to feature f, and $Z(x)$ is a term that represents the systematic bias due to the limitations of the basis functions.

The kriging meta-model is based on the idea that the bias function $Z(x)$ is a realization of a stationary Gaussian stochastic process. If x is a d-dimensional vector, we assume that the covariance between $Z(x)$ and $Z(x')$ can be written as

$$Cov(Z(x), Z(x')) = \beta \exp\left[-\sum_{i=1}^{d} \alpha_i (x_i - x_i')^2\right],$$

where, as before, β is the variance of the stochastic process while α_i scales the covariance function for each dimension.

Samples of one-dimensional Gaussian surfaces are illustrated in Figure 16.3, which shows curves generated from different values of the activity variable α. Smaller values of α produces slower, undulating surfaces, while larger values of α (which reduces the correlations between nearby points) produces surfaces that undulate with higher frequencies. If this is the bias, it means that the true surface is likely to be reasonably smooth. It is important to realize that $Z(x)$ is not a random noise term with zero mean (which would imply that the approximation is unbiased). A sample realization of $Z(x)$ is the bias for a specific function $\mu(x)$ and a specific approximation represented by the set of basis functions ϕ and regression vector θ.

The best linear predictor $\bar{Y}^n(x)$ of $\mu(x)$ after n measurements is given by

$$
\bar{Y}^n(x) = \sum_{f \in \mathcal{F}} \theta^n_f \phi_f(x)
$$

$$
+ \sum_{i=1}^{n} Cov(Z(x_i), Z(x)) \sum_{j=1}^{n} Cov(Z(x_i), Z(x_j))(\hat{y}_i - \sum_{f \in \mathcal{F}} \theta^n_f \phi_f(x)),
$$

where θ^n is the least squares estimator of the regression parameters after n observations.

The idea of sequential kriging is to use the expected improvement $I(x)$ resulting from measuring the function at x. We start by illustrating this for a function $f(x)$ that can be observed deterministically. Let x^n be the current best solution given our current approximation. After n measurements, let $F(x)$ represent our belief about $f(x)$ which is normally distributed with mean $\bar{Y}^n(x)$ and variance $\sigma^{2,n}(x)$. The expected improvement, given the history of n observations, can be written as

$$
\mathbb{E}^n[I(x)] = \mathbb{E}^n \max(f(x^*) - F(x), 0). \tag{16.15}
$$

We need to build on this idea to handle the issue of noisy measurements. Sequential kriging optimization (SKO) uses the following expression for the expected improvement

$$
\mathbb{E}^n I(x) = \mathbb{E}^n\big[\max(\bar{Y}^n(x^{**}) - \mu(x), 0)\big] \left(1 - \frac{\sigma_\varepsilon}{\sqrt{\sigma^{2,n}(x) + \sigma^2_\varepsilon}}\right). \tag{16.16}
$$

The first term on the right hand side of (16.16) mimics the expected improvement term used if we could measure the function deterministically in equation (16.15). The second term in (16.16) represents a heuristic adjustment term that rewards uncertainty. If $\sigma^{2,n}(x) = 0$, then this adjustment term is zero and we would not place any value in measuring that point.

The expectation can be calculated analytically using

$$
\mathbb{E}^n\big[\max(\bar{Y}^n(x^{**}) - \mu(x))\big] = (\bar{Y}^n(x^{**}) - \bar{Y}^n(x))\Phi\left(\frac{\bar{Y}^n(x^{**}) - \bar{Y}^n(x)}{\sigma^n(x)}\right)
$$

$$
+ \sigma^n(x)\phi\left(\frac{\bar{Y}^n(x^{**}) - \bar{Y}^n(x)}{\sigma^n(x)}\right),
$$

where ϕ and Φ are the standard normal density and cumulative distribution functions.

Define a utility function that captures the uncertainty associated with a point x after n observations. For example, we might choose

$$
u^n(x) = -(\bar{Y}^n(x) + \sigma^n(x)).
$$

We choose our effective best solution, x^{**}, by maximizing the utility. We could try to search over the entire region $x \in \mathcal{X}$, but a common shortcut is to limit the search over previously sampled locations $x^1, \ldots, x^n$, giving us the rule

$$
x^{**} = \arg \max_{x^1, \ldots, x^n} [u^n(x)].
$$

16.3 THE KNOWLEDGE GRADIENT FOR CONTINUOUS PARAMETERS*

We now return to the knowledge gradient concept, and develop an algorithm that is adapted to problems with continuous parameters. It helps to remind ourselves of the definition of the knowledge gradient, which is given by

$$\nu^{KG,n}(x) = \mathbb{E}^n\left[\max_{x'\in\mathcal{X}}\theta^{n+1}(x')\Big|x^n = x\right] - \max_{x'\in\mathcal{X}}\theta^n(x'), \qquad (16.17)$$

where $\mathbb{E}^n$ refers to the conditional expectation given all prior measurements. Thus, $\mathbb{E}^n$ means the only random variable is the observation $\hat{y}^{n+1}$ from measuring x. The knowledge-gradient policy chooses the sampling decision at time n by maximizing the knowledge gradient,

$$x^n \in \operatorname{argmax}_{x\in\mathcal{X}}\nu^{KG,n}(x). \qquad (16.18)$$

When the set of choices $\mathcal{X}$ was a small, finite set of discrete measurements, solving the maximization problem in (16.18) was easy: We just evaluated the knowledge gradient at each of these discrete points and chose the best. Now that x is continuous (and probably a vector), solving (16.18) has become its own optimization problem. In this section, we describe a search algorithm for solving this optimization problem.

16.3.1 Maximizing the Knowledge Gradient

Recognizing that the knowledge gradient $\nu^{KG,n}(x)$ is a continuous surface, we can use a classical steepest ascent algorithm for finding the point x^n that maximizes $\nu^{KG,n}(x)$. The idea is to use a starting point $x^{n,0}$ and run a fixed number of iterations M of the algorithm

$$x^{n,m+1} = x^{n,m} + \alpha_m \nabla \nu^{KG,n}(x^{n,m}). \qquad (16.19)$$

If $\nu^{KG,n}(x)$ was a concave function, this algorithm should return the optimal solution. However, $\nu^{KG,n}(x)$ is highly nonconcave. The reason is that each time we measure a point x^n of our original function, the knowledge gradient tends to form a local minimum around this point, reflecting our higher level of confidence at this point and its neighboring region. This behavior is illustrated in Figure 16.4, which shows the knowledge gradient surface after we have evaluated four corner points and one interior point. It is not always the case, however, that a measured point produces a local minimum in the knowledge gradient surface, since high outcomes of $\hat{y}$ can produce local maxima. For this reason, we have to run the algorithm in equation (16.19) starting from a number of points, searching for local maxima. We then choose the best of these as our approximation of the global maximum.

We face three challenges. The first is that we have to approximate the expectation in equation (16.17), since we are no longer able to calculate this exactly as we were when the set $\mathcal{X}$ was discrete. The second and most difficult is the calculation of the

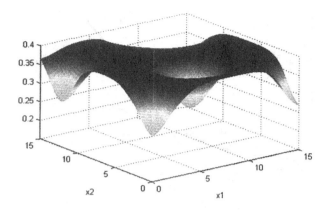

Figure 16.4   The knowledge gradient surface after measuring four corner points and one interior point.

gradient $\nu^{KG,n}(x)$ given the approximation of the expectation. Then, we describe the process of finding starting points in our search for a global maximum, and how we solve the issue of scaling the stepsize α_m.

16.3.2 Approximating the Knowledge Gradient

The first challenge we face when adapting the knowledge gradient to problems with continuous parameters is the imbedded maximization within the expectation in equation (16.17). We do this by replacing the maximum over all $x \in \mathcal{X}$ with the maximum over the previously sampled points $x^0, ..., x^n$, and the current sampling decision x,

$$\bar{\nu}^{KG,n}(x) = \mathbb{E}^n \left[\max_{i=0,..,n} \theta^{n+1}(x^i) \Big| x^n = x \right] - \max_{i=0,..,n} \theta^n(x^i) \Big|_{x^n=x}. \qquad (16.20)$$

Of course, we still have the problem of computing the expectation in equation (16.20), but we have already designed an algorithm for computing the expectation when our search region is limited to a finite set of points (as we do in equation (16.20)). This algorithm was presented in Section 5.3 under the name of the knowledge gradient for correlated beliefs.

We define the knowledge gradient for continuous parameters policy, π^{KGCP}, as the policy which selects the next sampling decision by maximizing the approximate knowledge gradient for continuous parameters,

$$x^n = \mathrm{argmax}_{x \in \mathcal{X}} \bar{\nu}^{KG,n}(x). \qquad (16.21)$$

The knowledge gradient for continuous parameters struggles with the same exploration vs. exploitation tradeoff that we have seen with discrete measurements. There

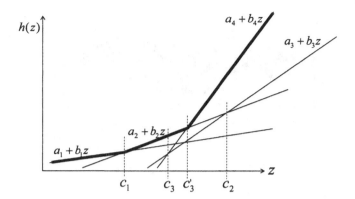

Figure 16.5   Regions of z over which different choices dominate.   Choice 3 is always dominated.

is a benefit, for example, from sampling near the current maximum of θ^n. At the same time, there is value in sampling points farther away from a previously sampled point because the uncertainty is higher. The SKO policy uses a heuristic adaptation to bias the search toward regions with more uncertainty, while KGCP uses a more formal expected value of information calculation. But as we see below, there is a computational price for this added elegance.

16.3.3   The Gradient of the Knowledge Gradient

The next step is finding the gradient of our approximate knowledge gradient, which we denote by $\nabla_x \bar{\nu}^{KG,n}(x)$. For this, we have to review the material from Section 5.3 on how we compute the expectation in our approximate knowledge gradient in (16.20). Recall from Section 5.3 that we can represent the maximization in the expectation in (16.20) as a series of cuts given by

$$\bar{\nu}^{KG,n}(x) \;=\; \arg\max_x \mathbb{E}^n\left[\max_i \theta_i^{n+1} \mid x^n = x\right]$$

$$\;=\; \arg\max_x \mathbb{E}^n\left[\max_i \theta_i^n + \tilde{\sigma}_i(\bar{\Sigma}^n, x^n)Z^{n+1} \mid x^n = x\right].$$

It is useful to reuse a figure first presented in Section 5.3, repeated again in Figure 16.5 for convenience, which illustrates the step of taking the maximum over the set of lines given by $\theta_i^n + \tilde{\sigma}_i(\bar{\Sigma}^n, x^n)z = a_i + b_i z$. We are going to assume that the lines have been sorted so that $b_{i+1} > b_i$, and we are going to further assume that we have removed any lines that are dominated by all the others. Throughout, while we will index lines over $0, \ldots, n$, we are going to assume that dominated lines are skipped. This means that if we compute the values of z that correspond to where the lines intersect, given by

$$c_i = \frac{a_i - a_{i+1}}{b_{i+1} - b_i},$$

then we are assured that $c_{i+1} > c_i$.

If we let

$$h(\theta^n, \tilde{\sigma}(x)) = \mathbb{E}^n\left[\max_i \theta_i^n + \tilde{\sigma}_i(\bar{\Sigma}^n, x^n)Z^{n+1} \mid x^n = x\right], \quad (16.22)$$

then we can substitute (16.22) into (16.22) to obtain

$$X^{KG}(s) = \arg\max_x h(\theta^n, \tilde{\sigma}(\bar{\Sigma}^n, x)).$$

Now let $h(a, b) = \mathbb{E}\max_i a_i + b_i Z$, where $a = \theta_i^n$, $b = \tilde{\sigma}_i(\bar{\Sigma}^n, x)$ and Z is our standard normal deviate. We can compute $h(a, b)$ using

$$h(a, b) = \sum_{i=0}^{n}(b_{i+1} - b_i)f(-|c_i|),$$

where $f(z) = z\Phi(z) + \phi(z)$.

We are now ready to talk about how to compute the derivative of the knowledge gradient. After some algebra, the gradient of the first term in equation (16.20) is given by

$$\nabla_x \mathbb{E}^n\left[\max_{i=0,..,n} \theta^{n+1}(x^i)\middle| x^n = x\right]$$

$$= \sum_{i=0}^{n}\left[\left(\nabla_{x^n}\theta^n(x^i)\right)(\Phi(c_{i+1}) - \Phi(c_i)) + \left(\nabla_{x^n}\tilde{\sigma}_i(\bar{\Sigma}^n, x^n)\right)(\phi(c_i) - \phi(c_{i+1}))\right]$$

$$+ \sum_{i=0}^{n}\left[\left(\theta^n(x^i) + \tilde{\sigma}_i(\bar{\Sigma}^n, x^n)c_{i+1}\right)\phi(c_{i+1})\nabla_{x^n}c_{i+1}\right.$$

$$\left. - \left(\theta^n(x^i) + \tilde{\sigma}_i(\bar{\Sigma}^n, x^n)c_i\right)\phi(c_i)\nabla_{x^n}c_i\right]. \quad (16.23)$$

The calculation of $\nabla_{x^n}c_i$ for $i = 0, ..., n + 1$ is relatively straightforward. An equivalent equation for the c_i's which are output from Algorithm 1 is $c_i = \frac{a_{i-1}-a_i}{b_i-b_{i-1}}$ for $i = 1, ..., n$ with $c_0 = -\infty$ and $c_{n+1} = +\infty$. Then using the quotient rule we can calculate the following:

$$\nabla_{x^n}c_i = \begin{cases} \frac{(b_i-b_{i-1})(\nabla a_{i-1}-\nabla a_i)-(a_{i-1}-a_i)(\nabla b_i-\nabla b_{i-1})}{(b_i-b_{i-1})^2}, & \text{for } i = 1, ..., n, \\ \vec{0}, & \text{for } i = 0, n+1. \end{cases} \quad (16.24)$$

As long as we can calculate $\nabla_{x^n}\theta^n(x^i)$ and $\nabla_{x^n}\tilde{\sigma}_i(\bar{\Sigma}^n, x^n)$ for $i = 0, ..., n$, we can calculate the expression in equation (16.23). We calculate $\nabla_{x^n}\theta^n(x^i)$ using

$$\nabla_{x^n}\theta^n(x^i) = \begin{cases} \vec{0}, & \text{if } i < n, \\ \nabla_{x^n}\theta^0(x^n) + J^n[S^n]^{-1}\tilde{y}^n, & \text{if } i = n, \end{cases}$$

where we let J^n be the matrix of first-order partial derivatives

$$
\begin{aligned}
J^n &= \left[\nabla_{x^n} \Sigma^0(x^0, x^n) \quad , \cdots , \quad \nabla_{x^n} \Sigma^0(x^{n-1}, x^n) \right] \\
&= 2 \begin{bmatrix} \alpha_1(x_1^0 - x_1^n)\Sigma^0(x^0, x^n) & \cdots & \alpha_1(x_1^{n-1} - x_1^n)\Sigma^0(x^{n-1}, x^n) \\ \vdots & \ddots & \vdots \\ \alpha_p(x_p^0 - x_p^n)\Sigma^0(x^0, x^n) & \cdots & \alpha_p(x_p^{n-1} - x_p^n)\Sigma^0(x^{n-1}, x^n) \end{bmatrix}.
\end{aligned}
$$

We calculate $\nabla_{x^n} \tilde{\sigma}_i(\bar{\Sigma}^n, x^n)$ using

$$
\nabla_{x^n} \tilde{\sigma}_i(\bar{\Sigma}^n, x^n) = \frac{B \nabla_{x^n} e_{x^i}^T \bar{\Sigma}^n e_{x^n} - e_{x^i}^T \bar{\Sigma}^n e_{x^n} \nabla_{x^n} B}{B^2},
$$

where $B = \sqrt{\lambda(x^n) + e_{x^n}^T \bar{\Sigma}^n e_{x^n}}$ and

$$
\nabla_{x^n} e_{x^i}^T \bar{\Sigma}^n e_{x^n} = \begin{cases} 2\mathrm{diag}(\alpha)(x^i - x^n)\Sigma^0(x^i, x^n) - J^n[S^n]^{-1}\Sigma^0 e_{x^i}, & \text{if } i < n, \\ -2J^n[S^n]^{-1} \begin{bmatrix} \Sigma^0(x^0, x^n) \\ \vdots \\ \Sigma^0(x^{n-1}, x^n) \end{bmatrix}, & \text{if } i = n \end{cases}
$$

and

$$
\nabla_{x^n} B = \frac{1}{2}(\lambda(x^n) + \Sigma^n(x^n, x^n))^{-\frac{1}{2}} \left(\nabla_{x^n} \lambda(x^n) - 2J^n[S^n]^{-1} \begin{bmatrix} \Sigma^0(x^0, x^n) \\ \vdots \\ \Sigma^0(x^{n-1}, x^n) \end{bmatrix} \right).
$$

These equations are a bit tedious, but this is a price we pay to handle more complex problems.

16.3.4 Maximizing the Knowledge Gradient

We are now ready to use our steepest descent algorithm, but we have to deal with a separate issue. The problem is that our knowledge gradient function $\bar{\nu}^{KG,n}(x)$ is highly nonconcave in x. The situation is illustrated in Figure 16.6(a), where we show the knowledge gradient for a scalar function after four noisy observations. Figure 16.6(b) shows the knowledge gradient approximation along with the exact knowledge gradient calculated on a finely discretized function. The figure illustrates the presence of local minima at or near previously sampled points. It also shows that the knowledge gradient tends to be largest at points farthest from previously sampled points, which hints at how we can run our steepest ascent algorithm.

Using the gradient, we can design a steepest ascent algorithm to find the maximum of $\bar{\nu}^{KG,n}(x)$. The biggest challenge here is that the knowledge gradient is nonconcave, but there is a lot of structure to its shape. Specifically, the local minima are very close to points that have been previously measured, and it is reasonable to expect that local maxima will be located halfway between local minima. It is also a good

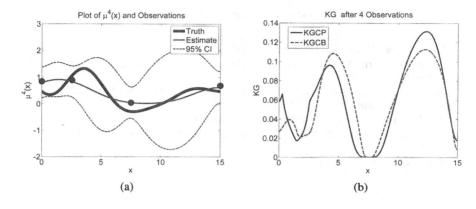

Figure 16.6   (a) The estimate of the function along with the 95 percent confidence intervals of the estimate after four observations. (b) The knowledge gradient for continuous parameters (KGCP) and exact knowledge gradient over a finely discretized set of decisions (KGCB) after four observations.

strategy to do restarts that start at previously sampled points, since local maxima are sometimes nearby (and not necessarily in between other previously sampled points). This suggests a restart procedure that starts with points that are halfway between previously sampled points.

This structure also gives us a good estimate of a starting stepsize, since we know the distance between pairs of local minima, and we can scale the stepsize accordingly. For example, a good rule is to set the stepsize so that the first step is one fourth the distance between the two previously sampled points.

After starting the search algorithm at all these points, we choose the best among all the optima returned by the steepest ascent algorithm. Of course, this process grows with the square of previously sampled points, but the idea is that this method would be used for expensive functions where the number of samples is not expected to grow very large. If this is not the case, then it is possible to limit the number of pairs of points that are used for restarts.

16.3.5 The KGCP Policy

We close by summarizing all the steps of the KGCP policy, outlined in Figure 16.1. We start in line 1 by choosing a set of initial measurement points x so that we can use a statistical procedure such as maximum likelihood to form estimates of the covariance parameters α, β and λ. In line 3 we choose the sampling decision by maximizing the knowledge gradient for continuous parameters defined in (16.20). This maximization should be approximated by using the algorithm in Section 16.3.4.

Table 16.1   The KGCP Policy.

---

(1) Choose an initial sample of points x (either randomly or using a Latin hypercube design) to form the basis of initial statistical estimates of α, β and λ (and possibly θ^0).

(2) **For** $n = 0, ..., N - 1$

(3)   Choose sampling decision: $x^n = \arg\max_{x \in \mathcal{X}} \bar{\nu}^{\mathrm{KG},n}(x)$ using Section 16.3.4.

(4)   Get noisy observation $\hat{y}^{n+1}$ of function at x^n.

(5)   Update θ^{n+1} and Σ^{n+1} using (16.8) and (16.9).

(6)   Update the estimates of α, β and λ using maximum likelihood estimation.

(7) **end**

(8) Implement $x^\star \in \arg\max_{x \in \mathcal{X}} \theta^N(x)$.

16.4   EFFICIENT GLOBAL OPTIMIZATION

While our interest is in solving problems where we can only obtain noisy observations of our function $\mu(x)$, an important contribution is a procedure known as *efficient global optimization* (EGO) which was developed for the case where there is no noise. EGO is similar to the knowledge gradient in that it uses the same idea of choosing to measure the point that maximizes the expected value of information. We have already seen this concept in Section 5.6, where it was known as the expected improvement (EI) procedure. In the literature, the names EGO and EI are used interchangeably to refer to the same technique.

Following our standard notation, we assume that we choose a measurement x^n and then observe $\hat{y}^{n+1} = \mu(x^n)$. Our indexing system follows the style of the rest of the book, and is designed to handle uncertainty. Below, $\hat{y}^{n+1}$ will be a random variable when we choose x^n, but in this section it is deterministic.

We start by defining

$$I^{n+1}(x) = \max\left(\theta^{n+1}(x) - \max_{i=1,..,n} \hat{y}^i, 0\right). \tag{16.25}$$

If we have no observation noise, $\bar{\nu}^{KG,n}(x) \leq \mathbb{E}^n[I^{n+1}(x)]$. Furthermore, $\mathbb{E}^n[I^{n+1}(x)] = \mathbb{E}^n[\max_{i=0,..,n} \theta^{n+1}(x^i)|x^n = x] - \max_{i=0,..,n-1} \theta^n(x^i)$. We show this using

$$\begin{aligned}
\bar{\nu}^{KG,n}(x) &= \mathbb{E}^n\left[\max_{i=0,..,n} \theta^{n+1}(x^i)|x^n = x\right] - \max_{i=0,..,n} \theta^n(x^i)\Bigg|_{x^n=x} \\
&\leq \mathbb{E}^n\left[\max_{i=0,..,n} \theta^{n+1}(x^i)\bigg|x^n = x\right] - \max_{i=0,..,n-1} \theta^n(x^i) \\
&= \mathbb{E}^n\left[\max\left(\theta^{n+1}(x^n), \max_{i=0,..,n-1} \theta^n(x^i)\right)\bigg|x^n = x\right] - \max_{i=0,..,n-1} \theta^n(x^i)
\end{aligned}$$

$$
\begin{aligned}
&= \quad \mathbb{E}^n \left[\max \left(\theta^{n+1}(x^n), \max_{i=1,..,n} \hat{y}^i \right) \Big| x^n = x \right] - \max_{i=1,..,n} \hat{y}^i \\
&= \quad \mathbb{E}^n \left[\max \left(\theta^{n+1}(x^n) - \max_{i=1,..,n} \hat{y}^i, 0 \right) \Big| x^n = x \right] \\
&= \quad \mathbb{E}^n [I^{n+1}(x)].
\end{aligned}
\tag{16.26}
$$

In the third line, we used the fact that, given what we know at time n, $\hat{y}^{i+1} = \theta^n(x^i) = \theta^{n+1}(x^i)$ for $i = 0, ..., n-1$ since there is no observation noise. The EGO algorithm maximizes the expected improvement given by equation (16.26) at each iteration. If we assume that there is no measurement noise, the expectation of (16.25) has a closed-form solution

$$
\nu^{EGO,n}(x) = \tilde{\sigma}_x(\Sigma^n, x) f \left(\frac{\theta^n(x) - \max_{i=1,...,n} \hat{y}^i}{\tilde{\sigma}_x(\Sigma^n, x)} \right),
$$

much like its analog in (5.28). This policy closely parallels the knowledge gradient for the case where there is no observation noise, and does not account for correlations in the belief structure.

16.5 EXPERIMENTS

It will always be hard to draw firm conclusions about the performance of algorithms in real experiments. In this section, we draw on a series of experiments reported in Scott et al. (2011), where SKO and KGCP were compared using a series of standard test problems as well as new test problems that were generated directly from the Gaussian process model. The standard test problems are known as Ackley, Branin, Hartman3 and the Six Hump Camelback. The Gaussian process datasets generated scalar functions using $\alpha = .1, 1.0, 10.0$, which captures the structural deviation between the true surface and the approximation. Recall that for smaller values of α, these deviations are described by smooth undulations, while larger values of α produce high frequency ripples. All of the test problems were run with three different levels of measurement noise, where we used $\lambda = .1, 1.0$, and 10.0.

Table 16.2 summarizes seven test problems, along with the results of comparisons of SKO and KGCP using the expected opportunity cost as the performance metric. These results suggest that the knowledge gradient algorithm consistently outperforms SKO, sometimes dramatically so. There was only one dataset where SKO outperformed KGCP in a statistically significant way (for the Six Hump Camelback, with $\lambda = 10$), but the expected opportunity cost for KGCP was 1.0264 versus .8488 for SKO, which is a small difference, especially compared to the relative performance for some of the other problems. As a general pattern, the relative improvement of KGCP over SKO was largest for problems where the measurement noise was the smallest. We suspect that as a general rule, differences in algorithms will become less noticeable as the measurement noise becomes larger.

Table 16.2 Comparison of the knowledge gradient algorithm for continuous parameters to sequential kriging optimization for noisy measurements. Source: Scott et al. (2011).

		KGCP		SKO	
Test Function	$\sqrt{\lambda}$	$\mathbb{E}(OC)$	$\sigma(OC)$	$\mathbb{E}(OC)$	$\sigma(OC)$
Ackley 5 ($\mathcal{X} = [-15, 30]^5$)	$\sqrt{.1}$	**5.7304**	.1874	7.8130	.1802
	$\sqrt{1.0}$	**10.8315**	.2413	12.6346	.2088
$p = 5, \sigma = 1.126$	$\sqrt{10.0}$	**17.3670**	.1477	18.1126	.1156
Branin	$\sqrt{.1}$	**.0141**	.0044	.0460	.0023
	$\sqrt{1.0}$	**.0462**	.0039	.1284	.0218
$p = 2, \sigma = 51.885$	$\sqrt{10.0}$	**.2827**	.0186	.4396	.0248
Hartman3	$\sqrt{.1}$	**.0690**	.0063	.1079	.0075
	$\sqrt{1.0}$	.5336	.0296	.5012	.0216
$p = 3, \sigma = .938$	$\sqrt{10.0}$	1.8200	.0541	1.8370	.0510
Six Hump Camelback	$\sqrt{.1}$	**.0714**	.0087	.1112	.0059
	$\sqrt{1.0}$	.3208	.0192	.3597	.0156
$p = 2, \sigma = 3.181$	$\sqrt{10.0}$	1.0264	.0391	**.8488**	.0370
GP ($\alpha = .1, \beta = 100$)	$\sqrt{.1}$	.0076	.0057	.0195	.0041
	$\sqrt{1.0}$	.0454	.0243	.0888	.0226
$p = 1, \sigma = 8.417$	$\sqrt{10.0}$	.3518	.0587	.2426	.0216
GP ($\alpha = 1, \beta = 100$)	$\sqrt{.1}$	**.0077**	.0022	.0765	.0311
	$\sqrt{1.0}$	**.0270**	.0045	.1993	.0486
$p = 1, \sigma = 9.909$	$\sqrt{10.0}$	**.4605**	.1028	.6225	.0669
GP ($\alpha = 10, \beta = 100$)	$\sqrt{.1}$	**.1074**	.0259	.5302	.0799
	$\sqrt{1.0}$	**.1846**	.0286	.6638	.0839
$p = 1, \sigma = 10.269$	$\sqrt{10.0}$	**1.0239**	.1021	1.8273	.1450

16.6 EXTENSION TO HIGHER-DIMENSIONAL PROBLEMS

The algorithms we have presented in this chapter can be used for multidimensional parameter vectors, but care has to be used when optimizing over higher-dimensional parameter spaces. It is not unusual to see algorithms which work in multiple dimensions being tested on problems with one or two dimensions. Transitioning to as few as five dimensions can actually be quite difficult for certain algorithms. Searching a parameter space with 10 or 20 dimensions is dramatically harder than five dimensions without making suitable approximations.

There are several strategies that can be used to search higher-dimensional parameter spaces. Some of these include

- Make a random sample of the parameter space $\mathcal{X}$, producing a set of possible parameters $x_1, x_2, \ldots, x_M$. Now, use traditional optimal learning policies for discrete alternatives.

- Assume that the function $\mu(x)$ is approximately separable in x, and use this property to search each dimension separately (possibly using the techniques in this chapter for continuous parameters).

- Stitch together a solution by optimizing over small subsets of dimensions. This can be done in parallel for different subsets of dimensions, after which new, overlapping subsets can be chosen.

The challenge of dimensionality exists even with classical stochastic search algorithms, where efficient learning may not be an issue. For example, imagine that we have a stochastic function $F(x, W)$ that depends on a controllable vector x and a random vector W. Let $W(\omega)$ be a sample realization of $F(x, W)$, and assume that we are able to compute the gradient $\nabla_x F(x, W(\omega))$. We might be able to solve the optimization problem

$$\max_x \mathbb{E} F(x, W)$$

using a classical stochastic gradient algorithm of the form

$$x^n = x^{n-1} + \alpha_{n-1} \nabla_x F(x^{n-1}, W(\omega^n)). \tag{16.27}$$

Stochastic gradient algorithms such as (16.27) can handle problems with thousands of dimensions, but can exhibit very slow convergence, even when we assume that we can compute the gradient $\nabla_x F(x^{n-1}, W(\omega^n))$. These algorithms work by basically linearizing the function and dealing with each dimension individually. Although we have not seen this in the research literature, it is possible to envision the use of optimal learning techniques applied in each dimension individually. However, for high dimensional problems, we would be limited to very simple policies. Needless to say, optimal learning for high dimensional problems is an interesting area of research.

16.7 BIBLIOGRAPHIC NOTES

There is an extensive literature on stochastic search for continuous variables which we have not covered in this chapter because they do not explicitly address the issue of value of information. An excellent review of the literature as of 2003 is given by Spall (2003). This literature can be largely divided between algorithms that assume that we have access to gradient information (or at least stochastic gradient), which is our focus.

There is a number of papers tackling the problem of optimizing noisy functions without derivative information. These algorithms primarily depend on methods for Monte Carlo sampling of the search region $\mathcal{X}$, and while everyone is looking for algorithms with fast convergence, most of the theory focuses on asymptotic convergence analysis (assuming an off-line application) while convergence rates are studied empirically. Benveniste et al. (1990) and Kushner & Yin (2003) are important references for the theory behind stochastic search algorithms. Some recent contributions include adaptive search with resampling (Andradóttir & Prudius 2010) and model reference adaptive search (Hu et al. 2007).

Sections 16.1 - Our belief model is based on material presented in Sacks et al. (1989) and Frazier et al. (2009). Our presentation of Gaussian process regression is based on Rasmussen & Williams (2006); the material on regression kriging in Forrester et al. (2008). The recursive updating equations for the means and variances are based on Frazier et al. (2009). The maximum likelihood estimation method for the Gaussian process regression model is based on Rasmussen & Williams (2006).

Sections 16.2 - Sequential kriging optimization was proposed by Huang et al. (2006). Stein (1999) provides a thorough introduction to the field of kriging, which evolved from the field of spatial statistics. This field continues to be very active; see Ankenman et al. (2010) for some recent developments.

Sections 16.3 - The adaptation of the knowledge gradient for continuous measurements was developed by Scott et al. (2011).

Sections 16.4 - Efficient global optimization was proposed by Jones et al. (1998).

Sections 16.5 - We draw on a series of experiments reported in Scott et al. (2011). Our test functions are culled from Frazier et al. (2009), Huang et al. (2006), and Jones et al. (1998). See also Scott et al. (2010) for additional empirical work on calibration of an airline business simulator.

CHAPTER 17

LEARNING WITH A PHYSICAL STATE

All of the problems that we have considered in earlier chapters have one characteristic in common. In each of these problems, the decision we make depends only on our state of knowledge about the problem. The decision itself can be discrete (as in Chapter 4), a subset (Chapter 9), a scalar (Chapter 10), or continuous (Chapter 16), and we have considered many different types of objective functions (online, offline, or linked to a complex implementation decision) and different belief structures (lookup table, parametric and nonparametric). However, within each specific problem class, the set of possible decisions has always stayed the same over time, and every policy we have looked at has made decisions based purely on the available information.

There are many problems where this is not the case, and our decision also depends on a *physical state* as well as a state of knowledge. A simple example arises when we are trying to learn the travel times on links in a transportation network. We can traverse from node i to node j to learn more about the time τ_{ij} to get from i to j, but doing so puts us at node j, which changes the decisions we can consider next. If we decide to go from i to j, we have to balance the information we gain about τ_{ij} against the impact of now being at location j.

Another application arises in medical treatments. Consider the problem of treating a diabetes patient. The doctor has certain beliefs about the effectiveness of different

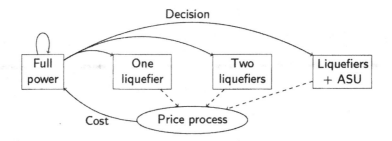

Figure 17.1   An operational flexibility problem, where the state of the power plant affects the energy consumed and the cost of resuming operations.

treatments. At the same time, the doctor also observes the results of blood sugar tests that are regularly undergone by the patient. The decision to recommend a particular type of treatment is based on the doctor's beliefs about the effectiveness of that treatment, but also on the patient's physical condition. For example, a class of diabetes drugs known as "secretagogues" works by stimulating pancreatic cells to release insulin. Secretagogues can work well in reducing blood sugar, but may cause weight gain as a side effect. The decision to try a treatment can change the physical state of the patient (captured by his weight), which can then change the decisions we can make in the future.

A second example involves the management of a sensor moving around to collect information. This might be a robot sensing radiation near an accident, or a medical technician collecting information about the prevalence of a viral outbreak in the population. The information can be used to help us design response strategies. If we collect information at one location, it reduces the cost of collecting information at nearby locations. Our decision of what information to collect depends on the physical location of the sensor.

We use the term "physical state" somewhat loosely, but we always mean a state variable that changes the decisions that we are allowed to make. For example, imagine that we are selling a product where we can change the price and then learn how the market might respond to the price. We may feel that the market will respond badly to rapid changes in the price, and as a result we impose a restriction that we cannot change the price by more than one dollar. If we charge $20 during one week, we cannot then try charging $28 or $15 the next week. While we may not think of price as a physical state, this falls within the problem class that we wish to consider.

There are many practical problems where learning arises which also have a physical state. Below are some additional examples of problems that have a physical state:

- Operational flexibility - A manufacturer of industrial gases buys energy on the real-time market to run its cryogenic plants. The spot price of energy is subject to very sudden spikes, known as "coincident peaks" in the energy literature. Each plant runs an air separation unit (ASU) which breaks up air into component gases, as well as one or two recycle compressors or liquefiers. Shutting down all of these components allows us to save more money during a coincident peak, but if the peak never arrives, we will waste a lot of energy ramping back up to normal operations (which can take upwards of 24 hours). We have a state of knowledge that helps us predict coincident peaks, but the physical state of the power plant also affects the costs incurred.

- Competitive games - Imagine a game of chess. Each player has a certain belief about the opponent's strategy, and uses this belief to decide on the next move. However, the decisions a player can make depends on the state of the board.

- Pricing with finite inventory - A store sets a price for a product with the goal of maximizing revenue. As in Chapters 8 and 11, we may wish to experiment with different prices and observe the effect on revenues. However, if we have finite inventory, our decision will depend on the amount we have in stock as well as on our beliefs about the revenue curve. Experimentation is less useful when there is less inventory remaining.

- Mutual fund cash balance - A mutual fund needs cash to meet shareholder redemptions. At the same time, keeping more cash on hand means having less money to invest, leading to lower profits. Over time, we gradually learn about shareholder behavior and the rate at which redemptions arise. Our ability to meet redemptions at any given time, however, is constrained by the amount of cash on hand.

The physical state introduces a whole new level of challenge. The question of how to make decisions based on both physical and knowledge states is at the very frontier of optimal learning. In this chapter, we give a framework for where to begin thinking about this problem. We also suggest a few approaches that integrate the optimal learning concepts we have developed throughout this book into the world of physical states, and close with a discussion of how some of these concepts might be taken further.

17.1 INTRODUCTION TO DYNAMIC PROGRAMMING

Sequential, stochastic decision problems have long been approached using dynamic programming. Assume that we have a process that may be in a physical state $S_t = s$ at time t. (Throughout this book, we have denoted time by n; the reason why we switch to t here will become clear shortly.) If we choose a (discrete) action x, we may land in a state $S_{t+1} = s'$ with probability $p(s'|s, x)$. Assume that we earn a contribution $C(S_t, x)$ from taking action x when we are in state S_t, and that contributions are discounted by $0 < \gamma < 1$ with each passing time period. If $V_t(S_t)$ is the value of being in state S_t at time t, then we would like to choose our action x_t by solving

$$x_t = \arg\max_x \left(C(S_t, x) + \gamma \sum_{s'} p(s'|S_t, x) V_{t+1}(s') \right). \qquad (17.1)$$

The problem is that we do not know $V_{t+1}(s')$. If we have a finite horizon problem with T as the last time period, we might assume $V_T(s) = 0$ then compute the remaining value functions using

$$V_t(S_t) = \max_x \left(C(S_t, x) + \gamma \sum_{s'} p(s'|S_t, x) V_{t+1}(s') \right).$$ (17.2)

Equation (17.2) needs to be computed by looping over all states S_t. If S_t is discrete and very low dimensional (as in, three or less), this approach can work quite well. But there are many problems where the state S_t has more than three dimensions, or it may be continuous. For these problems (which are quite common), we encounter what is widely known as the "curse of dimensionality."

17.1.1 Approximate Dynamic Programming

The dynamic programming community has developed an algorithmic strategy known broadly as *approximate dynamic programming* (ADP) that is designed to circumvent the curse of dimensionality. ADP is actually an umbrella for a variety of algorithms, but its most popular form works as follows. Let's begin by assuming that the state S_t is discrete. Further assume that we are given an approximation $\bar{V}_t^0(s)$ for all states s and times $t = 1, \ldots, T$ (for example, we might start with $\bar{V}_t^0(s) = 0$).

If we use equation (17.2), we are solving the problem by proceeding backward in time. It is this process that requires that we loop over all possible states, because we do not know in advance which states that we might actually visit. With approximate dynamic programming, we are going to progress *forward* in time, starting with a given initial state S_0. If we are in state S_t at time t and choose action x_t (according to some rule that we discuss below), we are then going to observe a sample realization of a random variable W_{t+1} (which was unknown at time t). Finally, we are going to use a *transition function* (also known as the state transition model), which gives us the next state using

$$S_{t+1} = S^M(S_t, x_t, W_{t+1}).$$

We are going to simulate our way from time 0 to time T iteratively. To firm up our notation, let n the iteration index. In iteration n, let S_t^n be the state that we visit at time t. Also let ω^n index the sample path of observations of W_t, which means that $W_{t+1}(\omega^n)$ is the sample realization of what we observe between t and $t + 1$ while following sample path ω^n.

We have to figure out how we are going to make decisions. The most natural policy is to mimic equation (17.1) and solve

$$x_t^n = \arg\max_x \left(C(S_t^n, x) + \gamma \sum_{s'} p(s'|S_t^n, x) \bar{V}_{t+1}^{n-1}(s') \right).$$ (17.3)

Not surprisingly, we are going to refer to this as a pure exploitation policy, because it means to choose the best action based on our current belief about the downstream values captured by $\bar{V}_{t+1}^{n-1}(s')$. Before we start criticizing this policy (as we will), we need to close the loop by mentioning how we are coming up with these value function approximations. First compute

$$\hat{v}_t^n = C(S_t^n, x_t^n) + \gamma \sum_{s'} p(s'|S_t^n, x_t^n)\bar{V}_{t+1}^{n-1}(s'). \tag{17.4}$$

We can think of $\hat{v}_t^n$ as a sample estimate of the value of being in state S_t^n. This is the value that we experienced while we were following our sample path through the states $S_0^n, S_1^n, \ldots, S_T^n$. It would be nice to mimic (17.2) and just set $V_t(S_t^n) = \hat{v}_t^n$, but $\hat{v}_t^n$ is a noisy sample estimate. For this reason, we have to do smoothing, which we can do using

$$\bar{V}_t^n(S_t^n) = (1 - \alpha_{n-1})\bar{V}_t^{n-1}(S_t^n) + \alpha_{n-1}\hat{v}_t^n.$$

Here, α_{n-1} is known as a stepsize, which is a parameter between 0 and 1. It might be a constant (such as $\alpha_{n-1} = .001$), but more often it is assigned a declining sequence such as $\alpha_n = a/(a + n - 1)$, where a is a tunable parameter.

Notice that, in the above presentation, we have two time indices, n and t. This is because we considered a finite-horizon problem ending at time T. The index t denotes a stage of the problem, whereas the index n denotes the number of times we have updated our approximation $\bar{V}$ of the value function. The algorithm works by making multiple passes through the problem, so that we solve (17.3) backward for $t = T - 1, ..., 1$ for each fixed value of n.

We can also accommodate infinite-horizon problems ($T \to \infty$), for which the optimality equation (17.2) becomes

$$V(S) = \max_x \left(C(S, x) + \gamma \sum_{s'} p(s'|S, x)V(s') \right). \tag{17.5}$$

for all states S. In this case, the ADP algorithm merges the two time indices into one, such that (17.3) becomes

$$x^n = \arg\max_x \left(C(S^n, x) + \gamma \sum_{s'} p(s'|S^n, x)\bar{V}^{n-1}(s') \right). \tag{17.6}$$

and (17.4) becomes

$$\hat{v}^n = C(S^n, x^n) + \gamma \sum_{s'} p(s'|S^n, x^n)\bar{V}^{n-1}(s'). \tag{17.7}$$

The approximation is updated using

$$\bar{V}^n(S^n) = (1 - \alpha_{n-1})\bar{V}^{n-1}(S^n) + \alpha_{n-1}\hat{v}^n. \tag{17.8}$$

We still run the algorithm by progressing forward in time starting from S^0. Now, a single iteration consists of a visit to a single state S^n, and n is our only time index, as in the rest of the book. We hope that, after a large number of iterations, $\bar{V}^n$ will

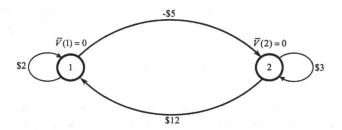

Figure 17.2   A two-state dynamic program, where the estimated value of being in each state is currently zero.

be close to the true solution of (17.5). To simplify our presentation, we will focus on infinite-horizon problems when developing our algorithms in this chapter, though it is important to keep in mind the wealth of finite-horizon applications.

We now have the beginnings of an algorithm that seems to have considerable appeal. We can handle arbitrarily complex state variables S_t, and even the random information W_t can be quite complex (problems have been solved with thousands of dimensions), because at any point in time we are working with a single state and a single sample realization of W_t. It almost seems like magic, so you know that there has to be something wrong. Indeed, the problem is known as the exploration vs. exploitation problem, something that should be now be quite familiar to readers of this book.

17.1.2   The Exploration vs. Exploitation Problem

Unfortunately, the problem with our elementary approximate dynamic programming algorithm is not just that it may not work well but that it can, and generally will, work terribly. To see why, we only need the very simple two-state dynamic program illustrated in Figure 17.2. Assume that we are initially in state 1, and we have to choose an action. We can choose to stay in state 1, in which case we earn a $2 contribution plus the approximate value of being in state 1, where our initial approximation is $\bar{V}(1) = 0$. Alternatively, we can choose to go to state 2, where we incur a loss of $5, but then we receive the value of being in state 2, where again we have an approximate value of $\bar{V}(2) = 0$. Looking at these two alternatives, we would naturally decide to stay in state 1. As a result, we never discover that the transition from state 2 back to state 1 would earn us $12.

This is hardly an artificial example. A more real-world setting is illustrated with an interstate truck driver who arrives in a city and is offered a set of loads of freight. He has to choose one of the loads, which has him moving from his current location to the destination of the load. Let S_t be the current location of the driver (typically a city or three-digit zip code), let x_t be his choice of load from a set of offered loads (these arrive randomly and only become known when he arrives at a location), and

Figure 17.3   The effect of a pure exploitation policy for our nomadic trucker.

let $C(S_t, x)$ be the net profit he receives from accepting the load and moving to its destination. For this simple problem, $S_{t+1} = S^M(S_t, x_t, W_{t+1}) = S^M(S_t, x_t)$ captures the destination of the load (which is deterministic).

A new truck driver might not have any information about the value of being in a city, and as a result might reasonably start with $\bar{V}^0(s) = 1$ for each state s. This means that he would initially take the load paying the most. Over time, he will learn about the value of being in states that he visits, but this will lead to situations where he is choosing between a load that returns him to a city which he has already visited (in which case $\bar{V}^{n-1}(s') > 0$), and a load that takes him to a city he has never visited (which means $\bar{V}^{n-1}(s') = 0$). As a result, there is a natural bias to take loads to cities he has visited before. Sound familiar? This is like returning to restaurants that you know, rather than trying new restaurants. Needless to say, this is fairly intuitive behavior.

The problem with this policy is that, as we would expect, there is a tendency to get caught visiting a small number of cities. Figure 17.3 shows the results of a simulation of this policy, and out of a set of 30 possible locations, the driver finds that he is constantly returning to the same seven cities. This might seem familiar, but it is hardly optimal.

17.1.3 Discussion

Our ADP algorithm has not really solved the curse of dimensionality. While we have eliminated the need to loop over all states to calculate the value of being in a state, we still need good approximations of the value of each state that we *might* visit, as shown in equation (17.3). While this may be much smaller than the size of the full state space, this is still a large number of states.

There are two changes we have to consider to overcome this problem. The first is that we have to introduce *exploration* steps, where we visit a state just to collect information about the state. We have seen this idea throughout this volume, so there should be no surprise that it arises in this setting as well. The second change is that we have to use some form of *generalized learning*. That is, we need to learn more than just the value of being in a single state. Again, this has been a common theme that we have seen before, first through the mechanism of correlated beliefs, and later when we made the transition to parametric belief models as we did in Chapter 8.

First, we present some simple heuristics for exploration. Then, we progress to more formal methods that address both of the issues raised in this discussion. As a rule, the exploration strategies in this chapter are designed for infinite-horizon problems, so we use a single time index n.

17.2 SOME HEURISTIC LEARNING POLICIES

One very simple policy that is frequently used in practice is our old acquaintance epsilon-greedy: At time n, we make a random decision with probability ε, or follow the pure exploitation policy (17.6) with probability $1 - \varepsilon$. We face the usual problem of tuning ε, but otherwise, it will be possible to achieve good performance as long as $\mathcal{X}$ is fairly small.

Two more sophisticated exploration strategies developed within the computer science community are known as R-max and E^3. Both of these policies are based on the idea of categorizing the states according to whether we have "enough" or "not enough" information about them.

The R-max policy makes decisions according to the rule

$$X^{Rmax,n}(S^n; \alpha^n) = \arg \max_x \big(C(S^n, x) + \gamma \sum_s \rho_s^n(S^n, x) F(s) \big),$$

where

$$F(s) = \begin{cases} R^{\max} & \text{if } s \text{ has been visited fewer than } m \text{ times,} \\ V^n(s) & \text{if } s \text{ has been visited at least } m \text{ times.} \end{cases}$$

The integer m is a tunable parameter representing the number of times we need to visit a state in order to obtain "enough" information. The value R^{max} is deliberately chosen to be very large, such that we are more likely to choose an action x if it is more likely to lead us to states for which we have little information. Eventually, once we visit every state enough times, R^{max} reduces to the pure exploitation policy . We are still, however, left with the issue of tuning m and R^{max}.

The E^3 policy ("Explicit Explore or Exploit") can be viewed as a modification of epsilon-greedy. As in R-max, we have a parameter m representing the amount of information that is "sufficient" for us to know the value of being in a state. Upon reaching state S^n, our decision is made as follows. If S^n is a state that we have never visited before, we make a random decision. If we have visited S^n before, but fewer than m times, we make the decision that we have tried the fewest number of times among all our previous visits to the state. Finally, if we have visited S^n more than m times, we follow the pure exploitation policy and make our decision according to (17.6). Once again, the policy reduces to pure exploitation once we have sufficiently explored the state space. In the early stages, the policy encourages us to make decisions with which we are unfamiliar, in order to learn about them.

17.3 THE LOCAL BANDIT APPROXIMATION

Our first formal technique, known as the local bandit approximation or LBA, attempts to convert the dynamic program into a multi-armed bandit problem at each time step, and then uses a Gittins-like calculation to make the next decision. The multi-armed bandit problem, which we covered in Chapter 6, has no physical state. We make our decision based purely on our knowledge about the rewards obtainable from M different arms. To apply bandit-style thinking to our current problem, we need a way to remove the physical state.

Let us start by defining a policy π_n that always makes decisions according to (17.6) for $\bar{V}^{n-1}$ fixed at the current value of n. At first glance, this sounds like the pure exploitation policy. However, pure exploitation updates the approximation $\bar{V}^{n-1}$ using (17.8) after $\hat{v}^n$ is observed. That is, pure exploitation assumes that the approximation is fixed when we make a decision, but we continue to update the approximation in every iteration. This is exactly the concept of experiential learning discussed back in Section 5.4.

By contrast, the policy π_n *always* makes decisions according to the *fixed* approximation $\bar{V}^{n-1}$. It is analogous to the "Stop" policy discussed in Section 6.4. If we use this policy, we will commit to our current value function approximation and stop learning altogether. The dynamic program is thus reduced to a Markov chain (Y_n) whose transition probabilities are given by

$$P(Y_{n+1} = s' \mid Y_n = s) = p(s' \mid s, X^{\pi,n}(s)),$$

where $X^{\pi,n}(s)$ is the decision that solves (17.6) for $S^n = s$. Our reward for visiting state s is thus $\tilde{C}(s) = C(s, X^{\pi,n}(s))$.

We use this policy much the same way that we did in Section 6.4: We assume that we will make only one more decision before switching to this naive policy. Suppose that S^n is our state at time n. Let us adopt a perspective that is centered around the state S^n. In this time step, we will make a decision x^n and transition to a different state. Suppose that, from that point onward, we will only make decisions according to the policy π_n. This policy will lead us to visit many different states, before we eventually return to S^n. We use the word *sojourn* to describe the period of time in which we travel around states other than S^n. The sojourn ends as soon as we return to this state. Figure 17.4 provides a visual illustration, with states represented by circles and decisions represented by squares.

As long as we use the policy π_n to make decisions, we can study the sojourn via the Markov chain (Y_n). Let

$$\tau_{s,s'} = \min\{n \geq 0 \mid Y_n = s', Y_0 = s\}$$

be the number of transitions needed for the Markov chain (Y_n) to reach state s' for the first time, provided that it started in state s. The quantities

$$R(s,s') = \mathbb{E}\left(\sum_{n=0}^{\tau_{s,s'}-1} \gamma^n f(Y_n) \mid Y_0 = s\right),$$

$$T(s,s') = \mathbb{E}\left(\sum_{n=0}^{\tau_{s,s'}-1} \gamma^n \mid Y_0 = s\right),$$

represent the average discount reward collected and the average discounted time elapsed, respectively, before our first visit to state s'. We can compute these quantities using first-transition analysis on Y_n. First,

$$R(s,s') = \begin{cases} f(s) + \gamma \sum_{s'' \in \mathcal{S}} P(Y_{n+1} = s'' \mid Y_n = s)R(s'',s') & \text{if } s \neq s', \\ 0 & \text{if } s = s'. \end{cases}$$

Similarly,

$$T(s,s') = \begin{cases} 1 + \gamma \sum_{s'' \in \mathcal{S}} P(Y_{n+1} = s'' \mid Y_n = s)T(s'',s') & \text{if } s \neq s', \\ 0 & \text{if } s = s'. \end{cases}$$

It follows that, if we visit state S^n and choose action x, the expected reward collected during the ensuing sojourn, and the expected length of the sojourn, are given by

$$R^{sojourn}(S^n, x) = C(S^n, x) + \gamma \sum_{s' \in \mathcal{S}} p(s' \mid S^n, x)R(s', S^n),$$

$$T^{sojourn}(S^n, x) = 1 + \gamma \sum_{s' \in \mathcal{S}} p(s' \mid S^n, x)T(s', S^n).$$

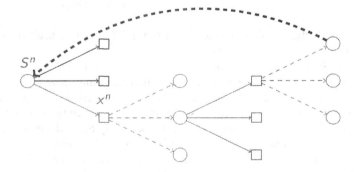

Figure 17.4   If we visit a state S^n and make decision x^n, we enter a sojourn that takes us to other states and actions. The sojourn ends as soon as we return to S^n.

The quantity

$$\nu^{LBA,n}(S^n, x) = \frac{R^{sojourn}(S^n, x)}{T^{sojourn}(S^n, x)} \tag{17.9}$$

represents the expected reward per expected unit time that we receive during the sojourn.

It turns out that the ratio in (17.9) resembles a Gittins index. In Chapter 6, we defined the Gittins index of a bandit arm as a risk-free reward that would make us indifferent between playing the arm repeatedly or just collecting the reward. An equivalent definition describes the Gittins index as the long-term expected reward per play that we can obtain from playing the arm over and over.

Thus, in a sense, this analysis is converting the dynamic program into a bandit problem. When we visit state S^n, we behave as if S^n were the only state in the dynamic program. All other states are grouped together into the sojourn. The different decisions available in state S^n are viewed as "arms" of a bandit, and the expected reward that we collect during the sojourn after pulling the "arm" x is treated as the "reward" of that arm. Thus, (17.9) is viewed as the Gittins index of decision x. We then make decisions according to the simple rule

$$X^{LBA,n}(S^n) = \arg\max_x \nu_x^{LBA,n},$$

analogous to the Gittins index policy for multi-armed bandits.

Unlike the original Gittins index policy, LBA is not optimal, because we are using the policy π_n instead of the optimal policy in the analysis of the sojourn. However, when this approach was first introduced, it was a pioneering example of an attempt to bring optimal learning ideas over to learning with a physical state. It also illustrates some of the difficulties inherent in carrying optimal learning concepts over to the physical state setting.

17.4 THE KNOWLEDGE GRADIENT IN DYNAMIC PROGRAMMING

We are going to develop the concept of the knowledge gradient in the context of dynamic programs with a physical state. We do this in two steps. First, we are going to introduce the idea of approximating a value function using a linear regression model. This makes it possible to generalize what we learn from a visit to a single state to what we know about all states. Then, we show how to compute the knowledge gradient in this setting.

17.4.1 Generalized Learning Using Basis Functions

A limitation of all the policies we have reviewed up to now is that they are limited by the use of a lookup table belief model for value functions. A much more powerful way of learning uses some form of generalization. While there are several ways of doing this, by far the most popular involves using linear regression to approximate the value of being in a state.

Before we do this, we have to solve a small technical difficulty. Our pure exploitation policy would have chosen an action using

$$x_t^n = \arg \max_x \left(C(S_t^n, x) + \gamma \sum_{s'} p(s'|S_t^n, x) \bar{V}_{t+1}^{n-1}(s') \right).$$

If we have a large state space (which is the only time where learning is an issue), then we would never be able to compute the one-step transition matrix $p(s'|S_t^n, x_t^n)$. A more natural way to write this equation is using its expectation form, where we would write

$$x_t^n = \arg \max_x (C(S_t^n, x) + \gamma \mathbb{E}\{\bar{V}_{t+1}^{n-1}(S')|S_t^n, x\}),$$

with S' denoting the randomly determined next state. We have not solved anything yet, because we can only compute the expectation for special cases. We can get around this by using the concept of the *post-decision state variable* which we designate S_t^x. The post-decision state is the state immediately after we have made a decision, but before we have learned any new information. For example, imagine we have a simple inventory problem where at time t we need to serve a random demand $\hat{D}_t$ with some resource, where R_t is the amount of resource (medicine, money, vaccines, water, power plants...) currently available. The (pre-decision) state is given by $S_t = (R_t, \hat{D}_t)$. Unsatisfied demands are lost. The resource variable evolves according to

$$R_{t+1} = R_t - x_t + \hat{R}_{t+1},$$

where $0 \leq x_t \leq R_t$ is our decision of how much to withdraw to satisfy demands, and $\hat{R}_{t+1}$ represents random additions (blood donation, cash deposits). The post decision state $S_t^x = R_t^x$, where

$$R_t^x = R_t - x_t.$$

The key idea with the post-decision state is that it is a deterministic function of the state S_t and the action x_t. More generally, we assume we have a function $S^{M,x}(S_t, x_t)$ that returns the post-decision state S_t^x.

Using this concept, our exploitation policy would be computed using

$$x_t^n = \arg \max_x \left(C(S_t^n, x) + \gamma \bar{V}_t^{x,n-1}(S^{M,x}(S_t^n, x)) \right), \qquad (17.10)$$

where $\bar{V}_t^{x,n-1}(S_t^x)$ is our value function approximation around the post-decision state S_t^x. Note that this is a deterministic optimization problem, which is much easier to solve. Below, we will again focus on the infinite-horizon formulation, in which (17.10) becomes

$$x^n = \arg \max_x (C(S^n, x) + \gamma \bar{V}^{x,n-1}(S^{M,x}(S^n, x))), \qquad (17.11)$$

We now just have to resolve the problem of approximating the value function. For this purpose, we propose using a simple linear model of the form

$$V(S^{x,n}) = \sum_{f \in \mathcal{F}} \theta_f \phi_f(S^{x,n}) = (\phi(S^{x,n}))^T \theta.$$

Here, $\phi_f(S)$ is known as a *feature* or *basis function*. These are functions that capture what an expert feels are the important characteristics of the state variable. Instead of having to estimate the value of being in each state, we now have to find the regression vector θ.

In Chapter 8, we showed how we could use the knowledge gradient to estimate the parameters of a linear regression model. We are going to do the same here, but now we have to deal with the issue of a physical state. We are also going to adopt another convention that we have used widely in our presentation which is the idea that we have a prior for our regression model. That is, we are going to assume that we have an initial estimate θ^0 as a starting point that is an unbiased estimate of the true value. This is like saying that we have a rough idea of what the value function looks like. This assumption allows us to claim that $\hat{v}^n$ is an unbiased observation of the true value $V(S^{x,n-1})$ of being in the post-decision state $S^{x,n-1}$. We are going to further assume that the error $\epsilon = \hat{v}^n - V(S^{x,n-1})$, in addition to having mean zero (since we assume that our prior is unbiased) also has a known error with variance σ_ϵ^2.

With this foundation, we can quickly describe the equations needed to update our regression vector θ. We start by defining the matrix

$$X^n = \begin{bmatrix} \phi_1(S^{x,0}) & \cdots & \phi_F(S^{x,0}) \\ \vdots & \ddots & \vdots \\ \phi_1(S^{x,n-1}) & \cdots & \phi_F(S^{x,n-1}) \end{bmatrix}.$$

This matrix consists of the value of each basis function evaluated at every state that we have visited. Also define the vector of observations

$$y^n = \begin{bmatrix} \hat{v}^1 \\ \vdots \\ \hat{v}^n \end{bmatrix}.$$

If we were using traditional linear regression, we could compute the best regression vector θ^n using

$$\theta^n = \left[(X^n)^T X^n \right]^{-1} (X^n)^T y^n.$$

We can compute θ^n recursively without having to perform matrix inversion. First let $B^n = \left[(X^n)^T X^n \right]^{-1}$. The updating equation for θ^n is given by

$$\theta^n = \theta^{n-1} + \frac{\hat{v}^n - \phi(S^{x,n-1})^T \theta^{n-1}}{1 + \phi(S^{x,n-1})^T B^{n-1} \phi(S^{x,n-1})} B^{n-1} \phi(S^{x,n-1}). \quad (17.12)$$

We can also compute the matrix B^n recursively using

$$B^n = B^{n-1} - \frac{B^{n-1}\phi(S^{x,n-1})\phi(S^{x,n-1})^T B^{n-1}}{1 + \phi(S^{x,n-1})^T B^{n-1}\phi(S^{x,n-1})}. \qquad (17.13)$$

We are going to make the somewhat heroic assumption that the observations $\hat{v}^n$ are independent and identically distributed with variance σ_ϵ^2. This means that the covariance matrix for the observation vector Y^n is $I\sigma_\epsilon^2$, where I is the identity matrix. Our estimate of the regression vector is given by

$$\theta^n = B^n (X^n)^T Y^n.$$

That is, our approximation of the value of state $S^{x,n}$ at time n is $\bar{V}^n(S^n) = \phi(S^{x,n})^T\theta^n$. It is also possible to show (as we did in Chapter 8) that the covariance matrix for θ is given by

$$\Sigma^{\theta,n} = \sigma_\epsilon^2 B^n.$$

We can now rewrite equation (17.12) as

$$\theta^n = \theta^{n-1} - \frac{\hat{v}^n - \phi(S^{x,n-1})^T\theta^{n-1}}{\sigma_\epsilon^2 + \phi(S^{x,n-1})^T\Sigma^{\theta,n-1}\phi(S^{x,n-1})}\Sigma^{\theta,n-1}\phi(S^{x,n-1}),$$

and

$$\Sigma^{\theta,n} = \Sigma^{\theta,n-1} - \frac{\Sigma^{\theta,n-1}\phi(S^{x,n-1})\phi(S^{x,n-1})^T\Sigma^{\theta,n-1}}{\sigma_\epsilon^2 + \phi(S^{x,n-1})^T\Sigma^{\theta,n-1}\phi(S^{x,n-1})}.$$

17.4.2 The Knowledge Gradient

We are now ready to compute a knowledge gradient for each potential action in a dynamic program. We are going to build on the principles we have developed earlier in this volume, but care has to be used in the presence of a physical state.

We start by writing our exploitation policy (doing the best given what we know now) in the form

$$X^{Exp,n}(S^n) = \arg\max_x Q^n(S^n, x) \qquad (17.14)$$

where

$$Q^n(S^n, x) = C(S^n, x) + \gamma\phi(S^{x,n})^T\theta^n.$$

When we use the knowledge gradient, we capture the fact that our decision will generate information (presumably from an exogenous distribution) that would allow us to update our regression vector, giving us an updated estimate θ^{n+1} (which is random at time n, when we choose our action). We can write the knowledge gradient policy as

$$X^{KG,n}(S^n) = \arg\max_x Q^{KG,n}(S^n, x), \qquad (17.15)$$

where

$$Q^{KG,n}(S^n, x) = C(S^n, x) + \gamma \mathbb{E}_x^n \bar{V}^{n+1}(S^{x,n}). \tag{17.16}$$

To understand this, it is useful to talk through the steps. If we are in state S^n and choose action x, we would then observe a random quantity W^{n+1} that would determine the next pre-decision state S^{n+1}. Once in state S^{n+1}, we would observe $\hat{v}^{n+1}$, which would then be used to update our parameter vector θ^n to give us an updated estimate θ^{n+1}. However, we are still in state S^n and W^{n+1} (and therefore S^{n+1} and $\hat{v}^{n+1}$) is a random variable. For this reason, we have to take the expectation $\mathbb{E}_x^n$ of $\bar{V}^{n+1}$ given what we know now (given by S^n, which captures both our physical state as well as our state of knowledge), and the action x that we are considering.

Note that while we are in state S^n, the information we gain from a transition from S^n to S^{n+1} from taking action x updates our belief about the value function that may benefit us even though our next decision is from state S^{n+1} (which is random when we are still in state S^n). We did not enjoy this property when we used a lookup table representation. If we learned something about state s' from a decision to move from s to s', this new information was of little or no value while we were in state s'. The information did not add value until we were in some other state where we might consider a transition back to state s. We could apply the same principle using a lookup table representation if we exploited the idea of correlated beliefs.

We start by expanding equation (17.16) using

$$Q^{KG,n}(S^n, x) \approx C(S^n, x)$$
$$+ \gamma \sum_{S^{n+1}} p(S^{n+1}|S^n, x) \mathbb{E}_x^n \max_{x'} Q^{n+1}(S^{n+1}, x'). \tag{17.17}$$

To compute the expectation in (17.17) we build on ideas we first presented in Chapter 5 and write θ^{n+1} using

$$\theta^{n+1} \sim \theta^n + \frac{\Sigma^{\theta,n+1} \phi(S^{x,n})}{\sqrt{\sigma_\varepsilon^2 + \phi(S^{x,n})^T \Sigma^{\theta,n} \phi(S^{x,n})}} Z, \tag{17.18}$$

where Z is a standard normal random variable with mean 0 and variance 1. Multiplying both sides in (17.18) by $\phi(S^y)$, for some decision y not necessarily equal to x, gives us the relationship between $\bar{V}^n$ and $\bar{V}^{n+1}$ which we can write as

$$\bar{V}^{n+1}(S^y) \sim \bar{V}^n(S^y) + \frac{\phi(S^y)^T \Sigma^{\theta,n-1} \phi(S^{x,n})}{\sqrt{\sigma_\varepsilon^2 + \phi(S^{x,n})^T \Sigma^{\theta,n} \phi(S^{x,n})}} Z. \tag{17.19}$$

The quantity $\phi(S^y)^T \Sigma^{\theta,n-1} \phi(S^{x,n})$ is precisely $Cov^n(S^y, S^{x,n})$.

Using equation (17.19), the last term in (17.17) can be written as

$$\mathbb{E}_x^n \max_{x'} Q^{n+1}(S^{n+1}, x') = \mathbb{E} \max_{x'} (a_{x'}^n + b_{x'}^n Z), \tag{17.20}$$

where

$$a_{x'}^n = C(S^{n+1}, x') + \gamma \bar{V}^n(S^{M,x}(S^{n+1}, x')),$$
$$b_{x'}^n = \gamma \frac{\phi(S^{M,x}(S^{n+1}, x'))^T \Sigma^{\theta,n-1} \phi(S^{x,n})}{\sqrt{\sigma_\varepsilon^2 + \phi(S^{x,n})^T \Sigma^{\theta,n} \phi(S^{x,n})}}.$$

Now we have to use the methods for computing the knowledge gradient with correlated beliefs (see Section 5.3) to compute (17.20) using

$$\mathbb{E} \max_{x'} (a_{x'}^n + b_{x'}^n Z) = \left(\max_{x'} a_{x'}^n \right) + \sum_{x'} (b_{x'+1}^n - b_{x'}^n) f(-|c_{x'}|),$$

where f is defined as $f(z) = z\Phi(z) + \phi(z)$, where ϕ and Φ denote the standard Gaussian pdf and cdf. Readers familiar with Section 5.3 will remember that it is important that the actions be sorted so that the slopes b_y^n are sorted in increasing order, and that dominated lines have been eliminated.

We can now compute the knowledge gradient using

$$\nu^{KG,n}(S^{x,n}, S^{n+1}) = \mathbb{E} \max_{x'} (a_{x'}^n + b_{x'}^n Z) - \max_{x'} a_{x'}^n$$

$$= \sum_{x'} (b_{x'+1}^n - b_{x'}^n) f(-|c_{x'}|).$$

Since $a_{x'}^n = Q^n(S^{n+1}, x')$, the quantity $\nu^{KG,n}(S^{x,n}, S^{n+1})$ can be viewed as the expected improvement in our estimate of the value of being in state S^{n+1}, obtained as a result of making the random transition from $S^{x,n}$ (which depends on our action) to S^{n+1}. We can substitute the definition of the knowledge gradient back into (17.17) to obtain

$$\sum_{S^{n+1}} p(S^{n+1}|S^n, x) \mathbb{E}_x^n \max_{x'} Q^{n+1}(S^{n+1}, x')$$

$$= \sum_{S^{n+1}} p(S^{n+1}|S^n, x) \max_{x'} Q^n(S^{n+1}, x')$$

$$+ \sum_{S^{n+1}} p(S^{n+1}|S^n, x) \nu^{KG,n}(S^{x,n}, S^{n+1}).$$

The value of being in the post-decision state is the expected value of being in the next pre-decision state. We can write this expectation using

$$\sum_{S^{n+1}} p(S^{n+1}|S^n, x) \max_{x'} Q^n(S^{n+1}, x') \approx \bar{V}^n(S^{x,n}),$$

and (17.15) reduces to

$$X^{KG,n}(S^n) = \max_x C(S^n, x) + \gamma \bar{V}^n(S^{x,n})$$

$$+ \sum_{S^{n+1}} p(S^{n+1}|S^n, x) \nu^{KG,n}(S^{x,n}, S^{n+1}). \quad (17.21)$$

We see that our knowledge gradient policy looks very similar to the pure exploitation policy in equation (17.14), with the exception that we now have one more term that captures the value of information.

One potential problem with equation (17.21) is that we may not be able to compute the conditional probability $p(S^{n+1}|S^n, x)$, especially if the random variables are continuous. In this case, a quick work-around is to to approximate the expectation in the value of information term using Monte Carlo simulation. In this case, we would use

$$\sum_{S^{n+1}} p(S^{n+1}|S^n, x)\nu^{KG,n}(S^{x,n}, S^{n+1}) \;\approx\; \frac{1}{K}\sum_{k=1}^{K} \nu^{KG,n}(S^{x,n}, S_k^{n+1}),$$

where $S_k^{n+1} = S^{M,W}(S^{x,n}, W^{n+1}(\omega_k))$ for the kth sample path generated. This technique works well for relatively low values of K, such as $K \approx 20$, as long as we have some sort of simulation model from which we can generate transitions.

The structure of the policy in (17.21) is suited for online learning, as we saw in Section 6.4. If we are trying to learn the value functions as quickly as possible in an offline setting, we would focus purely on the value of information, giving us a policy of the form

$$X^{Off,n}(S^n) = \arg\max_x \sum_{S^{n+1}} p(S^{n+1}|S^n, x)\nu^{KG,n}(S^{x,n}, S^{n+1}). \qquad (17.22)$$

This policy will look odd to readers familiar with approximate dynamic programming, since it appears to ignore the contribution from an action and the downstream value. However, in an offline setting, our goal is to just learn the value function approximation, with the hope that if we can learn this function quickly, then a pure exploitation policy (holding the value function fixed) will return good results.

17.4.3 Experiments

Experience with this use of the knowledge gradient is quite limited, but we report on experimental work that is available as this book went to press. We are going to use these ideas to optimize the performance of an energy storage device such as a battery. Let R^n be the amount of energy in the battery as a percentage of total capacity, which is taken to be 35 megawatt-hours. We allow R^n to take integer values between 0 and 100. We can buy energy on the spot market at a price P^n which evolves according to the model

$$\log \frac{P^{n+1}}{P^n} = -\alpha \log \frac{P^n}{P^0} + \sigma Z^{n+1}, \qquad (17.23)$$

where Z, as before, is a standard normal random variable with mean 0 and variance 1. We use the values $\alpha = 0.0633$ and $\sigma = 0.2$, along with an initial price $P^0 = 30$. The decision x^n is an integer from -50 to 50 representing how much to charge ($x^n \geq 0$) or discharge ($x^n < 0$) the storage device. The decisions x^n are constrained so that R^n does not go negative or above the maximum capacity of the battery. The post-decision state is given by

$$
\begin{aligned}
R^{x,n} &= R^n + x^n, \\
P^{x,n} &= P^n.
\end{aligned}
$$

The next pre-decision state is obtained by letting $R^{n+1} = R^{x,n}$ (we assume no exogenous changes to the amount stored in the battery) and generating P^{n+1} using equation (17.23). The single-period reward is given by $C(S^n, x^n) = -P^n x^n$, the cost incurred or revenue obtained as a result of our decision. Our goal is to determine a storage and withdrawal policy that maximizes total discounted profits over time, using a discount factor of $\gamma = .99$.

The spot price P^n is continuous, and thus we cannot solve the problem exactly. We use a parametric value function approximation with six polynomial basis functions given by $\phi(S^{x,n}) = \left(1, R^{x,n}, (R^{x,n})^2, P^{x,n}, (P^{x,n})^2, R^{x,n}P^{x,n}\right)$. We run the KG policy with the Monte Carlo approximation from equation (17.22), with $K = 20$.

For each sample path, the algorithm started with a prior $\theta_1^0 = 15000, \theta_2^0, ..., \theta_6^0 = 0$ and a diagonal covariance matrix $\Sigma^{\theta,0}$ with all diagonal elements equal to 500^2. An optimistic value for θ^0 was chosen heuristically to reduce the likelihood of getting stuck in a subset of the state space. The measurement noise was chosen to be $\sigma_\varepsilon^2 = 2000^2$.

Most experiments with approximate dynamic programming use a series of training iterations to estimate the value function approximation, after which θ is fixed and a series of simulations are run to determine how well the policy works. This is classical offline learning, and yet it is most common to use the exploitation policy given in equation (17.10) which has more of an online structure. For this reason, we report the results of four comparisons: We are going to use the online policy $X^{KG,n}(S^n)$ in equation (17.21) and the offline policy $X^{Off,n}(S^n)$ in equation (17.22), tested in two different objective functions. The first is an online objective function, where we assume that we are accumulating contributions in a real setting as we are learning the value function. The second is a more traditional offline setting, where we use a budget to fit the value function approximation and then evaluate the policy using a series of testing iterations. Naturally, we expect the online policy $X^{KG,n}(S^n)$ to do better on the online objective function, and we expect the offline policy $X^{Off,n}(S^n)$ to do better on the offline objective function.

Table 17.1 compares our two learning policies using basis functions against an algorithm that uses a lookup table representation, and an epsilon-greedy learning policy that we introduced in Section 17.2. The average performance of each algorithm, averaged over several hundred problems, shows that as we hoped, the offline KG policy (using either basis functions or a lookup table representation) does better on the offline objective function, and the online KG policy does better on the online objective function. Interestingly, the basis functions worked best for the offline objective function while the lookup table worked best for the online objective function. The epsilon-greedy policy performed poorly on both objective functions.

A separate but important issue involves the computational effort to support a policy. For this problem, we can safely say that the epsilon-greedy policy requires virtually no computational effort, and for this reason could perhaps be run for more iterations. The knowledge gradient policy when using a lookup table representation required 0.025 seconds for each calculation. The KG policy when using basis functions required .205 seconds to determine the best action which is small, but not negligible. In a true online problem, these execution times are probably negligible (perhaps we are making decisions once a day or once an hour). However, in many approximate dynamic

Table 17.1 Means and standard errors for the storage problem.

	Offline objective		Online objective	
	Mean	Avg. SE	Mean	Avg. SE
Offline KG (basis functions)	1136.20	3.54	-342.23	19.96
Online KG (basis functions)	871.13	3.15	44.58	27.71
Offline KG (lookup)	210.43	0.33	-277.38	15.90
Online KG (lookup)	79.36	0.23	160.28	5.43
Epsilon-greedy (param.)	-475.54	2.30	-329.03	25.31

programming applications, this may be considered a fairly significant amount of overhead. The value of the policy would have to be judged based on the degree to which it accelerated convergence.

One final note of caution has to do with the basis function model itself. If the vector ϕ of basis functions is poorly chosen (for instance, if the features do not adequately describe the value function, or if V is nonlinear in the features), then ADP can perform very poorly. The literature has found cases in which the parameters θ^n will never converge, regardless of how long we run the algorithm. This issue has nothing to do with our particular choice of exploration strategy, but rather is intrinsic to the linear model. Notwithstanding, basis functions continue to be a very popular algorithmic strategy in ADP, due to their remarkable ease of use. In the above model, performance can often be vastly improved by tuning σ_ε^2 or the starting covariance matrix $\Sigma^{\theta,0}$. Unfortunately, we are not able to avoid tunable parameters altogether in our ADP algorithm, the way we did in Chapter 5. It is important to bear in mind, however, that other policies such as epsilon-greedy would add even more tunable parameters if we were to use them together with basis functions.

17.5 AN EXPECTED IMPROVEMENT POLICY

As in Sections 5.6 and 16.4, our version of KG for learning with a physical state also has a close relative in the form of an expected improvement policy. This time, EI appears under the name "value of perfect information" or VPI. The difference in name is due to the fact that this method was developed independently in the computer science community. Recalling that

$$Q^n(S^n, x) = C(S^n, x) + \gamma \phi(S^{x,n})^T \theta^n,$$

and defining

$$\sigma_x^{2,n} = \phi(S^{x,n})^T \Sigma^{\theta,n} \phi(S^{x,n}),$$

we calculate

$$\nu^{VPI,n}(S^{x,n}) = \sigma_x^{2,n} f\left(-\frac{\left|\phi\left(S^{x,n}\right)^T \theta^n - \max_{x' \neq x} \phi\left(S^{x',n}\right)^T \theta^n\right|}{\sigma_x^{2,n}}\right).$$

We then make our decision according to the formula

$$X^{VPI,n}\left(S^n\right) = \arg\max_x C(S^n, x) + \gamma V^n(S^{x,n}) + \nu^{VPI,n}(S^{x,n}).$$

Just as the other variants of EI, this policy implicitly assumes that we will learn the true value of $S^{x,n}$ immediately after choosing the decision x. Furthermore, just like the KG policy, VPI makes a decision by solving Bellman's equation, plus a value of information term.

17.6 BIBLIOGRAPHIC NOTES

In addition to the models discussed in this chapter, there is a separate stream of literature within the computer science community on learning unknown transition probabilities in a Markov decision process. This particular learning problem uses the Dirichlet distribution that we briefly touched on in Section 2.3 to model the unknown probabilities; the resulting algorithms tend to have a high degree of computational complexity. The LBA method of Duff & Barto (1996) also grew out of this literature. An example of an early approach is Silver (1963). For additional work on this topic, see also Dearden et al. (1999), Steele (2000) and Duff (2003).

Section 17.1 - We give a very streamlined introduction to ADP based on Chapters 3 and 4 of Powell (2011). Other good references include Puterman (1994), Bertsekas & Tsitsiklis (1996), and Si et al. (2005).

Section 17.2 - The R-max method was developed by Brafman & Tennenholtz (2003), whereas E^3 is due to Kearns & Singh (2002). Our heuristics from Chapter 4 can also be used in this setting; implementation is described, for example, in Sutton & Barto (1998) and Kaelbling (1993).

Section 17.3 - The LBA policy was originally proposed by Duff & Barto (1996) as a conceptual algorithm. The full implementation was worked out by Ryzhov et al. (2010), where some experimental results are given. The expression of the Gittins index as expected reward per expected unit time is due to Katehakis & Veinott (1987); see also Dupacova (1995). There have also been other crossovers of bandit methods into dynamic programming and reinforcement learning; see Szepesvári (2010) for a UCB-based approach.

Section 17.4 - Basis functions are a standard model for generalized learning; the work by Tesauro (1992) is an example of an early treatment. The model continues to see widespread use; see Sutton et al. (2009) for some recent advances. The first derivation of the knowledge gradient in a dynamic programming setting (that is, in the presence of a physical state) was given in Ryzhov & Powell (2011*b*).

Section 17.5 - The VPI policy was proposed by Dearden et al. (1998).

Bibliography

Abernethy, J., Hazan, E. & Rakhlin, A. (2008), Competing in the dark: An efficient algorithm for bandit linear optimization, *in* 'Proceedings of the 21st Annual Conference on Learning Theory', pp. 263–274.

Aghion, P., Bolton, P., Harris, C. & Jullien, B. (1991), 'Optimal learning by experimentation', *The Review of Economic Studies* **58**, 621–654.

Agrawal, R. (1995), 'Sample mean based index policies with O (log n) regret for the multi-armed bandit problem', *Advances in Applied Probability* **27**(4), 1054–1078.

Alrefaei, M. H. & Andradóttir, S. (1999), 'A simulated annealing algorithm with constant temperature for discrete stochastic optimization', *Management Science* **45**, 748–764.

Andradóttir, S. & Prudius, A. A. (2010), 'Adaptive random search for continuous simulation optimization', *Naval Research Logistics* **57**(6), 583–604.

Ankenman, B., Nelson, B. L. & Staum, J. (2010), 'Stochastic kriging for simulation metamodeling', *Operations Research* **58**(2), 371–382.

Araman, V. & Caldentey, R. (2009), 'Dynamic pricing for non-perishable products with demand learning', *Operations Research* **57**(5), 1169–1188.

Audibert, J. Y., Bubeck, S. & Munos, R. (2010), Best arm identification in multi-armed bandits, *in* 'Proceedings of the 23rd Annual Conference on Learning Theory (COLT)', pp. 1–14.

Auer, P., Cesa-Bianchi, N. & Fischer, P. (2002), 'Finite-time analysis of the multi-armed bandit problem', *Machine Learning* **47**(2), 235–256.

Auer, P., Jaksch, T. & Ortner, R. (2008), Near-optimal regret bounds for reinforcement learning, *in* D. Koller, Y. Bengio, D. Schuurmans, L. Bottou & A. Culotta, eds, 'Advances in Neural Information Processing Systems', Vol. 21, pp. 89–96.

Aviv, Y. & Pazgal, A. (2005), 'A partially observed Markov decision process for dynamic pricing', *Management Science* **51**(9), 1400–1416.

Avriel, M. & Wilde, D. (1966), 'Optimality proof for the symmetric Fibonacci search technique', *Fibonacci Quarterly* **4**, 265–269.

Azadivar, F. & Tompkins, G. (1999), 'Simulation optimization with qualitative variables and structural model changes: A genetic algorithm approach', *European Journal of Operational Research* **113**, 169–182.

Azadivar, F. & Wang, J. (2000), 'Facility Layout Optimization Using Simulation and Genetic Algorithms', *International Journal of Production Research* **38**(17), 4369–4383.

Banks, J., Nelson, B. L. & J. S. Carson, I. I. (1996), *Discrete-Event System Simulation*, Prentice-Hall, Inc., Englewood Cliffs, N.J.

Barr, D. R. & Rizvi, M. H. (1966), 'An introduction to ranking and selection procedures', *J. Amer. Statist. Assoc.* **61**(315), 640– 646.

Bartlett, P., Dani, V., Hayes, T., Kakade, S., Rakhlin, A. & Tewari, A. (2008), High-probability regret bounds for bandit online linear optimization, *in* 'Proceedings of the 21st Annual Conference on Learning Theory', pp. 335–342.

Bechhofer, R. E. (1954), 'A Single-Sample Multiple Decision Procedure for Ranking Means of Normal Populations with known Variances', *The Annals of Mathematical Statistics* **25**, 16–39.

Bechhofer, R. E., Kiefer, J. & Sobel, M. (1968), *Sequential Identification and Ranking Procedures*, University of Chicago Press, Chicago.

Bechhofer, R., Santner, T. & Goldsman, D. (1995), *Design and Analysis of Experiments for Statistical Selection, Screening and Multiple Comparisons*, J.Wiley & Sons, New York.

Bellman, R. & Kalaba, R. (1959), 'On adaptive control processes', *IRE Trans.* **4**, 1–9.

Ben-Tal, A., Ghaoui, L. E. & Nemirovski, A. (2009), *Robust Optimization*, Princeton University Press, Princeton NJ.

Benveniste, A., Metivier, M. & Priouret, P. (1990), *Adaptive Algorithms and Stochastic Approximations*, Springer-Verlag, New York.

Bergemann, D. & Välimäki, J. (1996), 'Learning and strategic pricing', *Econometrica* **64**(5), 1125–1149.

Berry, D. A. & Fristedt, B. (1985), *Bandit Problems*, Chapman and Hall, London.

Bertsekas, D. P. & Tsitsiklis, J. N. (1996), *Neuro-dynamic programming*, Athena Scientific, Belmont, MA.

Bertsimas, D. J. & Nino-Mora, J. (2000), 'Restless bandits, linear programming relaxations, and a primal-dual index heuristic', *Operations Research* **48**(1), 80–90.

Better, M., Glover, F. W., Kochenberger, G. & Wang, H. (2008), 'Simulation optimization: applications in risk management', *International Journal of Information Technology and Decision Making* **7**(4), 571–587.

Beyer, H. & Sendhoff, B. (2007), 'Robust optimization: A comprehensive survey', *Computer Methods in Applied Mechanics and Engineering* **196**(33-34), 3190–3218.

Birchler, U. & Butler, M. (2007), *Information Economics*, Routledge, London.

Boyen, X. & Koller, D. (1998), Tractable inference for complex stochastic processes, *in* 'Proceedings of the 14th Conference on Uncertainty in Artificial Intelligence', pp. 33–42.

Brafman, R. I. & Tennenholtz, M. (2003), 'R-max – a general polynomial time algorithm for near-optimal reinforcement learning', *Journal of Machine Learning Research* **3**, 213–231.

Branke, J., Chick, S. E. & Schmidt, C. (2005), New developments in ranking and selection: an empirical comparison of the three main approaches, *in* M. Kuhl, N. Steiger, F. Argstrong & J. Joines, eds, 'Proc. 2005 Winter Simulation Conference', IEEE, Inc., Piscataway, NJ, pp. 708–717.

Branke, J., Chick, S. E. & Schmidt, C. (2007), 'Selecting a Selection Procedure', *Management Science* **53**, 1916–1932.

Brezzi, M. & Lai, T. L. (2002), 'Optimal learning and experimentation in bandit problems', *Journal of Economic Dynamics and Control* **27**(1), 87–108.

Broder, J. & Rusmevichientong, P. (2010*a*), Dynamic pricing under a general parametric choice model, Technical report, Cornell University.

Broder, J. & Rusmevichientong, P. (2010*b*), Dynamic pricing under a logit choice model, Technical report, Cornell University.

Bruss, F. (1984), 'A unified approach to a class of best choice problems with an unknown number of options', *The Annals of Probability* **12**(3), 882–889.

Bubeck, S. (2010), Bandits Games and Clustering Foundations, PhD thesis, Universite Lille.

Bubeck, S., Munos, R., Stoltz, G. & Szepesvari, C. (2011), 'X-Armed Bandits', *Journal of Machine Learning Research* **12**, 1655–1695.

Bull, A. D. (2011), 'Convergence rates of efficient global optimization algorithms', *Submitted for publication*.

Can, B., Beham, A. & Heavey, C. (2008), A comparative study of genetic algorithm components in simulation-based optimization, *in* S. Mason, R. Hill, L. Mönch, O. Rose, T. Jefferson & J. Fowler, eds, 'Proceedings of the 2008 Winter Simulation Conference', pp. 1829–1837.

Carvalho, A. & Puterman, M. (2003), Dynamic pricing and reinforcement learning, *in* 'Proceedings of the 2003 International Joint Conference on Neural Networks', Vol. 4, pp. 2916–2921.

Carvalho, A. & Puterman, M. (2005), 'Learning and pricing in an Internet environment with binomial demands', *Journal of Revenue and Pricing Management* **3**(4), 320–336.

Cayley, A. (1875), 'Mathematical questions with their solutions, No. 4528', *Educational Times*.

Chade, H. & Schlee, E. E. (2002), 'Another Look at the Radner-Stiglitz Nonconcavity in the Value of Information', *Journal of Economic Theory* **107**(2), 421–452.

Chang, H. S., Fu, M. C., Hu, J. & Marcus, S. I. (2007), *Simulation-Based Algorithms for Markov Decision Processes*, Springer, New York.

Chaudhry, S. & Luo, W. (2005), 'Application of genetic algorithms in production and operations management: a review', *International Journal of Production Research* **43**(19), 4083–4101.

Chen, C.-H. & Lee, L. H. (2010), *Stochastic Simulation Optimization: An Optimal Computing Budget Allocation*, World Scientific.

Chen, C.-H., Fu, M. C. & Shi, L. (2008), Simulation and optimization, *in* P. Gray, Z.-L. Chen & S. Raghavan, eds, '2008 TutORials in Operations Research', pp. 247–260.

Chen, C.-H., He, D. & Fu, M. C. (2006), 'Efficient Dynamic Simulation Allocation in Ordinal Optimization', *IEEE Transactions Automatic Control* **51**, 2005–2009.

Chen, C.-H., Lin, J., Yücesan, E. & Chick, S. E. (2000), 'Simulation budget allocation for further enhancing the efficiency of ordinal optimization', *Discrete Event Dynamic Systems* **10**(3), 251–270.

Chhabra, M. & Das, S. (2011), Learning the Demand Curve in Posted-Price Digital Goods Auctions, *in* 'Proceedings of the 10th International Conference on Autonomous Agents and Multi-Agent Systems', pp. 63–70.

Chhabra, M. & Das, S. (2012), Dynamic pricing with non-conjugate Pareto priors, Technical report, Rensselaer Polytechnic Institute.

Chick, S. E. (2003), Expected opportunity cost guarantees and indifference zone selection procedures, *in* S. E. Chick, P. J. Sánchez, D. Ferrin & D. J. Morrice, eds, 'Proceedings of the 2003 Winter Simulation Conference', pp. 465–473.

Chick, S. E. (2006), Subjective Probability and Bayesian Methodology, *in* S. Henderson & B. Nelson, eds, 'Handbooks of Operations Research and Management Science, vol. 13: Simulation', North-Holland Publishing, Amsterdam, pp. 225–258.

Chick, S. E. & Gans, N. (2009), 'Economic analysis of simulation selection problems', *Management Science* **55**(3), 421–437.

Chick, S. E. & Inoue, K. (2001), 'New two-stage and sequential procedures for selecting the best simulated system', *Operations Research* **49**(5), 732—-743.

Chick, S. E. & Wu, Y. (2005), 'Selection procedures with frequentist expected opportunity cost bounds', *Operations Research* **53**(5), 867–878.

Chick, S. E., Branke, J. & Schmidt, C. (2010), 'Sequential Sampling to Myopically Maximize the Expected Value of Information', *INFORMS Journal on Computing* **22**(1), 71–80.

Chick, S. E., He, D. H. & Chen, C.-H. (2007), 'Opportunity cost and OCBA selection procedures in ordinal optimization for a fixed number of alternative systems', *IEEE Transactions on Systems Man and Cybernetics* **C37**, 951–961.

Chvátal, V. (1983), *Linear programming*, WH Freeman.

Cinlar, E. (2011), *Probability and Stochastics*, Springer.

Cohen, J. D., McClure, S. M. & Yu, A. J. (2007), 'Should I stay or should I go? How the human brain manages the trade-off between exploitation and exploration.', *Philosophical transactions of the Royal Society of London* **B362**(1481), 933–42.

Cohn, D., Atlas, L. & Ladner, R. (1994), 'Improving generalization with active learning', *Machine Learning* **5**(2201), 221.

Cohn, D., Ghahramani, Z. & Jordan, M. (1996), 'Active learning with statistical models', *Journal of Artificial Intelligence Research* **4**, 129–145.

Cozzolino, J., Gonzalez-Zubieta, R. & Miller, R. (1965), Markov decision processes with uncertain transition probabilities, Technical Report 11, Operations Research Center, MIT.

Crooks, G. (2009), Logistic approximation to the logistic-normal integral, Technical Report 002v2, Lawrence Berkeley National Laboratory.

Davis, L. & Mitchell, M. (1991), *Handbook of genetic algorithms*, Van Nostrand Reinhold.

Dayanik, S., Powell, W. B. & Yamazaki, K. (2008), 'Index policies for discounted bandit problems with availability constraints', *Adv. in Appl. Probab* **40**, 377–400.

Dearden, R., Friedman, N. & Andre, D. (1999), Model-based Bayesian Exploration, *in* 'Proceedings of the 15th Conference on Uncertainty in Artificial Intelligence', pp. 150–159.

Dearden, R., Friedman, N. & Russell, S. (1998), Bayesian Q-learning, *in* 'Proceedings of the 15th National Conference on Artificial Intelligence', pp. 761–768.

DeGroot, M. H. (1970), *Optimal Statistical Decisions*, John Wiley and Sons.

Delara, M. & Gilotte, L. (2007), 'A tight sufficient condition for Radner-Stiglitz non-concavity in the value of information', *Journal of Economic Theory* **137**(1), 696–708.

Duff, M. (2003), Design for an optimal probe, *in* 'Proceedings of the 20th International Conference on Machine Learning', pp. 131–138.

Duff, M. & Barto, A. (1996), Local bandit approximation for optimal learning problems, *in* M. Mozer, M. Jordan & T. Pesche, eds, 'Advances in Neural Information Processing Systems', Vol. 9, Cambridge, MA: MIT Press, pp. 1019–1025.

Dupacova, J. (1995), 'Multistage Stochastic Programs - The State of the Art and Selected Bibliography', *Kybernetica* **31**, 151–174.

Farias, V. & Van Roy, B. (2010), 'Dynamic pricing with a prior on market response', *Operations Research* **58**(1), 16–29.

Forrester, A. I. J., Sobester, A. & Keane, A. J. (2008), *Engineering design via surrogate modelling: a practical guide*, John Wiley and Sons.

Frazier, P. I. & Powell, W. B. (2010), 'Paradoxes in Learning and the Marginal Value of Information', *Decision Analysis* **7**(4), 378–403.

Frazier, P. I. & Powell, W. B. (2011), 'Consistency of Sequential Bayesian Sampling Policies', *SIAM Journal on Control and Optimization* **49**(2), 712–731.

Frazier, P. I., Powell, W. B. & Dayanik, S. (2008), 'A Knowledge Gradient Policy for Sequential Information Collection', *SIAM Journal on Control and Optimization* **47**(5), 2410—2439.

Frazier, P. I., Powell, W. B. & Dayanik, S. (2009), 'The Knowledge-Gradient Policy for Correlated Normal Beliefs', *INFORMS Journal on Computing* **21**(4), 599–613.

Fu, M. C. (2002), 'Optimization for simulation: Theory vs. practice', *INFORMS Journal on Computing* **14**(3), 192–215.

Fu, M. C., Glover, F. & April, J. (2005), Simulation optimization: a review, new developments, and applications, *in* M. E. Kuhl, N. M. Steiger, F. B. Armstrong & J. A. Joines, eds, 'Proceedings of the 2005 Winter Simulation Conference', pp. 83–95.

Fu, M. C., Hu, J. Q., Chen, C.-H. & Xiong, X. (2004), Optimal computing budget allocation under correlated sampling, *in* R. G. Ingalls, M. D. Rossetti, J. S. Smith & B. A. Peters, eds, 'Proceedings of the 2004 Winter Simulation Conference', pp. 595–603.

Gallego, G. & Van Ryzin, G. (1994), 'Optimal dynamic pricing of inventories with stochastic demand over finite horizons', *Management Science* **40**(8), 999–1020.

Gelman, A., Carlin, J. B., Stern, H. S. & Rubin, D. B. (2004), 'Bayesian Data Analysis, 2nd ed', *Chapman & Hall, New York* p. 63.

Geman, S., Bienenstock, E. & Doursat, R. (1992), 'Neural networks and the bias/variance dilemma', *Neural computation* **4**(1), 1–58.

Ghaffari-Hadigheh, A. & Terlaky, T. (2006), 'Sensitivity analysis in linear optimization: Invariant support set intervals', *European Journal of Operational Research* **169**(3), 1158–1175.

Ginebra, J. & Clayton, M. K. (1995), 'Response Surface Bandits', *Journal of the Royal Statistical Society* **B57**, 771–784.

Gittins, J. C. (1979), 'Bandit processes and dynamic allocation indices', *Journal of the Royal Statistical Society* **B41**(2), 148–177.

Gittins, J. C. (1989), 'Multi-armed Bandit Allocation Indices', *Wiley and Sons: New York*.

Gittins, J. C. & Jones, D. M. (1974), A dynamic allocation index for the sequential design of experiments, *in* J. Gani, ed., 'Progress in statistics', North Holland, Amsterdam, pp. 241––266.

Gittins, J. C., Glazebrook, K. D. & Weber, R. R. (2011), *Multi-Armed Bandit Allocation Indices*, John Wiley & Sons, New York.

Glasserman, P. (2004), *Monte Carlo Methods in Financial Engineering*, Springer-Verlag, New York.

Glazebrook, K. D. (1982), 'On the evaluation of suboptimal strategies for families of alternative bandit processes', *Journal of Applied Probability* **19**(3), 716–722.

Glazebrook, K. D. & Minty, R. (2009), 'A Generalized Gittins Index for a Class of Multiarmed Bandits with General Resource Requirements', *Mathematics of Operations Research*.

Goldberg, D. E. & Deb, K. (1991), A comparative analysis of selection schemes used in genetic algorithms, *in* G. Rawlings, ed., 'Foundations of genetic algorithms', Morgan Kaufmann Publishers, San Mateo, CA, pp. 69–93.

Gosavi, A. (2003), *Simulation-Based Optimization*, Kluwer Academic Publishers, Norwell, MA.

Gramacy, R. B. & Lee, H. K. H. (2011), 'Optimization under unknown constraints', *Arxiv preprint arXiv:1004.4027*.

Gupta, S. S. & Miescke, K. J. (1994), 'Bayesian look ahead one stage sampling allocations for selecting the largest normal mean', *Statistical Papers* **35**, 169–177.

Gupta, S. S. & Miescke, K. J. (1996), 'Bayesian look ahead one-stage sampling allocations for selection of the best population', *Journal of Statistical Planning and Inference* **54**, 229–244.

Harrison, J., Keskin, N. & Zeevi, A. (2010), Bayesian dynamic pricing policies: Learning and earning under a binary prior distribution, Technical report, Working paper, Columbia and Stanford University.

Hastie, T., Tibshirani, R. & Friedman, J. (2009), *The elements of statistical learning: data mining, inference and prediction*, Springer, New York.

Hastie, T., Tibshirani, R., Friedman, J. & Franklin, J. (2005), *The elements of statistical learning: data mining, inference and prediction*, Vol. 27, Springer, New York.

Hedlund, H. E. & Mollaghasemi, M. (2001), A Genetic Algorithm and an Indifference-Zone Ranking and Selection Framework for Simulation Optimization, *in* B. A. Peters, J. S. Smith, D. J. Medeiros & M. W. Rohrer, eds, 'Proceedings of the 2001 Winter Simulation Conference', pp. 417–421.

Hong, L. J. & Nelson, B. L. (2006), 'Discrete Optimization via Simulation Using COMPASS', *Operations Research* **54**(1), 115–129.

Hong, L. J. & Nelson, B. L. (2009), A Brief Introduction To Optimization Via Simulation, *in* M. Rosetti, R. Hill, B. Johansson, A. Dunkin & R. Ingalls, eds, 'Proceedings of the 2009 Winter Simulation Conference', pp. 75–85.

Horrace, W., Marchand, J. & Smeeding, T. (2008), 'Ranking inequality: Applications of multivariate subset selection', *Journal of Economic Inequality* **6**(1), 5–32.

Horstein, M. (1963), 'Sequential transmission using noiseless feedback', *IEEE Transactions on Information Theory* **9**(3), 136–143.

Howard, R. A. (1966), 'Information value theory', *IEEE Transactions on systems science and cybernetics* **2**(1), 22—-26.

Hu, J. & Hu, P. (2010), An Approximate Annealing Search Algorithm to Global Optimization and its Connection to Stochastic Approximation, *in* B. Johansson, S. Jain, J. Montoya-Torres, J. Hugan & E. Yücesan, eds, 'Proceedings of the 2010 Winter Simulation Conference', pp. 1223–1234.

Hu, J. & Hu, P. (2011), 'Annealing Adaptive Search, Cross-Entropy, and Stochastic Approximation in Global Optimization', *Naval Research Logistics (to appear)*.

Hu, J. & Wang, C. (2011), Discrete Optimization Via Approximate Annealing Adaptive Search With Stochastic Averaging, *in* S. Jain, R. R. Creasey, J. Himmelspach, K. P. White & M. C. Fu, eds, 'Proceedings of the 2011 Winter Simulation Conference'.

Hu, J., Fu, M. & Marcus, S. (2007), 'A model reference adaptive search method for global optimization', *Operations Research* **55**(3), 549–568.

Huang, D., Allen, T. T., Notz, W. I. & Zeng, N. (2006), 'Global Optimization of Stochastic Black-Box Systems via Sequential Kriging Meta-Models', *Journal of Global Optimization* **34**(3), 441–466.

Inoue, K., Chick, S. E. & Chen, C.-H. (1999), 'An empirical evaluation of several methods to select the best system', *ACM Transactions on Modeling and Computer Simulation* **9**, 381–407.

Jaakkola, T. & Jordan, M. (2000), 'Bayesian parameter estimation via variational methods', *Statistics and Computing* **10**(1), 25–37.

Jedynak, B., Frazier, P. I. & Sznitman, R. (2011), 'Questions with noise: Bayes optimal policies for entropy loss', *Journal of Applied Probability (to appear)*.

Jones, D., Schonlau, M. & Welch, W. (1998), 'Efficient global optimization of expensive black-box functions', *Journal of Global Optimization* **13**(4), 455–492.

Kaelbling, L. P. (1993), *Learning in embedded systems*, MIT Press, Cambridge, MA.

Kalyanam, K. (1996), 'Pricing decisions under demand uncertainty: A Bayesian mixture model approach', *Marketing Science* **15**(3), 207–221.

Kaminski, P., Bryson, A. & Schmidt, S. (1971), 'Discrete square root filtering: A survey of current techniques', *IEEE Transactions on Automatic Control* **16**(6), 727–736.

Katehakis, M. & Veinott, A. (1987), 'The Multi-Armed Bandit Problem: Decomposition and Computation', *Mathematics of Operations Research* **12**(2), 262–268.

Katz, R. & Ionescu, F. (1977), 'Application of the Free-Wilson Technique to Structurally Related Series of Homologues. Quantitative S truct ure-Activity Relations hip Studies of Narcotic Analgetics', **20**(11), 1413–1419.

Kearns, M. & Singh, S. (2002), 'Near-optimal reinforcement learning in polynomial time.', *Machine Learning* **49**, 209–232.

Kim, S.-H. & Nelson, B. L. (2001), 'A fully sequential procedure for indifference-zone selection in simulation', *ACM Trans. Model. Comput. Simul.* **11**, 251–273.

Kim, S.-H. & Nelson, B. L. (2006), 'On the asymptotic validity of fully sequential selection procedures for steady-state simulation', *Operations Research* **54**, 475–488.

Kleinberg, R. (2004), Nearly Tight Bounds for the Continuum-Armed Bandit Problem, *in* L. Saul, Y. Weiss & L. Bottou, eds, 'Advances in Neural Information Processing Systems', MIT Press, Cambridge, MA, pp. 697–704.

Kleinberg, R., Niculescu-Mizil, A. & Sharma, Y. (2010), 'Regret bounds for sleeping experts and bandits', *Machine Learning* **80**(2), 245–272.

Kulkarni, V. (1986), 'Shortest paths in networks with exponentially distributed arc lengths', *Networks* **16**, 255–274.

Kushner, H. J. & Yin, G. G. (2003), *Stochastic Approximation and Recursive Algorithms and Applications*, Springer.

Lai, T. L. (1987), 'Adaptive treatment allocation and the multi-armed bandit problem', *The Annals of Statistics* **15**(3), 1091–1114.

Lai, T. L. & Robbins, H. (1985), 'Asymptotically Efficient Adaptive Allocation Rules', *Advances in Applied Mathematics* **6**, 4–22.

Mandelbaum, A. (1987), 'Continuous multi-armed bandits and multiparameter processes', *The Annals of Probability* **15**(4), 1527–1556.

Martin, J. (1967), *Bayesian Decision Problems and Markov Chains*, John Wiley and Sons.

Matejcik, F. & Nelson, B. (1995), 'Two-stage multiple comparisons with the best for computer simulation', *Operations Research* **43**(4), 633–640.

McKinsey & Company (2007), 'Reducing U.S. Greenhouse Gas Emissions: How Much at What Cost?', U.S. Greenhouse Gas Abatement Mapping Initiative, Executive Report.

Miller, A. (2002), *Subset selection in regression*, CRC Press.

Minka, T. (2001), A family of algorithms for approximate Bayesian inference, PhD thesis, Massachusetts Institute of Technology.

Montgomery, D. C. (2008), *Design and analysis of experiments (7th ed.)*, John Wiley and Sons.

Negoescu, D. M., Frazier, P. I. & Powell, W. B. (2011), 'The Knowledge-Gradient Algorithm for Sequencing Experiments in Drug Discovery', *INFORMS Journal on Computing* **23**(3), 346–363.

Nelson, B. L. & Matejcik, F. J. (1995), 'Using common random numbers for indifference-zone selection and multiple comparisons in simulation', *Management Science* **41**(12), 1935–1945.

Nino-Mora, J. (2010), 'Computing a Classic Index for Finite-Horizon Bandits', *INFORMS Journal on Computing* **23**(2), 254–267.

Pasupathy, R. & Kim, S. (2011), 'The stochastic root-finding problem: Overview, solutions, and open questions', *ACM Transactions on Modeling and Computer Simulation* **21**(3), 19:1–19:23.

Paulson, E. (1964), 'A Sequential Procedure for Selecting the Population with the Largest Mean from k Normal Populations', *The Annals of Mathematical Statistics* **35**, 174–180.

Peer, S. & Sharma, D. (2007), 'Finding the shortest path in stochastic networks', *Computers and Mathematics with Applications* **53**, 729–740.

Powell, W. B. (2007), *Approximate Dynamic Programming: Solving the curses of dimensionality*, John Wiley & Sons, Hoboken, NJ.

Powell, W. B. (2011), *Approximate Dynamic Programming: Solving the curses of dimensionality*, 2nd. edn, John Wiley & Sons, Hoboken, NJ.

Puterman, M. L. (1994), *Markov Decision Processes*, 1st edn, John Wiley and Sons, Hoboken.

Radner, R. & Stiglitz, J. (1984), A Nonconcavity in the Value of Information, *in* M. Boyer & R. Kihlstrom, eds, 'Bayesian models in economic theory', Vol. 5, North-Holland, Amsterdam, chapter 3, pp. 33–52.

Rasmussen, C. E. & Williams, C. K. I. (2006), 'Gaussian Processes for Machine Learning', *The MIT Press*.

Rinott, Y. (1978), 'On two-stage selection procedures and related probability inequalities', *Communications in Statistics A7* pp. 799–811.

Robbins, H. & Monro, S. (1951), 'A stochastic approximation method', *The Annals of Mathematical Statistics* **22**(3), 400–407.

Roberts, C. P. & Casella, G. (2004), *Monte Carlo Statistical Methods*, Springer-Verlag, New York.

Romeijn, H. & Smith, R. (1994), 'Simulated annealing and adaptive search in global optimization', *Probability in the Engineering and Informational Sciences* **8**(4), 571–590.

Rothschild, M. (1974), 'A two-armed bandit theory of market pricing', *Journal of Economic Theory* **9**(2), 185–202.

Rubinstein, R. Y. & Kroese, D. P. (2008), *Simulation and the Monte Carlo Method*, Wiley-Interscience, New York.

Rusmevichientong, P. & Tsitsiklis, J. N. (2010), 'Linearly Parameterized Bandits', *Mathematics of Operations Research* **35**(2), 395–411.

Ruszczynski, A. & Shapiro, A. (2003), *Handbooks in Operations Research and Management Science: Stochastic Programming*, Vol. 10, Elsevier, Amsterdam.

Ryzhov, I. O. & Powell, W. B. (2009*a*), A Monte-Carlo Knowledge Gradient Method For Learning Abatement Potential Of Emissions Reduction Technologies, *in* M. D. Rossetti, R. R. Hill, B. Johansson, A. Dunkin & R. G. Ingalls, eds, 'Proceedings of the 2009 Winter Simulation Conference'.

Ryzhov, I. O. & Powell, W. B. (2009*b*), The knowledge gradient algorithm for on-line subset selection, *in* 'Proceedings of the 2009 IEEE Symposium on Adaptive Dynamic Programming and Reinforcement Learning', pp. 137–144.

Ryzhov, I. O. & Powell, W. B. (2011*a*), 'Information collection for linear programs with unknown objective coefficients', *Submitted for publication*.

Ryzhov, I. O. & Powell, W. B. (2011*b*), 'Information Collection on a Graph', *Operations Research* **59**(1), 188–201.

Ryzhov, I. O. & Powell, W. B. (2011*c*), The value of information in multi-armed bandits with exponentially distributed rewards, *in* 'Proceedings of the 2011 International Conference on Computational Science', pp. 1363–1372.

Ryzhov, I. O., Powell, W. B. & Frazier, P. I. (2011), 'The knowledge gradient algorithm for a general class of online learning problems', *Operations Research (to appear)*.

Ryzhov, I. O., Valdez-Vivas, M. R. & Powell, W. B. (2010), Optimal Learning of Transition Probabilities in the Two-Agent Newsvendor Problem, *in* B. Johansson, S. Jain, J. Montoya-Torres, J. Hugan & E. Yücesan, eds, 'Proceedings of the 2010 Winter Simulation Conference', pp. 1088–1098.

Sacks, J., Welch, W., Mitchell, T. J. & Wynn, H. P. (1989), 'Design and analysis of computer experiments', *Statistical Science* **4**(4), 409–423.

Satia, J. & Lave, R. (1973), 'Markov decision processes with imprecise transition probabilities', *Operations Research* **21**(3), 755–763.

Scott, W. R., Frazier, P. I. & Powell, W. B. (2011), 'The Correlated Knowledge Gradient for Simulation Optimization of Continuous Parameters using Gaussian Process Regression', *SIAM Journal on Optimization* **21**(3), 996–1026.

Scott, W. R., Powell, W. B. & Simão, H. P. (2010), Calibrating simulation models using the knowledge gradient with continuous parameters, *in* B. Johansson, S. Jain, J. Montoya-Torres, J. Hugan & E. Yücesan, eds, 'Proceedings of the 2010 Winter Simulation Conference', pp. 1099–1109.

Settles, B. (2009), Active learning literature survey, Computer Sciences Technical Report 1648, University of Wisconsin–Madison.

Si, J., Barto, A. G., Powell, W. B. & Wunsch, D. (2005), *Learning and Approximate Dynamic Programming*.

Silver, E. (1963), Markovian decision processes with uncertain transition probabilities or rewards, Technical Report 1, Operations Research Center, MIT.

Singh, S. P., Jaakkola, T., Szepesvari, C. & Littman, M. (2000), 'Convergence results for single-step on-policy reinforcement-learning algorithms', *Machine Learning* **38**(3), 287–308.

Snyder, T. & Steele, J. (1995), Probabilistic networks and network algorithms, *in* M. Ball, T. Magnanti & C. Monma, eds, 'Handbooks of Operations Research and Management Science, vol. 7: Networks', North-Holland Publishing, Amsterdam, pp. 401–424.

Spall, J. C. (2003), *Introduction to Stochastic Search and Optimization: Estimation, Simulation and Control*, John Wiley & Sons, Hoboken, NJ.

Srinivas, N., Krause, A., Kakade, S. M. & Seeger, M. (2010), Gaussian process optimization in the bandit setting: No regret and experimental design, *in* 'Proceedings of the 27th International Conference on Machine Learning', pp. 1015–1022.

Steele, J. M. (2000), *Stochastic Calculus and Financial Applications*, Springer, New York.

Stein, M. (1999), *Interpolation of Spatial Data: Some theory for kriging*, Springer Verlag.

Streeter, M. & Smith, S. (2006), A simple distribution-free approach to the max k-armed bandit problem, *in* 'Principles and Practice of Constraint Programming', Vol. 4204 of *Lecture Notes in Computer Science*, pp. 560–574.

Sutton, R. S. & Barto, A. G. (1998), *Reinforcement Learning*, Vol. 35, MIT Press, Cambridge, MA.

Sutton, R. S., Maei, H. R., Precup, D., Bhatnagar, S., Silver, D., Szepesvári, C. & Wiewiora, E. (2009), Fast gradient-descent methods for temporal-difference learning with linear function approximation, *in* 'Proceedings of the 26th International Conference on Machine Learning', pp. 993–1000.

Szepesvári, C. (2010), *Algorithms for reinforcement learning*, Vol. 4 of *Synthesis Lectures on Artificial Intelligence and Machine Learning*, Morgan & Claypool Publishers.

Takimoto, E. & Warmuth, M. K. (2003), 'Path kernels and multiplicative updates', *Journal of Machine Learning Research* **4**, 773–818.

Tesauro, G. (1992), 'Practical Issues in Temporal Difference Learning', *Machine Learning* **8**(3-4), 257–277.

Vanderbei, R. J. (1980), 'The optimal choice of a subset of a population', *Mathematics of Operations Research* **5**(4), 481–486.

Vanderbei, R. J. (2008), *Linear programming: foundations and extensions (3rd ed.)*, Springer.

Vermorel, J. & Mohri, M. (2005), 'Multi-armed bandit algorithms and empirical evaluation', *Proceedings of the 16th European Conference on Machine Learning* pp. 437–448.

Wald, A. & Wolfowitz, J. (1948), 'Optimum Character of the Sequential Probability Ratio Test', *The Annals of Mathematical Statistics* **19**, 326–339.

Weber, R. R. & Weiss, G. (1990), 'On an index policy for restless bandits', *J. Appl. Prob.* **27**(3), 637–648.

Weibull, J. W., Mattsson, L.-G. & Voorneveld, M. (2007), 'Better May be Worse: Some Monotonicity Results and Paradoxes in Discrete Choice Under Uncertainty', *Theory and Decision* **63**(2), 121–151.

Wetherill, G. B. & Glazebrook, K. D. (1986), *Sequential methods in statistics*, Chapman & Hall.

Whittle, P. (1980), 'Multi-armed bandits and the Gittins index', *Journal of the Royal Statistical Society* **B42**(2), 143–149.

Whittle, P. (1988), 'Restless bandits: Activity Allocation in a Changing World', *J. Appl. Prob.* **25**(1988), 287–298.

INDEX

WILEY SERIES IN PROBABILITY AND STATISTICS

ESTABLISHED BY WALTER A. SHEWHART AND SAMUEL S. WILKS

Editors: *David J. Balding, Noel A. C. Cressie, Garrett M. Fitzmaurice,
Harvey Goldstein, Iain M. Johnstone, Geert Molenberghs, David W. Scott,
Adrian F. M. Smith, Ruey S. Tsay, Sanford Weisberg*
Editors Emeriti: *Vic Barnett, J. Stuart Hunter, Joseph B. Kadane, Jozef L. Teugels*

The *Wiley Series in Probability and Statistics* is well established and authoritative. It covers
many topics of current research interest in both pure and applied statistics and probability
theory. Written by leading statisticians and institutions, the titles span both state-of-the-art
developments in the field and classical methods.

Reflecting the wide range of current research in statistics, the series encompasses applied,
methodological and theoretical statistics, ranging from applications and new techniques
made possible by advances in computerized practice to rigorous treatment of theoretical
approaches.

This series provides essential and invaluable reading for all statisticians, whether in academia, industry, government, or research.

† ABRAHAM and LEDOLTER · Statistical Methods for Forecasting
AGRESTI · Analysis of Ordinal Categorical Data, *Second Edition*
AGRESTI · An Introduction to Categorical Data Analysis, *Second Edition*
AGRESTI · Categorical Data Analysis, *Second Edition*
ALTMAN, GILL, and McDONALD · Numerical Issues in Statistical Computing for the
    Social Scientist
AMARATUNGA and CABRERA · Exploration and Analysis of DNA Microarray and
    Protein Array Data
ANDĚL · Mathematics of Chance
ANDERSON · An Introduction to Multivariate Statistical Analysis, *Third Edition*
* ANDERSON · The Statistical Analysis of Time Series
ANDERSON, AUQUIER, HAUCK, OAKES, VANDAELE, and WEISBERG ·
    Statistical Methods for Comparative Studies
ANDERSON and LOYNES · The Teaching of Practical Statistics
ARMITAGE and DAVID (editors) · Advances in Biometry
ARNOLD, BALAKRISHNAN, and NAGARAJA · Records
* ARTHANARI and DODGE · Mathematical Programming in Statistics
* BAILEY · The Elements of Stochastic Processes with Applications to the Natural
    Sciences
BAJORSKI · Statistics for Imaging, Optics, and Photonics
BALAKRISHNAN and KOUTRAS · Runs and Scans with Applications
BALAKRISHNAN and NG · Precedence-Type Tests and Applications
BARNETT · Comparative Statistical Inference, *Third Edition*
BARNETT · Environmental Statistics
BARNETT and LEWIS · Outliers in Statistical Data, *Third Edition*
BARTOSZYNSKI and NIEWIADOMSKA-BUGAJ · Probability and Statistical Inference
BASILEVSKY · Statistical Factor Analysis and Related Methods: Theory and
    Applications
BASU and RIGDON · Statistical Methods for the Reliability of Repairable Systems
BATES and WATTS · Nonlinear Regression Analysis and Its Applications
BECHHOFER, SANTNER, and GOLDSMAN · Design and Analysis of Experiments for
    Statistical Selection, Screening, and Multiple Comparisons
BEIRLANT, GOEGEBEUR, SEGERS, TEUGELS, and DE WAAL · Statistics of
    Extremes: Theory and Applications

*Now available in a lower priced paperback edition in the Wiley Classics Library.
†Now available in a lower priced paperback edition in the Wiley–Interscience Paperback Series.

*Now available in a lower priced paperback edition in the Wiley Classics Library.
†Now available in a lower priced paperback edition in the Wiley–Interscience Paperback Series.

CHILÈS and DELFINER · Geostatistics: Modeling Spatial Uncertainty
CHOW and LIU · Design and Analysis of Clinical Trials: Concepts and Methodologies, *Second Edition*
CLARKE · Linear Models: The Theory and Application of Analysis of Variance
CLARKE and DISNEY · Probability and Random Processes: A First Course with Applications, *Second Edition*
* COCHRAN and COX · Experimental Designs, *Second Edition*
COLLINS and LANZA · Latent Class and Latent Transition Analysis: With Applications in the Social, Behavioral, and Health Sciences
CONGDON · Applied Bayesian Modelling
CONGDON · Bayesian Models for Categorical Data
CONGDON · Bayesian Statistical Modelling, *Second Edition*
CONOVER · Practical Nonparametric Statistics, *Third Edition*
COOK · Regression Graphics
COOK and WEISBERG · An Introduction to Regression Graphics
COOK and WEISBERG · Applied Regression Including Computing and Graphics
CORNELL · A Primer on Experiments with Mixtures
CORNELL · Experiments with Mixtures, Designs, Models, and the Analysis of Mixture Data, *Third Edition*
COVER and THOMAS · Elements of Information Theory
COX · A Handbook of Introductory Statistical Methods
* COX · Planning of Experiments
CRESSIE · Statistics for Spatial Data, *Revised Edition*
CRESSIE and WIKLE · Statistics for Spatio-Temporal Data
CSÖRGŐ and HORVÁTH · Limit Theorems in Change Point Analysis
DANIEL · Applications of Statistics to Industrial Experimentation
DANIEL · Biostatistics: A Foundation for Analysis in the Health Sciences, *Eighth Edition*
* DANIEL · Fitting Equations to Data: Computer Analysis of Multifactor Data, *Second Edition*
DASU and JOHNSON · Exploratory Data Mining and Data Cleaning
DAVID and NAGARAJA · Order Statistics, *Third Edition*
* DEGROOT, FIENBERG, and KADANE · Statistics and the Law
DEL CASTILLO · Statistical Process Adjustment for Quality Control
DeMARIS · Regression with Social Data: Modeling Continuous and Limited Response Variables
DEMIDENKO · Mixed Models: Theory and Applications
DENISON, HOLMES, MALLICK and SMITH · Bayesian Methods for Nonlinear Classification and Regression
DETTE and STUDDEN · The Theory of Canonical Moments with Applications in Statistics, Probability, and Analysis
DEY and MUKERJEE · Fractional Factorial Plans
DILLON and GOLDSTEIN · Multivariate Analysis: Methods and Applications
DODGE · Alternative Methods of Regression
* DODGE and ROMIG · Sampling Inspection Tables, *Second Edition*
* DOOB · Stochastic Processes
DOWDY, WEARDEN, and CHILKO · Statistics for Research, *Third Edition*
DRAPER and SMITH · Applied Regression Analysis, *Third Edition*
DRYDEN and MARDIA · Statistical Shape Analysis
DUDEWICZ and MISHRA · Modern Mathematical Statistics
DUNN and CLARK · Basic Statistics: A Primer for the Biomedical Sciences, *Third Edition*
DUPUIS and ELLIS · A Weak Convergence Approach to the Theory of Large Deviations

*Now available in a lower priced paperback edition in the Wiley Classics Library.
†Now available in a lower priced paperback edition in the Wiley–Interscience Paperback Series.

---

*Now available in a lower priced paperback edition in the Wiley Classics Library.

†Now available in a lower priced paperback edition in the Wiley–Interscience Paperback Series.

HINKELMANN and KEMPTHORNE · Design and Analysis of Experiments, Volume 1: Introduction to Experimental Design, *Second Edition*

HINKELMANN and KEMPTHORNE · Design and Analysis of Experiments, Volume 2: Advanced Experimental Design

HINKELMANN (editor) · Design and Analysis of Experiments, Volume 3: Special Designs and Applications

HOAGLIN, MOSTELLER, and TUKEY · Fundamentals of Exploratory Analysis of Variance

* HOAGLIN, MOSTELLER, and TUKEY · Exploring Data Tables, Trends and Shapes

* HOAGLIN, MOSTELLER, and TUKEY · Understanding Robust and Exploratory Data Analysis

HOCHBERG and TAMHANE · Multiple Comparison Procedures

HOCKING · Methods and Applications of Linear Models: Regression and the Analysis of Variance, *Second Edition*

HOEL · Introduction to Mathematical Statistics, *Fifth Edition*

HOGG and KLUGMAN · Loss Distributions

HOLLANDER and WOLFE · Nonparametric Statistical Methods, *Second Edition*

HOSMER and LEMESHOW · Applied Logistic Regression, *Second Edition*

HOSMER, LEMESHOW, and MAY · Applied Survival Analysis: Regression Modeling of Time-to-Event Data, *Second Edition*

HUBER · Data Analysis: What Can Be Learned From the Past 50 Years

HUBER · Robust Statistics

† HUBER and RONCHETTI · Robust Statistics, *Second Edition*

HUBERTY · Applied Discriminant Analysis, *Second Edition*

HUBERTY and OLEJNIK · Applied MANOVA and Discriminant Analysis, *Second Edition*

HUITEMA · The Analysis of Covariance and Alternatives: Statistical Methods for Experiments, Quasi-Experiments, and Single-Case Studies, *Second Edition*

HUNT and KENNEDY · Financial Derivatives in Theory and Practice, *Revised Edition*

HURD and MIAMEE · Periodically Correlated Random Sequences: Spectral Theory and Practice

HUSKOVA, BERAN, and DUPAC · Collected Works of Jaroslav Hajek— with Commentary

HUZURBAZAR · Flowgraph Models for Multistate Time-to-Event Data

IMAN and CONOVER · A Modern Approach to Statistics

JACKMAN · Bayesian Analysis for the Social Sciences

† JACKSON · A User's Guide to Principle Components

JOHN · Statistical Methods in Engineering and Quality Assurance

JOHNSON · Multivariate Statistical Simulation

JOHNSON and BALAKRISHNAN · Advances in the Theory and Practice of Statistics: A Volume in Honor of Samuel Kotz

JOHNSON and BHATTACHARYYA · Statistics: Principles and Methods, *Fifth Edition*

JOHNSON, KEMP, and KOTZ · Univariate Discrete Distributions, *Third Edition*

JOHNSON and KOTZ · Distributions in Statistics

JOHNSON and KOTZ (editors) · Leading Personalities in Statistical Sciences: From the Seventeenth Century to the Present

JOHNSON, KOTZ, and BALAKRISHNAN · Continuous Univariate Distributions, Volume 1, *Second Edition*

JOHNSON, KOTZ, and BALAKRISHNAN · Continuous Univariate Distributions, Volume 2, *Second Edition*

JOHNSON, KOTZ, and BALAKRISHNAN · Discrete Multivariate Distributions

JUDGE, GRIFFITHS, HILL, LÜTKEPOHL, and LEE · The Theory and Practice of Econometrics, *Second Edition*

JUREČKOVÁ and SEN · Robust Statistical Procedures: Aymptotics and Interrelations

---

*Now available in a lower priced paperback edition in the Wiley Classics Library.
†Now available in a lower priced paperback edition in the Wiley–Interscience Paperback Series.

LE · Applied Survival Analysis

LEE · Structural Equation Modeling: A Bayesian Approach

LEE and WANG · Statistical Methods for Survival Data Analysis, *Third Edition*

LePAGE and BILLARD · Exploring the Limits of Bootstrap

LEYLAND and GOLDSTEIN (editors) · Multilevel Modelling of Health Statistics

LIAO · Statistical Group Comparison

LINDVALL · Lectures on the Coupling Method

LIN · Introductory Stochastic Analysis for Finance and Insurance

LINHART and ZUCCHINI · Model Selection

LITTLE and RUBIN · Statistical Analysis with Missing Data, *Second Edition*

LLOYD · The Statistical Analysis of Categorical Data

LOWEN and TEICH · Fractal-Based Point Processes

MAGNUS and NEUDECKER · Matrix Differential Calculus with Applications in
Statistics and Econometrics, *Revised Edition*

MALLER and ZHOU · Survival Analysis with Long Term Survivors

MALLOWS · Design, Data, and Analysis by Some Friends of Cuthbert Daniel

MANN, SCHAFER, and SINGPURWALLA · Methods for Statistical Analysis of
Reliability and Life Data

MANTON, WOODBURY, and TOLLEY · Statistical Applications Using Fuzzy Sets

MARCHETTE · Random Graphs for Statistical Pattern Recognition

MARDIA and JUPP · Directional Statistics

MARKOVICH · Nonparametric Analysis of Univariate Heavy-Tailed Data: Research and
Practice

MARONNA, MARTIN and YOHAI · Robust Statistics: Theory and Methods

MASON, GUNST, and HESS · Statistical Design and Analysis of Experiments with
Applications to Engineering and Science, *Second Edition*

McCULLOCH, SEARLE, and NEUHAUS · Generalized, Linear, and Mixed Models,
Second Edition

McFADDEN · Management of Data in Clinical Trials, *Second Edition*

* McLACHLAN · Discriminant Analysis and Statistical Pattern Recognition

McLACHLAN, DO, and AMBROISE · Analyzing Microarray Gene Expression Data

McLACHLAN and KRISHNAN · The EM Algorithm and Extensions, *Second Edition*

McLACHLAN and PEEL · Finite Mixture Models

McNEIL · Epidemiological Research Methods

MEEKER and ESCOBAR · Statistical Methods for Reliability Data

MEERSCHAERT and SCHEFFLER · Limit Distributions for Sums of Independent
Random Vectors: Heavy Tails in Theory and Practice

MENGERSEN, ROBERT, and TITTERINGTON · Mixtures: Estimation and
Applications

MICKEY, DUNN, and CLARK · Applied Statistics: Analysis of Variance and
Regression, *Third Edition*

* MILLER · Survival Analysis, *Second Edition*

MONTGOMERY, JENNINGS, and KULAHCI · Introduction to Time Series Analysis
and Forecasting

MONTGOMERY, PECK, and VINING · Introduction to Linear Regression Analysis,
Fourth Edition

MORGENTHALER and TUKEY · Configural Polysampling: A Route to Practical
Robustness

MUIRHEAD · Aspects of Multivariate Statistical Theory

MULLER and STOYAN · Comparison Methods for Stochastic Models and Risks

MURRAY · X-STAT 2.0 Statistical Experimentation, Design Data Analysis, and
Nonlinear Optimization

MURTHY, XIE, and JIANG · Weibull Models

MYERS, MONTGOMERY, and ANDERSON-COOK · Response Surface Methodology:

---

*Now available in a lower priced paperback edition in the Wiley Classics Library.

†Now available in a lower priced paperback edition in the Wiley–Interscience Paperback Series.

Process and Product Optimization Using Designed Experiments, *Third Edition*
MYERS, MONTGOMERY, VINING, and ROBINSON · Generalized Linear Models.
    With Applications in Engineering and the Sciences, *Second Edition*
† NELSON · Accelerated Testing, Statistical Models, Test Plans, and Data Analyses
† NELSON · Applied Life Data Analysis
NEWMAN · Biostatistical Methods in Epidemiology
OCHI · Applied Probability and Stochastic Processes in Engineering and Physical
    Sciences
OKABE, BOOTS, SUGIHARA, and CHIU · Spatial Tesselations: Concepts and
    Applications of Voronoi Diagrams, *Second Edition*
OLIVER and SMITH · Influence Diagrams, Belief Nets and Decision Analysis
PALTA · Quantitative Methods in Population Health: Extensions of Ordinary Regressions
PANJER · Operational Risk: Modeling and Analytics
PANKRATZ · Forecasting with Dynamic Regression Models
PANKRATZ · Forecasting with Univariate Box-Jenkins Models: Concepts and Cases
PARDOUX · Markov Processes and Applications: Algorithms, Networks, Genome and
    Finance
* PARZEN · Modern Probability Theory and Its Applications
PEÑA, TIAO, and TSAY · A Course in Time Series Analysis
PIANTADOSI · Clinical Trials: A Methodologic Perspective
PORT · Theoretical Probability for Applications
POURAHMADI · Foundations of Time Series Analysis and Prediction Theory
POWELL · Approximate Dynamic Programming: Solving the Curses of Dimensionality,
    Second Edition
POWELL and RYZHOV · Optimal Learning
PRESS · Bayesian Statistics: Principles, Models, and Applications
PRESS · Subjective and Objective Bayesian Statistics, *Second Edition*
PRESS and TANUR · The Subjectivity of Scientists and the Bayesian Approach
PUKELSHEIM · Optimal Experimental Design
PURI, VILAPLANA, and WERTZ · New Perspectives in Theoretical and Applied
    Statistics
† PUTERMAN · Markov Decision Processes: Discrete Stochastic Dynamic Programming
QIU · Image Processing and Jump Regression Analysis
* RAO · Linear Statistical Inference and Its Applications, *Second Edition*
RAO · Statistical Inference for Fractional Diffusion Processes
RAUSAND and HØYLAND · System Reliability Theory: Models, Statistical Methods,
    and Applications, *Second Edition*
RAYNER · Smooth Tests of Goodnes of Fit: Using R, *Second Edition*
RENCHER · Linear Models in Statistics
RENCHER · Methods of Multivariate Analysis, *Second Edition*
RENCHER · Multivariate Statistical Inference with Applications
* RIPLEY · Spatial Statistics
* RIPLEY · Stochastic Simulation
ROBINSON · Practical Strategies for Experimenting
ROHATGI and SALEH · An Introduction to Probability and Statistics, *Second Edition*
ROLSKI, SCHMIDLI, SCHMIDT, and TEUGELS · Stochastic Processes for Insurance
    and Finance
ROSENBERGER and LACHIN · Randomization in Clinical Trials: Theory and Practice
ROSS · Introduction to Probability and Statistics for Engineers and Scientists
ROSSI, ALLENBY, and McCULLOCH · Bayesian Statistics and Marketing
† ROUSSEEUW and LEROY · Robust Regression and Outlier Detection
ROYSTON and SAUERBREI · Multivariate Model Building: A Pragmatic Approach to
    Regression Analysis Based on Fractional Polynomials for Modeling Continuous
    Variables

*Now available in a lower priced paperback edition in the Wiley Classics Library.
†Now available in a lower priced paperback edition in the Wiley–Interscience Paperback Series.

* RUBIN · Multiple Imputation for Nonresponse in Surveys
RUBINSTEIN and KROESE · Simulation and the Monte Carlo Method, *Second Edition*
RUBINSTEIN and MELAMED · Modern Simulation and Modeling
RYAN · Modern Engineering Statistics
RYAN · Modern Experimental Design
RYAN · Modern Regression Methods, *Second Edition*
RYAN · Statistical Methods for Quality Improvement, *Third Edition*
SALEH · Theory of Preliminary Test and Stein-Type Estimation with Applications
SALTELLI, CHAN, and SCOTT (editors) · Sensitivity Analysis
* SCHEFFE · The Analysis of Variance
SCHIMEK · Smoothing and Regression: Approaches, Computation, and Application
SCHOTT · Matrix Analysis for Statistics, *Second Edition*
SCHOUTENS · Levy Processes in Finance: Pricing Financial Derivatives
SCHUSS · Theory and Applications of Stochastic Differential Equations
SCOTT · Multivariate Density Estimation: Theory, Practice, and Visualization
* SEARLE · Linear Models
† SEARLE · Linear Models for Unbalanced Data
† SEARLE · Matrix Algebra Useful for Statistics
† SEARLE, CASELLA, and McCULLOCH · Variance Components
SEARLE and WILLETT · Matrix Algebra for Applied Economics
SEBER · A Matrix Handbook For Statisticians
† SEBER · Multivariate Observations
SEBER and LEE · Linear Regression Analysis, *Second Edition*
† SEBER and WILD · Nonlinear Regression
SENNOTT · Stochastic Dynamic Programming and the Control of Queueing Systems
* SERFLING · Approximation Theorems of Mathematical Statistics
SHAFER and VOVK · Probability and Finance: It's Only a Game!
SHERMAN · Spatial Statistics and Spatio-Temporal Data: Covariance Functions and
    Directional Properties
SILVAPULLE and SEN · Constrained Statistical Inference: Inequality, Order, and Shape
    Restrictions
SINGPURWALLA · Reliability and Risk: A Bayesian Perspective
SMALL and McLEISH · Hilbert Space Methods in Probability and Statistical Inference
SRIVASTAVA · Methods of Multivariate Statistics
STAPLETON · Linear Statistical Models, *Second Edition*
STAPLETON · Models for Probability and Statistical Inference: Theory and Applications
STAUDTE and SHEATHER · Robust Estimation and Testing
STOYAN, KENDALL, and MECKE · Stochastic Geometry and Its Applications, *Second
    Edition*
STOYAN and STOYAN · Fractals, Random Shapes and Point Fields: Methods of
    Geometrical Statistics
STREET and BURGESS · The Construction of Optimal Stated Choice Experiments:
    Theory and Methods
STYAN · The Collected Papers of T. W. Anderson: 1943–1985
SUTTON, ABRAMS, JONES, SHELDON, and SONG · Methods for Meta-Analysis in
    Medical Research
TAKEZAWA · Introduction to Nonparametric Regression
TAMHANE · Statistical Analysis of Designed Experiments: Theory and Applications
TANAKA · Time Series Analysis: Nonstationary and Noninvertible Distribution Theory
THOMPSON · Empirical Model Building: Data, Models, and Reality, *Second Edition*
THOMPSON · Sampling, *Third Edition*
THOMPSON · Simulation: A Modeler's Approach
THOMPSON and SEBER · Adaptive Sampling
THOMPSON, WILLIAMS, and FINDLAY · Models for Investors in Real World Markets

*Now available in a lower priced paperback edition in the Wiley Classics Library.
†Now available in a lower priced paperback edition in the Wiley–Interscience Paperback Series.

*Now available in a lower priced paperback edition in the Wiley Classics Library.
†Now available in a lower priced paperback edition in the Wiley–Interscience Paperback Series.

Printed in the United States
By Bookmasters